Philosophy and Medicine

Founding Editors
H. Tristram Engelhardt Jr.
Stuart F. Spicker

Volume 151

The Philosophy and Medicine series is dedicated to publishing monographs and collections of essays that contribute importantly to scholarship in bioethics and the philosophy of medicine. The series addresses the full scope of issues in bioethics and philosophy of medicine, from euthanasia to justice and solidarity in health care, and from the concept of disease to the phenomenology of illness. The Philosophy and Medicine series places the scholarship of bioethics within studies of basic problems in the epistemology, ethics, and metaphysics of medicine. The series seeks to publish the best of philosophical work from around the world and from all philosophical traditions directed to health care and the biomedical sciences. Since its appearance in 1975, the series has created an intellectual and scholarly focal point that frames the field of the philosophy of medicine and bioethics. From its inception, the series has recognized the breadth of philosophical concerns made salient by the biomedical sciences and the health care professions. With over one hundred and twenty five volumes in print, no other series offers as substantial and significant a resource for philosophical scholarship regarding issues raised by medicine and the biomedical sciences.

* * *

Maartje Schermer • Nicholas Binney

Editors

A Pragmatic Approach to Conceptualization of Health and Disease

 Springer

Editors
Maartje Schermer
Section Medical Ethics, Philosophy and
History of Medicine
Erasmus MC University Medical Center
Rotterdam, The Netherlands

Nicholas Binney
Section Medical Ethics, Philosophy and History
of Medicine
Erasmus MC University Medical Center
Rotterdam, The Netherlands

ISSN 0376-7418   ISSN 2215-0080   (electronic)
Philosophy and Medicine
ISBN 978-3-031-62240-3   ISBN 978-3-031-62241-0   (eBook)
https://doi.org/10.1007/978-3-031-62241-0

This work was supported by Erasmus MC Rotterdam.

This Springer imprint is published by the registered company Springer Nature Switzerland AG
The registered company address is: Gewerbestrasse 11, 6330 Cham, Switzerland

If disposing of this product, please recycle the paper.

Acknowledgement

The research project 'Health and Disease as Practical Concepts' was supported by the Dutch Research Council (NWO), project number 406.18.FT.002.

The editors also want to acknowledge the editorial support of Sybren Siebesma.

Contents

About the Editors

Maartje Schermer studied medicine and philosophy at the University of Amsterdam, and is professor of philosophy of medicine and head of the Department of Medical Ethics, Philosophy and History of Medicine at the Erasmus MC, Rotterdam. Her research mainly concerns philosophical and ethical questions surrounding new (bio)-medical technologies. She is the former chair of the Dutch standing committee on Ethics and Law of the Health Council and of the Center for Ethics and Health (CEG).

Nicholas Binney is a philosopher and postdoctoral researcher at the Department of Medical Ethics, Philosophy and History of Medicine, at the Erasmus MC Rotterdam. His work focuses on developing pragmatic concepts of disease, integrating the history and philosophy of medicine, the history and philosophy of diagnosis, drawing especially on the philosophy of Ludwik Fleck.

About the Authors

Gemma Blok is a professor in the History of Mental Health and Culture at the Open Universiteit and professor in the History of Psychiatry at Utrecht University. Her areas of expertise are the histories of psychiatry, addiction treatment, and intoxication, focusing mostly on the Netherlands during the twentieth century. She was a principal investigator in the HERA project Governing the Narcotic City. Imaginaries, Practices and Discourses of Public Drug Cultures in European Cities from 1970 until today. She is currently leading the project Tales from the Dutch Drug Closet, on intoxicant use in the Netherlands since the 1960s.

Marianne Boenink graduated in health sciences and philosophy. She currently is a professor in Ethics of Healthcare at the Radboud University Medical Center in Nijmegen, the Netherlands. Her research focuses on philosophical and ethical questions raised by emerging technologies in healthcare, with a particular interest in predictive, diagnostic, and prognostic innovations. She is the PI of an interdisciplinary research project investigating the conceptual, ethical, and societal implications of the ongoing biomarkerisation of brain disease. In other projects, she explores ethical and societal conditions for responsible AI-based screening and digital monitoring.

Timo Bolt is associate professor of medical history at the Erasmus MC University Medical Center Rotterdam, the Netherlands. His work focuses on the modern and contemporary history of medicine and public health. His research interests include the history of clinical epidemiology and evidence-based medicine, the history of psychiatry and mental healthcare, and historical perspectives on concepts of health and disease. He is one of the initiators and researchers of the project 'Health and disease as practical concept', funded by the Dutch Research Council (NWO) within the Social Sciences and Humanities Open Competition programme.

Olaf Dekkers studied medicine and philosophy. He is professor of research medicine at the Leiden University Medical Center. Currently he is working as an internist and clinical epidemiologist.

Hans-Joerg Ehni is deputy director at the Institute for the Ethics and History of Medicine, University of Tuebingen. His research is focused on the ethics of biomedical research involving human subjects and on the ethics of ageing, particularly on the ethics of new biomedical interventions into the ageing process and increased longevity. He is also working on the theoretical understanding of the concept of health in a pluralistic society and its policy implications, in particular for policies promoting healthy ageing.

Heiner Fangerau is a historian of medicine. He is the director of the Department of the History, Philosophy and Ethics of Medicine, Heinrich-Heine-University Düsseldorf. His current research interests include the history of medical diagnostics, the history of the biomedical model (including medical technology), and the history and ethics of child and adolescent medicine.

Élodie Giroux is full professor in philosophy of science at Lyon 3 University and researcher at the Lyon Institute of Philosophical Researches. Her main works are on history and philosophy of epidemiology, causation in medicine and public health, health and disease concepts, precision medicine, and environmental health. Besides numerous papers and editions of special issues (on the history of risk factor epidemiology, on precision medicine, humanism in medicine, exposome research, etc.), her publications include several books such as *Naturalism in the Philosophy of Health: Issues and Implications* (2016, Springer) and *Integrative Approaches in Environmental Health and Exposome Research* (2023, Palgrave MacMillan).

Monica Greco is Professor of Social and Policy Sciences, University of Bath. Her research engages with the history, philosophy, and contemporary practices of psychosomatic medicine with a focus on conceptualisations of health, illness, explanation and classification, including their genealogies and ethico-political implications. Related interests include the history and philosophy of vitalism; healthism; and medical humanities as an epistemic project. In recent years Monica's work has focused on the problematic of so-called 'medically unexplained symptoms' and associated syndromes, also working in collaboration with clinical researchers on developing novel forms of intervention.

Bjørn Hofmann is a Norwegian professor in philosophy of medicine and bioethics with special interest for the relationship between epistemology and ethics. He is affiliated with the Centre for Medical Ethics at the University of Oslo in Norway and the Department of Health Science at the Norwegian University of Science and Technology (NTNU) at Gjøvik. Hofmann is trained in the natural sciences (electrical engineering and biomedical technology), history of ideas, and philosophy. His main subjects and interests are basic concepts for health care, norms of knowledge and evidence production, handling of technology, and (bio)medical ethics.

Lara Keuck is professor for history and philosophy of medicine at Bielefeld University, and directs the research group Practices of Validation in the Biomedical Sciences at the Max Planck Institute for the History of Science, Berlin. She has published widely on the epistemology of medical classification, the history of Alzheimer's disease, the validation of animal models of human diseases, and challenges to diagnostic validity in psychiatry.

Quill Kukla is professor of philosophy and director of disability studies at Georgetown University. From 2021 to 2023, they were also a Humboldt Stiftung prize winner at the Institut für Philosophie at Leibniz Universität Hannover. They received a PhD in philosophy from the University of Pittsburgh and completed a Greenwall Postdoctoral Fellowship at The Johns Hopkins School of Public Health. Their most recent book is *City Living: How Urban Dwellers and Urban Spaces Make One Another* (Oxford University Press 2021) and their forthcoming book is entitled *Sex Beyond 'Yes!'* (W. W. Norton & Co. 2024).

Martin Kusch is professor of philosophy of science and epistemology at the University of Vienna. He has published widely on the philosophy of relativism, the history and philosophy of the social sciences, and the sociology of science and technology. Between 2014 and 2019, he was principal investigator of an ERC Advanced Grant on "The Emergence of Relativism". His most recent book is *Relativism in the Philosophy of Science* (CUP, 2020).

Rik van der Linden works as a doctoral researcher at the Department of Medical Ethics, Philosophy and History of Medicine, at the Erasmus MC Rotterdam. His PhD research focuses on concepts of health and disease, which is part of the project 'Health and disease as practical concepts: a pragmatist approach to conceptualization of health and disease', funded by the Dutch Research Council. With a multidisciplinary background, including a BSc in health sciences, an MSc in clinical psychology, and an MA in philosophy and bioethics, Rik enjoys working on the intersection of medicine and philosophy.

Lennart van der Molen graduated in both biomedical sciences and philosophy. In his master's thesis, he examined the ethical feasibility of a preclinical diagnosis of Alzheimer's disease. He now works as a PhD student at the Ethics of Healthcare group of the Radboud University Medical Center in Nijmegen, the Netherlands. In his PhD project, he investigates the conceptual, ethical, and societal implications of the increasing use of biomarkers in the context of neurodegenerative diseases, currently focusing on the domain of Alzheimer's disease.

Wendy A. Rogers is a distinguished professor in the Philosophy Department and the School of Medicine at Macquarie University. She initially trained as a general practitioner, before moving into bioethics. Her current research interests include philosophy of medicine; evidence and ethics in surgery; ethics of new technologies (artificial intelligence, synthetic biology); feminist bioethics; and transplant abuse. She has an interest in health policy and has served two terms on the Australian Health Ethics Committee. She is co-Director of the Macquarie University Ethics and Agency Research Centre and a long-standing member of the International Network of Feminist Approaches to Bioethics (FAB).

Jenny Slatman is professor of medical and health humanities in the Department of Culture Studies at Tilburg University, the Netherlands. She has published widely on issues of embodiment in art, expression, and contemporary medical practices. www.jennyslatman.nl

Leen De Vreese is a postdoctoral researcher affiliated with the Centre for Logic and Philosophy of Science at Ghent University, Belgium. Her main research interests lie in the domain of general philosophy of science (causation, explanation, scientific understanding) and philosophy of medicine and psychiatry. Pluralism and pragmatism are recurring themes in all of her contributions to the field.

Mary Jean Walker is lecturer in philosophy at La Trobe University in Melbourne. Her research is in bioethics, philosophy of medicine, and personal identity. In bioethics, her work has focused on emerging technologies, including ethical issues related to overdiagnosis, personalised medicine, neurotechnologies, and artificial organs and prostheses. In philosophy of medicine, she has contributed to conceptual debates on definitions of health and disease, and epistemological investigation of personalised medicine. Her work on identity has examined narrative approaches to understanding persons, and their application to understanding ethical dimensions of self-change.

Frank J. Wolters (MD PhD) is assistant professor at the departments of Epidemiology and Radiology and Nuclear Medicine at Erasmus Medical Centre in Rotterdam, the Netherlands. Trained in clinical neurology and epidemiology at Oxford, Rotterdam, and Harvard, he developed expertise in research methodology and causal inference, applied notably to the prevention of cerebrovascular disease and dementia in clinical studies as well as large-scale datasets.

Gili Yaron trained as a social philosopher at the University of Groningen (thesis cum laude 2009). After completing her PhD in Medical Humanities (Maastricht University), she has held teaching and research positions with Zuyd University of Applied Sciences' 'What art knows' research group, the Netherlands Cancer institute-Antoni van Leeuwenhoek, and Maastricht University's 'Living Lab Sustainable Care'. Yaron's research interests include the embodiment of illness and disability, science-society interactions, and dementia care ethics. Yaron currently works as a senior researcher and ethics teacher at the research group 'Living Well with Dementia', Windesheim University of Applied Sciences.

Hub Zwart studied philosophy and psychology at Radboud University Nijmegen, worked as research associate at the Centre for Bioethics in Maastricht, and became professor of philosophy at the Faculty of Science of Radboud University in 2000. In 2018, he was appointed as Dean of Erasmus School of Philosophy (Erasmus University Rotterdam). The focus of his research is on philosophical and ethical issues in the emerging life sciences. He is editor-in-chief of the *Library for Ethics and Applied Philosophy* (Springer). Recently, he published *Continental Philosophy of Technoscience* (Springer Nature, 2022).

Chapter 1
Introduction

Nicholas Binney and Maartje Schermer ⓘ

The volume before you grew from an interdisciplinary research project "Health and disease as practical concepts", funded by the Dutch organization for scientific research (NWO), in which we aimed to develop a pragmatic approach to the conceptualization of health and disease.

The need for such a new approach springs from two observations. First, the actual landscape of health and disease is rapidly changing due to scientific, technological and societal developments. This creates new questions on how to conceptualize health and disease, for example because medical interventions are increasingly targeted at health risks and pre-diseases, or because the rising numbers of people with multiple chronic diseases in our ageing populations challenge our ideas of what it means to be healthy. Techno-scientific developments enable new understandings of 'pathology' while societal processes of medicalization change the boundaries of disease categories.

Second, within the philosophy of medicine, the decades old debate between naturalists and normativists concerning the definition of health and disease appears to be wearing thin. Moreover, this debate mainly focusses on the demarcation between 'the normal and the pathological', i.e. on health and disease as general concepts, but has little to say on defining, conceptualizing and classifying specific diseases. Also, the focus is solely on health and disease as theoretical concepts, while little attention is paid to the practical function these concepts have. Consequently, there appears to be a divide between the need for new ways of thinking about health and disease concepts, and the theoretical and analytical tools that traditional philosophy of medicine has to offer.

N. Binney (✉) · M. Schermer (✉)
Section Medical Ethics, Philosophy and History of Medicine, Erasmus MC University Medical Center, Rotterdam, The Netherlands
e-mail: n.binney@erasmusmc.nl; m.schermer@erasmusmc.nl

© The Author(s) 2024
M. Schermer, N. Binney (eds.), *A Pragmatic Approach to Conceptualization of Health and Disease*, Philosophy and Medicine 151,
https://doi.org/10.1007/978-3-031-62241-0_1

Luckily, we also observe some changes in the theoretical debate over the last years. Increasingly, attention is paid to plurality, complexity, contextuality and historicity of notions of health and disease, and methods other than classic conceptual analysis are being advocated. Some have explicitly called for a pragmatic turn. With this volume, we hope to contribute to this trend and develop the field further. We do not aim to provide a new definition or a new theory of health and disease. Instead our goal is to develop an outlook and an approach to identify, analyse and resolve concrete issues concerning the conceptualization of health and disease, as they arise in practice.

In an international, multidisciplinary workshop with philosophers, historians, social scientists, medical researchers and doctors, which we organized in Rotterdam, The Netherlands in the spring of 2023, this approach was discussed extensively and 'tried out', as it were, using different cases from different medical domains. This volume contains the contributions to this workshop. It consists of ten full papers, each followed by a shorter reflection. They can be read in consecutive order, but this is not necessary to get an idea of our proposed approach, and readers should of course feel free to pick out the topics they are most interested in.

The volume starts with a Prologue, in which we—Nicholas Binney, Timo Bolt, Rik van der Linden and Maartje Schermer—sketch the contours of our proposed pragmatic approach and formulate some challenging problems that we encountered in medical practice—broadly conceived—in relation to conceptualizations of health and disease. This chapter is based on a 'position paper' that was shared with all participants before the workshop. The prologue provides a background to place the separate chapters in the broader context of our research program and explain their coherence.

The next six chapters question the epistemic role of history and historical inquiry for our understanding of health and disease as concepts formed in medical practices through time, and discuss the related ontological questions about the kinds of realism or relativism that are compatible with a pragmatist position.

In Chap. 3, Heiner Fangerau compares concepts of disease and diagnosis found in the late nineteenth and early twentieth centuries. He explores the largely essentialist interpretations of diagnostic signs in the late nineteenth century, before comparing these with attempts to reconcile nominalism and essentialism in the early twentieth century. He pays particular attention to the philosophies of Richard Koch and Francis Crookshank, and how their views were linked to the popular 'As-If' philosophy of Hans Vaihinger. This historical work provides tools for contemporary thinking about how nature and culture are integrated.

In his reflections (Chap. 4), Hub Zwart focuses on the absence of the patient's voice from many of these historical discussions of diagnosis. He makes a plea for closer integration of the research laboratory and clinic.

In Chap. 5, Martin Kusch investigates medical relativism. The position we take in this book is that relativism need not be pernicious and can form a productive part of a pragmatic approach to health and disease. The types of relativism found in scholarly accounts of medical practice is under examined and Kusch takes important steps towards addressing this deficit. He describes a spectrum of relativist positions available to scholars and situates the work of three important scholars of medicine—Andrew Cunningham, Nicholas Jewson and Annemarie Mol—in this spectrum.

In his reflections (Chap. 6) Hans-Joerg Ehni focuses particularly on Kusch's comments on Cunningham, raising concerns about the viability of the forms of relativism on offer. In particular, he worries that the prospect of medical relativism opens the door to worrisome ethical relativism.

In Chap. 7, Nicholas Binney traces the historical development of thyroid tumour classification over the nineteenth and twentieth centuries. He makes a Fleckian analysis, following the development of a fluctuating network of active and passive elements of knowledge. Binney shows that this is a highly pragmatic process, with biological classification tailored to predict the clinical behaviour of tumours. Binney argues that understanding this history is important for the justification of medical knowledge in the present day.

In his comments on Binney (Chap. 8), Timo Bolt makes a plea for more history to supplement the approach taken by Binney. Bolt agrees that history does have a role to play in the justification of medical knowledge, and that historians should pay more attention to this role for medical history. Even so, he argues that more attention should also be paid to social and contextual factors neglected by Binney in order to fully realize his aims.

The next few chapters take their point of departure in medical research, and discuss how concepts of disease are shaped, changed and translated going from bench to bedside, while new technologies and methodologies also produce conceptual changes.

Lara Keuck (Chap. 9) proposes a new concept, *scope validity*, to improve the evaluation of biomedical knowledge. She argues that the scope of a disease operationalization in different experimental and clinical contexts is often under considered. Understanding disease entities as abstract constructs that can be applied uniformly obscures the need to tailor research to particular and concrete challenges found in practice. The concept of scope validity is designed to draw attention this need, and to promote a relational epistemology emphasizing an adequacy-for-purpose view of validity.

In his reflections Frank Wolters (Chap. 10) discusses whether scope validity is indeed an asset to the epidemiologist's armoury. He compares scope validity to other validity concepts, such as construct validity and external validity. He finds that there are subtle but important differences between scope validity and external validity and argues that scope validity is a valuable addition to validity concepts. For example, he argues that the explicit mention of a study's scope may help prevent confusion and crosstalk in discussions of Alzheimer's disease.

Marianne Boenink and Lennart van der Molen (Chap. 11) argue that biomarkers change our understanding of disease. Boenink and Van der Molen analyse how criteria for the diagnosis of Alzheimer's disease have changed following the introduction of biomarkers. They argue that the introduction of biomarkers has resulted in a gradual shift from an 'ontological' conception of disease to a 'physiological' conception of disease. They argue that biomarkers change the conceptualization of disease for people with and without symptoms, that they have the effect of increasing the recognition of patient heterogeneity, and that they shift medical attention from making a diagnosis to anticipating the future trajectory of disease.

Bjorn Hofmann (Chap. 12) considers whether the proliferation of biomarkers has encouraged a shift from and ontological to a physiological concept of disease, and whether they are responsible for a shift from diagnosis to anticipation. He argues that sometimes they do, and sometimes they don't. He suggests that biomarkers do more than detecting or anticipating disease. Biomarkers influence the concepts of illness and sickness as well, which he calls the biomarkerization of malady.

The main thesis in the Chap. (13) by Elodie Giroux is that an epidemiological risk approach represents a different way of modelling health phenomena than the binary and categorical approach of pathophysiology. Levels of biological variables used to define disease are standardly divided into 'normal', 'at risk' and 'pathological' ranges. The downward adjustment of these thresholds is associated with overdiagnosis and overtreatment, and risk factors are increasingly considered as diseases in their own right. Giroux reveals a number of conceptual confusions surrounding the concept of risk and argues that naturalist definitions of disease fail to distinguish risk and disease adequately. She argues that distinguishing risk-based concepts of disease and pathophysiology-based concepts of disease would address these conceptual confusions and benefit medical practice.

In Chap. 14, Olaf Dekker reflects on this and considers the interrelationship of risk-based approaches and diseased based approaches. Using examples such as the discovery of smoking as a cause of lung cancer and precision medicine, he wonders how easy it is to separate these two approaches.

The final eight chapters consider conceptualizations of specific conditions— addiction and medically unexplained symptoms—and of health.

Mary Walker and Wendy Rogers (Chap. 15) take a pragmatic approach to the disease status of addiction. They argue that addiction is profitably understood as a vague cluster concept. Such concepts have several elements that can be used in different combinations in different clinical and research contexts, including physiological addiction, loss of control and harm. Many of these elements can be understood as existing on a spectrum of severity, adding to the vagueness of the cluster concept. Walker and Rogers argue that conceptualising addiction solely as physiological dysfunction is not pragmatically valuable. They highlight that each possible combination has both strengths and weaknesses, meaning that adjusting concepts to address problems encountered in practice will also generate new problems that need to be taken into account.

Gemma Blok reflects on the disease status of addiction from a historical perspective (Chap. 16). She finds that the vague cluster concept suggested by Walker and Rogers resonates with her experience as an historian. Historical actors have had different attitudes to the disease status of their addictions, and the vague cluster concept allows for this flexibility. Blok also argues that thinking of addiction as a spectrum of substance abuse disorder may help break down the binary distinction between the diseased and the healthy, and that focusing on harms can help focus attention on the consequences of social policy decisions, rather than simply on physiology.

In Chap. 17, Monica Grecco investigates two different pragmatic approaches to 'medically unexplained symptoms'. One version, which Greco aligns with Richard Rorty's 'epistemological behaviourism', provides legitimacy to patients reinforcing the distinction between patients who have a disease and patients who only have symptoms. On this view, legitimate disease are problems of the body—disease is somatized. Another version, aligned with Willam James' radical empiricism and exemplified by a 'symptoms clinic', provides legitimacy to patients by integrating physiology, societal structures and patient experiences. With reference to the symptoms clinic, Greco argues that this is the more productive approach.

In her reflections (Chap. 18), Jenny Slatman points out the bodily deficit in contemporary healthcare. She locates the somatization of disease not with the humoral theories endorsed by Descartes, but with the anatomical lesion-oriented approach to disease adopted in the early 1800s. Slatman considers the benefits that might accrue to medicine if assumptions about the anatomical location of disease could be replaced with a vision of disease that integrates the patient with their wider environment.

A specific conceptualization of health, namely Positive Health, is analysed in Chap. 19. Rik van der Linden and Maartje Schermer evaluate the adequacy of the concept of Positive Health: a concept that was developed in response to problems experienced in medical practice, which has gained significant popularity within the Dutch healthcare system and beyond. They explore the reasons for re-engineering the concept of health, the kind of actors involved and the outcomes and effects of this re-engineering. They use this case study to exemplify a pragmatic approach to the philosophy of health and disease, in which current concepts are adjusted to address problems encountered in practice. This pragmatic approach stands in opposition to the analytic approach that uses conceptual analysis to try to arrive at the one, universally correct concept.

In her comments, Gili Yaron (Chap. 20) considers Positive Health as a response to increasing dissatisfaction with the biomedical model, drawing on her own research to complement van der Linden and Schermer. She considers the factors that have made this new concept so popular, highlighting that the concept does not only provide a psychological resource, but is also used to re-engineer material tools, such as those used in administration, project design and funding.

The chapter by Quill Kukla (Chap. 21) also concerns the notion of health. They urge caution when building a pragmatist, pluralist conception of health. Kukla argues that adopting such expansionist concepts opens the possibility that these will be weaponized by people and institutions in positions of power in ways that undermine social justice. They look at three case studies, healthy eating, healthy sex and healthy gender, which exemplify this tendency.

Leen de Vreese (Chap. 22) reflects on the possible misuses of pragmatic concepts of health and considers what humanities scholars could do to help address this problem. She argues that humanities scholars can develop reflective communities that participate in an ongoing culture of vigilance.

In the final chapter, the Epilogue (Chap. 23), we reflect on the lessons learned in the workshop and the further questions to be asked and steps to be taken for the further development and application of our pragmatist program. We end by making some suggestions for the development of a toolbox to be used in pragmatist attempts to improve conceptualizations of health and disease in order to help solve problematic situations encountered in medical practice.

Chapter 2
Prologue: A Pragmatist Approach to Conceptualization of Health and Disease

Nicholas Binney, Timo Bolt, Rik van der Linden, and Maartje Schermer (iD)

2.1 Introduction

In the research project "Health and disease as practical concepts" which ran from 2020–2024, we aimed to develop a new, pragmatist approach to the conceptualization of health and disease, starting from 'problematic situations' as they arise in medical practices, rather than from philosophical theorizing.[1]

In this chapter, we briefly outline the main tenets of our approach and their link to the philosophical tradition of pragmatism, and we discuss some of the problematic situations we identified during the first phase of our research project. The chapter is an adapted version of the 'position paper' that we distributed among the participants of the workshop from which this Volume has grown. The purpose of this position paper was to give participants in the workshop an idea of the approach we envisaged and give us some common ground for the discussion. The participants were invited to take one of the problematic situations that we identified as a starting point for their contributions and to relate to the theoretical starting points of this new approach. To be sure, what we offer here is a rough sketch, not of a new definition or set of definitions, nor of a new method, but of a way of looking and approaching the issue of conceptualizing health and disease. In the Epilogue, we will take stock of what the workshop and the various contributions collected in this volume, have rendered in

---

[1]With 'medical practices' we mean the whole broad range of practices that are related to medicine and healthcare, including biomedical and epidemiological research, clinical practice, preventive medicine, self-care and lifestyle practices, public health, healthcare systems and institutions, and health policy.

N. Binney · T. Bolt · R. van der Linden · M. Schermer (✉)
Section Medical Ethics, Philosophy and History of Medicine, Erasmus MC University Medical Center, Rotterdam, The Netherlands
e-mail: m.schermer@erasmusmc.nl

© The Author(s) 2024
M. Schermer, N. Binney (eds.), *A Pragmatic Approach to Conceptualization of Health and Disease*, Philosophy and Medicine 151,
https://doi.org/10.1007/978-3-031-62241-0_2

terms of new and fruitful insights, both with regard to concepts and the resolution of problematic situations, as to the approach itself.

2.2   Background Assumptions and Theoretical Starting Points

Point of departure for our pragmatist approach is that the way in which 'disease' and 'health' are conceptualized, as well as the way in which individual diseases are conceptualized, has *important practical consequences*. Medical research strategies, treatment regimens for patients, implementation of health-policy measures or eligibility for sickness benefits all depend on the concepts and definitions used. Disease and health concepts used—either explicitly or implicitly—in various medical and health-related practices ultimately have effects on the lives and well-being of people. How we conceptualize health and disease really matters.

In contrast to some of the classic medical-philosophical theories of health and disease, our project does not aim for a single definition or theory. Rather, *we believe a pluralist account will do better justice to the complexity of medicine and healthcare; we see concepts as 'tools' that can be more or less helpful or productive in different contexts.* We therefore do not suggest that one single concept of health and disease can address all of the problematic situations we encounter, as we do not see how one concept of health and disease could possibly operate effectively in all the different contexts relevant to medical practice. We propose to use several different concepts to address different problematic situations, championing a pluralist account of health and disease (cf van der Linden and Schermer 2022). Interestingly, contextual conceptual pluralism is visible in medical practice itself, and we believe we should try to learn from this.

One of the tenets of pragmatism is that human problems should be central to human inquiry. Hence, we take our starting point in actual problematic situations, as will be outlined in the next sections. Moreover, medicine itself is by nature a pragmatic discipline, centrally concerned with the problems of relieving human suffering.[2] *We believe the relief of human suffering should be central to medical inquiry and a touchstone to evaluating how conceptualizations play out.*

Another tenet of pragmatism is non-essentialism. In our project we explore the historical processes through which medical knowledge has developed and show how this is riddled with pragmatic choices and judgements. Hence, disease and health concepts are contingent and not as 'objective' as is often believed by medical

---

[2]We are aware there is a whole literature on the 'goals of medicine', as well as on the nature of suffering. We take suffering to include physical, mental, social and existential components as well as premature death, loss of function, abilities etc. The proximate goals of specific medical practices can differ, e.g. medical research may aim to understand a disease mechanism, or health policy may aim to improve public health, but we think the relief (or prevention) of suffering is the ultimate aim in all cases.

practitioners and others. Supposedly value-free knowledge of pathology and patho-physiology, and descriptions of what is happening in the diseased human body, are not discovered biological realities but rather human creations. We fear that the desire to make use of a supposedly entirely discovered knowledge of an ultimate biological reality will prevent consideration of what this knowledge was invented to do—and thus what it is useful for and what it is not. The possibility of accessing an ultimate reality is a seductive prize, and the utility of knowledge may be ignored if knowledge is understood as capturing an ultimate reality. On the other hand, framing concepts of health and disease as pragmatically constructed and historically contingent may raise concerns that these concepts are being reduced to whatever historical actors have found expedient to believe. Some may worry that defending a non-realist position will end in complete relativism. This concern is reasonable, and serious. Our aim, however, is *to carve out a position that rejects simplistic forms of realism and embraces contingency without slipping into pernicious and silly forms of relativism.*

We are of course aware that claims that knowledge of pathology and pathophysiology, and even of biology, are pragmatic inventions, made in local contexts for particular purposes, are not original. The pragmatist tradition—as we will discuss in the next section—is one of the sources of inspiration and of justification for this general claim. We find, however, that it is still a challenge *to explain how medical knowledge is at once an invention and a discovery* (cf Binney 2023). If we can meet this task, the goal of attaining objective knowledge of an ultimate biological reality will seem much less seductive, and the need to explicitly discuss the pragmatic interests and ethical intuitions of various stakeholders when doing medical science will be obvious.

Finally, as has become clear, we are convinced that understanding *the historical development of a field of medical practice has an important epistemic role to play* in understanding and therefore justifying medical knowledge and practice. Some of the chapters in this volume specifically deal with this question of how and in which ways it is possible for history to have an epistemic function.

2.3 Connections with Classical Pragmatism

Although the pragmatic attitudes to health and disease concepts explored in this volume owe no strict allegiance to the classical American pragmatists, such as Charles Peirce, William James and John Dewey, we do find much in common with their work and take inspiration from them. Dewey's famous plea for the need for a recovery of philosophy especially resonates with us. "Philosophy recovers itself when it ceases to be a device for dealing with the problems of philosophers and becomes a method, cultivated by philosophers, for dealing with the problems of men" (Dewey et al. 1917: 65). It is worth fleshing out the connections we see with classical pragmatist thought and our own philosophy.

Following Dewey, our philosophical inquiry starts with problematic situations (see the next section). Indeed, Dewey saw the problematic situation as central to inquiry. Peirce rejected Descartes' extreme doubt, arguing that it was impossible to doubt everything at once. Instead, Peirce focused on *genuine doubt*, that arises from activities in which something surprising happens. Such unexpected events trigger genuine doubt, and the purpose of inquiry for Peirce is to resolve its unpleasantness (Thayer 1970: 61–100). Dewey built on Peirce's insight, expanding the notion of genuine doubt to focus on the *problematic situation*, which Tom Burke describes as a "localized instance of disequilibrium" (Burke 1994: 29). For Dewey, problematic situations are the disturbed relations of an organism and its environment, which define doubt and trigger inquiry (Dewey 1938: 35).

Burke uses the beautiful metaphor of a sea anemone to clarify what problematic situations are. The anemone's circulatory system includes the surrounding sea water, which is therefore contiguous with the anemone's very body. The substance of this creature's body cannot be separated from its surroundings: changes to the surrounding water produce and indeed *are* changes to the creature's body. Analogously, the inquiring subject (the anemone) and the objects they experience (its environment) are not separated but integrated. This creature, integrated with its environment, pursues the goal-oriented activities of life. Whilst its goals are being achieved, equilibrium is maintained. However, should something shift within the creature or within its environment to produce "a proportionate excess or deficit in some factor", this equilibrium is disturbed (Dewey 1938: 27). To address the need to restore equilibrium, and reach a state of fulfilment, the anemone then engages in a form of "proto-inquiry" (Burke 1994: 28).

> A state of tension is set up which is an actual state (not mere feeling) of organic uneasiness and restlessness. This state of tension (which defines need) passes into search for material that will restore the condition of balance. In the lower organisms it is expressed in the bulgings and retractions of parts of the organism's periphery so that nutritive material is ingested. The matter ingested initiates activities throughout the rest of the animal that lead to a restoration of balance, which, as the outcome of the state of previous tension, is fulfilment (Dewey 1938: 27).

The resulting equilibrium need not be the same as it was before the problematic situation was encountered (Dewey 1938: 28). The anemone and the surrounding water are changed—life may be organized differently to resolve problematic situations. Inquiry is the process of acting to resolve the problematic situations encountered in the course of life.

For Dewey, higher organisms encounter problems and seek to resolve them in much the same way. We encounter problematic situations and act to try to resolve them through inquiry. The problems we encounter, and their solutions, only exist in the particular situations in which we live. Inquiry is not an attempt to grasp a knower-independent world. "Until it frees itself from identification with problems which are supposed to depend upon Reality as such, or its distinction from a world of Appearance, or its relation to a Knower as such, the hands of philosophy are tied" (Dewey et al. 1917: 65). The objects of experience themselves are the product of inquiry, not just its object. Problematic situations are the origin, the object and, ultimately, the outcome of inquiry (Dewey 1938: 35).

This talk of knowledge being the solutions to local, contextually bound problems, is gratifying. However, it smacks of problems long associated with pragmatism. Notoriously, pragmatism connects truth to things which are expedient to believe. James rejected the notion that our beliefs and ideas need to copy some underlying reality for them to be true, as he could make little sense of this notion: "it is hard to see exactly what your ideas can copy" (James 2014: 73). According to James, "'The true', to put it very briefly, is only the expedient in the way of our thinking, just as 'the right' is only the expedient in the way of our behaving" (James 2014: 80). James repeatedly equated truth with that which it is useful to believe. "You can say of it either that "it is useful because it is true" or that "it is true because it is useful." Both these phrases mean exactly the same thing, namely that there is an idea that is fulfilled and can be verified" (James 2014: 74). In light of this, it is forgivable to think that, for a pragmatist, to be true is to be useful, and to be useful is to be true.

It is little wonder, then, that many have objected to a pragmatist characterization of truth. Bertrand Russell, famously, utterly rejected it. "I find great intellectual difficulties in this doctrine. It assumes that a belief is" true "when its effects are good" (Russell 1967: 817). Clinging to the view that facts are ethically neutral, Russell objected to this characterization of truth, as it required an ethical evaluation of whether the effects of a belief were *good* to determine whether the belief was true. Russell held that neither the effects of a belief nor their ethical evaluation was relevant to whether a belief was true. Parodying James' arguments for the existence of God, Russell considered arguments for the existence of Santa Claus. Every year, hundreds of millions of adults engage in a world-wide conspiracy to convince young children that Santa Claus exists. They do this because this belief helps make Christmas a magical experience for their children, which they see as a good effect. Accepting that these effects are good, is this sufficient for the belief in Santa Claus to be true? Russell says the pragmatist must say yes, even though they should say no. Russell wrote: "I have always found that the hypothesis of Santa Claus 'works satisfactorily in the widest sense of the word'; therefore [accepting pragmatism] 'Santa Claus exists' is true, although [as we all know] Santa Claus does not exist" (Russell 1967: 818). For Russell, neither belief nor expedient belief is sufficient for truth. Consequently, he rejected pragmatism. "But this is only a form of the subjectivistic madness which is characteristic of most modern philosophy" (Russell 1967: 818).

The charge of reducing truth to whatever it is expedient to believe has dogged pragmatism from its earliest days. Concern that pragmatic theories of truth are insufficiently realist, and "violate basic intuitions about the nature and meaning of truth" have also been persistent (Capps 2023). Given the frequent appeals to utility as a determinant of truth made by pragmatists like James, and the calls for philosophy to free itself from Realism made by pragmatists like Dewey, these concerns are perhaps understandable. For our part, we agree with Russell that the belief in Santa Claus is not true—expedient belief is not sufficient for truth. However, we also disagree with Russell, as we hold that what is true is at least in part a human creation, dependent upon the effects of a belief and their ethical evaluation.

In recent years, philosophers have taken up the mantle of pragmatism, defending the early pragmatists from charges of subjectivistic madness. Hasok Chang, for example, has defended James. "The classical pragmatists' views on truth should not be caricatured as a notion that whatever pleases the believer is true" (Chang 2022: 197). Chang points out that James denied that this was a fair characterization of pragmatism. He also argues that pragmatism should be understood as a form of realism, albeit in a different sense to usually understood by philosophers.

Common sense, everyday, man-in-the-street intuitions about truth include at least two elements. The first is expressed beautifully in the final scene of the HBO series *Chernobyl*, which discusses what truth is and the consequences of being unfaithful to it:

> To be a scientist is to be naive. We are so focused on our search for truth, we fail to consider how few actually want us to find it. But it is always there, whether we can see it or not, whether we choose to or not. The truth doesn't care about our needs or wants. It doesn't care about our governments, our ideologies, our religions. It will lie in wait, for all time.

The truth, according to everyday intuitions, is something entirely independent of us. The truth is eternal, existing outside of history. This property of being entirely independent of us is connected to realistic intuitions that there is a world that is entirely independent of us and that out true beliefs correspond to the way this world is in itself.

The second element of everyday intuitions about truth is also expressed in the passage from *Chernobyl*—*the truth opposes our will*. We do not choose what the truth is. We may want, or even need, for certain things to be true, but find that they are not, regardless. This property of resisting our will is also connected to realistic intuitions about the truth corresponding to the way the world is in itself, as the character of such a world, being entirely independent of us, would not be for us to choose.

A key insight for us is that these two elements are *separable*. We can have the second without accepting the first. Pathologists may have to decide which histological structures are seen as cancer, but whether or not cancer (so defined) spreads around the body, or responds to a particular treatment, is not for the pathologists to decide. Facts about the prognosis and response to treatment of patients with cancer do not exist without people making contingent decisions about how to define things like cancer. Once these decisions have been made, however, the resulting facts about cancer are not for those people to decide. Facts are dependent upon human decisions, but not determined by them. As Chang puts it, "entities being mind-framed does not imply that they are mind-controlled...Even though real entities are concept-bound, they do not obey our wishes" (Chang 2022: 204). As facts are dependent upon human decisions, they are not independent of us, and do not correspond to how the world is in itself. We must reject the first element of truth. And yet, we do not determine or choose what the facts are. We can accept the second element of truth.

Chang argues that embracing this second element of truth is enough to qualify as a form of realism. "We can design a concept as we wish, but whether our concept can facilitate coherent activities is a matter that is quite outside our control. So I think my position does retain something very important in what many people value in realism" (Chang 2022: 204). We certainly agree with this but are ambivalent about whether a philosophy that rejects correspondence with the world in itself should qualify as realism. Philosophies that embrace the possibility of discovering facts that correspond with the world-in-itself form a long and honourable tradition. This tradition deserves a name, and 'realism' is widely used. 'Relativism' or 'pragmatism' might be used for philosophies that reject that facts correspond to the way the world is in itself, whilst accepting that the facts do not obey our wishes. The important point is to note that these elements are separable.

Having identified that which does not obey our wishes as an important element of common realistic intuitions, consider the role this element might play in a pragmatist philosophy. James, for instance, argued that whilst utility, or "satisfactions", was necessary for truth, it was not sufficient. Even if a belief was useful, a pragmatist would call it false if the putative reality did not obtain:

> The pragmatist calls satisfactions indispensable for truth building, but I have everywhere called them insufficient unless reality be also incidentally led to. If the reality assumed were cancelled from the pragmatist's universe of discourse, he would straightaway give the name falsehoods to the beliefs remaining, in spite of all their satisfactoriness. For him, as for his critic, there can be no truth if there is nothing to be true about (James 2014: 187).

But what might this "reality" be? It cannot be the independent world-in-itself, as for James "The trail of the human serpent is thus over everything" (James 2014: 27). Might this reality be that which does not obey our wishes? Perhaps so. Indeed, James says that we will not reach satisfaction unless we pay close attention to regularities that exist between the objects of our experience. Such regularities can resist our will, even if the objects that are related are invented by our wishes. If we are wayward in our attention to these obstinate regularities, then only woe waits for us.

> Our experience meanwhile is all shot through with regularities. One bit of it can warn us to get ready for another bit, can 'intend' or be 'significant of' that remoter object. The object's advent is the significance's verification. Truth, in these cases, meaning nothing but eventual verification, is manifestly incompatible with waywardness on our part. Woe to him whose beliefs play fast and loose with the order which realities follow in his experience; they will lead him nowhere or else make false connections (James 2014: 75).

Thus, James held that beliefs would not be useful if they misrepresented the regularities of our experience. The failure to recognize the possibility that beliefs that misrepresent regularities in our experience can sometimes be useful may have caused much confusion regarding the importance of these regularities to pragmatism. No one has ever seen a jolly fat man piloting a sledge drawn by flying reindeer. Even if you stay up all night on Christmas Eve, you will encounter no such stranger bearing gifts. Neither will he be found at the North Pole. If you get a job at the U.S. Postal Service, you will see that letters to Santa Claus are not delivered to him, and it is not he who replies. The person who behaves as Santa Claus, who is Santa Claus, does not exist. Consequently, according to the James of these passages, belief in Santa Claus is false.

Perhaps then there is a different pairing of elements that comprise the truth. Or perhaps there are two different concepts of truth that are used iteratively in pragmatism. According to the first, if those regularities that resist our wishes are as we say they are, then our claims are true. If they are not, then they are false. And yet, establishing that such regularities obtain is not sufficient for a pragmatic theory of truth (although it may be for some relativisms). To be true in this second sense, the regularities must not only obtain, but also provide satisfaction. *To be true, that which resists our will must also be useful.* If it is not, then we call the claim false, even if the regularity is as claimed.

For example, a pathologists might claim that half of the patients with a particular disease do not recover when they receive a particular treatment. This regularity resists their will. They would prefer it if all patients recovered, but the regularity is indifferent. The claim that "half of the patients with this disease do not recover when treated" is true, in the first sense. For the pragmatist, though, this still constitutes a problematic situation. Not enough people recover and this is not satisfactory. There is something *wrong* with how things are understood and this must be improved. There are many ways to respond. One way would be to search for a better treatment. Another way, however, would be to adopt different concept of the disease, such that patients would receive different, and perhaps superior, management. Perhaps there is a different treatment that would help some of these patients who do not recover. By seeing these patients as having a different disease, that needs a different treatment, their management might be improved. In Fleckian terms, active elements of knowledge could be adjusted to produce more productive passive resistance (Binney 2023, and Binney, Chap. 7, this volume).[3] Should a more satisfactory configuration of disease concepts and treatment options be found, then this would be considered true. The old configuration would be considered *false*, because there was something *wrong* with it, even though the regularities of the old configuration did obtain. As in the metaphor of the sea anemone above, the inquiring subject and the objects experienced would be thoroughly integrated, as the objects of experience would be produced by the activity of the inquiring subject, and the regularities between the objects so produced would condition the activity of the inquiring subject, in order to restore equilibrium and produce satisfaction. Understood like this, the truth does care about our needs and wants.

We propose this as a valuable, pragmatic theory of truth. It is relativist, as it rejects the notion that truth is a correspondence with the absolute reality of the world-in-itself. And yet, it is (perhaps) realist, as it embraces the notion that the truth should

---

[3] We do find Ludwik Fleck's distinction between active and passive elements of knowledge a useful tool for our pragmatism (Fleck 1979). Active elements are taken for granted by knowers, and are constitutive of the objects of experience. Once these objects have been brought into being, however, they do not necessarily relate to each other as knowers wish. This resistance to the knowers' will is the passive element of knowledge. Without the active elements the passive elements do not exist, as a river does not exist without its banks. As the active elements are the invention of human minds and culture, and the passive elements are dependent upon them but not determined by them, the passive elements are (in Chang's language) mind-framed but not mind-controlled. The passive resistance will then shape the active elements used to produce it, as a river shapes its banks. On this view, nature and culture are thoroughly integrated. "Does the river make its banks, or do the banks make the river? Does a man walk with his right leg or with his left leg more essentially? Just as impossible may it be to separate the real from the human factors in the growth of our cognitive experience" (James 2014: 90).

be obstinate and resist out wishes. At the very least, it is not a pernicious relativism. It is shot through with ethical considerations, as claims are not true unless they help people achieve their goals, and which goals to pursue requires ethical deliberation. For medicine, these goals will need to be connected to the relief of human suffering, as this is the ultimate legitimate purpose of the medical endeavour. Hence we claim that this needs to be a touchstone in evaluating how conceptualizations play out. This theory of truth allows for pluralism, as the same problem may have several solutions, or there may be many goals to pursue. Finally, it is a local theory of truth, as the problematic situations that emerge and what counts as their resolution are contingent upon time and place. In the next section, we give some concrete examples of problematic situations and how to approach them from a pragmatist perspective.

2.4   Problematic Situations Related to Health and Disease Concepts

Problematic situations, as we understand them, are instances of conflict between stakeholders that arise due to differing conceptions of disease, or due to disagreements about how disease should be conceptualized, or due to unsatisfactory outcomes of medical research or practice. They are situations in which suboptimal practices, or disagreement about how to practice, arise due to the concepts of health and disease employed by the parties involved (who could be doctors, medical researchers, philosophers, patients or other lay people). *Our aim is to consider how these problematic situations might be addressed by altering the concepts of health and disease employed in an effort to improve medical practice.*[4]

We have identified a number of problematic situations by studying concrete cases of conflicts, disagreements or (alleged) suboptimal situations related to medical practice. These were mostly taken from the medical and medical philosophical literature, but also inspired by conversations and more formal interviews with those involved in healthcare practices (van der Linden and Schermer, forthcoming). For example, we studied the evolution of the understanding of chronic pain over the last decades (van der Linden et al. 2022), the recent reconceptualization of Alzheimer's disease in medical research (Schermer and Richard 2019; Schermer 2023), the diagnostic criteria for osteoporosis in relation to over- and underdiagnosis (Binney 2022), and the different perception of health between diabetes patients and their doctors (Haalboom 2023). In all these cases we found various problematic situations related to the conceptualization of health or disease. By abstracting from

---

[4]Our aim can therefore be seen as a form of "conceptual engineering"—see also van der Linden and Schermer, Chap. 19, this volume. In Fleckian terms, we aim to adapt the active elements of knowledge to ultimately produce more useful resistance (passive elements of knowledge) and ultimately better outcomes for patients (Binney 2023).

the specifics of those cases we arrived at some tentative general formulations of types of problematic situations.[5]

2.4.1 Patients with Symptoms but No Pathology Are Not Understood as Diseased

Consider patients in chronic pain. Traditionally, pain was understood as a symptom of a pathological lesion that causes it (Raffaeli and Arnaudo 2017). However, many patients who report chronic pain have no identifiable injuries or other pathology which can explain why they are in pain. As pain is understood as a symptom, and not as a disease, patients are classified according to the lesions producing the pain. Patients without lesion cannot be classified, often making them invisible to the bureaucratic systems of patient management (Treede et al. 2019). As having a disease is identified with having a pathological lesion, patients with no lesions are often considered not to have a somatic disease at all. Instead, they are often understood as having a psychological problem, and are managed differently to patients with pathological lesions.

We identified this as a problematic situation for several reasons (Raffaeli and Arnaudo 2017; Nugraha et al. 2019; Treede et al. 2019). Firstly, many doctors, researchers and patients are concerned that patients in chronic pain but without lesions are not getting the attention, resources and recognition they deserve. Secondly, as these patients are often considered as having a psychological problem, and are treated differently to patients with observable lesions, many doctors feel that they may not be getting the care that they need. Thirdly, historical work reveals an evolving conversation about what pain is. Pain is not an easily explained phenomenon, reducible to the detection of pathological lesions by the nervous system. Some researchers think the time is ripe for pain to be fundamentally reconsidered. For example, instead of being understood as the conscious registration a nervous signal, pain might be understood as a conscious experience produced to modify the patient's behavior so that they act to protect themselves from injury. This may not be a popular way of thinking about pain amongst pain researchers, but it does show that patients suffering in pain need not be thought of as suffering from an entirely different condition depending on whether or not a lesion is present.

Although defining diseases in terms of pathological lesions provides a feeling of objectivity, in the sense of coming into contact with a culturally independent real

---

[5]Our list of problematic situations is not comprehensive or exhaustive. We realize that different stakeholders, engaging from different perspectives, may identify different problematic situations. The way in which disagreements or conflict relate to conceptualizations of health and disease may not always be directly obvious but require philosophical analysis. Moreover, conceptual issues may not be the only element making a situation problematic. However, we do believe that if philosophy of medicine aims to make a contribution to improving medical practice, it needs to start with actual problems (with its feet in the mud, so to speak).

world, our pragmatist perspective highlights that this feeling may not provide the security researchers seek. The decision to understand particular anatomical findings as lesions, and to use these lesions to define diseases, is contingent choice. This is an example of something which is mind-framed in the above sense, of an active element of knowledge in Fleckian terms. We would emphasize, however, that adopting such active elements of knowledge is a historically contingent choice, not a reflection of the mind and culture independent world. One could, just as reasonably, use the symptoms of pain to define disease, by adopting the association of pain and disease as an active element of knowledge. On this view, the relationship between pathological lesions and the disease 'chronic pain' would be a passive element of knowledge. Even though it is mind-framed, it is not mind controlled. Whether they want to or not, researchers would find that many patients with this disease would not have a lesion.

Understanding pain as the conscious registration of a nervous signal from damaged tissue has not turned out to be a satisfactory way of understanding pain, as there are so many anomalous observations regarding the relationship between the experience of pain and pathological lesions. In particular, this has not proved a particularly successful approach to relieving patient's pain. Consequently, it may well be that shifting to different ways of understanding pain, in which patients in pain are understood to have the same basic problem, which requires the same explanation, could be more profitable. Shifting the active elements of knowledge such that chronic pain is defined as a disease in its own right, in light of this history, may produce more productive passive elements of knowledge, such that human suffering may be reduced, and equilibrium restored.

Moreover, since the relief of human suffering is one of the pragmatic goals of medicine, it is difficult to see how defining disease so that patients who are suffering in pain are ignored in clinical practice and by the bureaucratic system serves those goals. Focusing on patients who are suffering by defining disease in terms of pain will help make such patients visible to the bureaucratic systems in which they are managed and will help such patients get the care that will hopefully benefit them. Consequently, we see the potential in the redefinition of chronic pain as a disease in its own right (as ICD 11 does). A similar argument might hold for other conditions in which patients are suffering, but no clear pathophysiology is found (Sharpe and Greco 2019; O'Leary 2020; Wilshire and Ward 2020; Tesio and Buzzoni 2021). At the same time, we emphasize that this reconfiguration of how disease is understood is a possible local solution to a particular type of problem—it is not intended as a universal claim on how we ought to view disease in general since other solutions may be fitting to other problems.

2.4.2 Patients with Pathology or 'Biomarkers' but No Symptoms Are Understood as Diseased

Consider patients with Alzheimer's disease. This disease is currently defined in terms of its associated pathologies: amyloid plaques and tau-tangles (Montine et al. 2012; Petersen 2018). By definition, patients with Alzheimer's disease have these pathologies and patients with these pathologies have Alzheimer's disease. However, patients with Alzheimer's disease, so defined, may never develop symptoms of dementia (Dubroff and Nasrallah 2015). For decades, research into Alzheimer's disease has focused on intervening in these pathologies, even though this has been unsuccessful in preventing or improving the symptoms of dementia.

Similarly, in the case of diabetes doctors sometimes become very focused on blood glucose levels and not on the suffering of patients. The aim of treatment may become to maintain a constant level of blood glucose, although there is evidence that maintaining a constant blood glucose levels does not lead to optimal care (Sleath 2015; Khunti and Davies 2018).

The abstract problematic situation is the focusing on some laboratory (or at least non-symptomatic) parameter, whilst forgetting that the symptom is the ultimate object of clinical interest. This is problematic for several reasons. First, in the case of Alzheimer, the pathological conception of the disease dominates the research agenda, making difficult to get funding for research without adopting this pathological concept of the disease. By contrast, research into treatments for and coping with the symptoms of dementia, is much less well funded. Hence, patients may not get proper support in living with the condition. Second, where symptom free patients with pathology may want to understand themselves as healthy, there may be negative consequences in labelling them as diseased (e.g. anxiety, stigma, bureaucratic consequences). Finally, it can also lead to overdiagnosis and overtreatment (see Sect. 2.4.4).

The origins of this problem appear at least partly conceptual. Medical science focuses on what is considered material and real, namely pathology and pathophysiology. These form the traditional conceptual basis of disease (Sharpe and Greco 2019). Defined in this way, to treat disease is to remove the pathology, or to correct the physiological disturbance. Pathological lesions are seen as solid and tangible, bolstering their claim to objectivity and reality, as opposed to the subjectivity of many symptoms. This view neglects the contingency of disease definitions and fails to see them as the product of an historical process that selected specific parameters and criteria for particular reasons in a particular context.

Another conceptual problem may be a monocausal conception of disease driving the definition and prognostic expectations, drawing attention away from the need to explore how pathologies or biomarkers relate to symptoms. In the case of Alzheimer, amyloid was suspected of playing a causal role in the development of Alzheimer's dementia, and then was promoted to serve as the defining feature of the disease. This makes sense in a monocausal aetiological model of disease, which assumes that a cause is always followed by an effect, and diseases are defined according to their causes. However, as many cases with amyloid never develop dementia, this conceptual model may be wrong.

As a direction to a solution of this problematic situation, there might be value in reconceiving of disease in a ways that focus more on symptoms and human suffering; and in reconsidering monocausal and linear models of disease in light of their historical developments. We believe there are many—and apparently rising—

numbers of instances of pathology without symptoms, and the ways in which we conceptualize these, will affect the ways in which they are handled.

2.4.3   Preventive Medicine Aimed at Preventing Pathology or Pathophysiology, as Opposed to Symptoms

Medicine is often divided into two projects: the treatment of disease and the prevention of disease. The treatment of disease tackles pathology and pathophysiology, whereas the prevention of disease prevents these from occurring. What should the aims of preventive medicine be? Should it be to prevent pathology and pathophysiology? Or should it be to prevent illness?

Many diseases, such as diabetes, have associated preconditions, such as prediabetes. Diabetes is defined as persistently high blood glucose.[6] Prediabetes is defined in the same way, but at lower thresholds. Treating diabetes is to treat a disease, whereas treating prediabetes is to prevent a disease. The stated rationale for identifying patients with prediabetes is to prevent morbidity and mortality from diabetes and cardiovascular disease in the long term (Viera 2011; Yudkin 2016). However, the thresholds for prediabetes are set (by the American Diabetes Association) according to the risk of developing diabetes. As diabetes is defined as persistently high blood glucose, it can be asymptomatic. Thus, the prevention of diabetes is not necessarily the prevention of symptoms. The thresholds for prediabetes are set at a certain risk of developing high blood glucose, and not at a certain risk of developing symptoms (Yudkin 2016). Indeed, the treatment of prediabetes is reportedly not particularly effective at preventing morbidity and mortality in the long run, although this is debated (Cefalu 2016; Yudkin 2016). Researchers have complained that the concept of prediabetes is too "glucocentric", highlighting that type 2 diabetes is a complex metabolic condition for which glucose is but one important causal factor (Yudkin 2016).

Even if it is useful to define diabetes in terms of blood sugar levels, it may not be optimal to define predisease in this glucocentric way. Doing so creates a condition (prediabetes) that puts a patient at risk of developing a disease (diabetes) that puts the patient at risk of developing pathology (e.g. retinopathy) that may or may not result in symptoms (e.g. loss of vison or heart attack). This is a very indirect way of assessing those risks. The therapeutic logic is equally indirect. Even if treating diabetics by lowering blood glucose does reduce the risk of symptoms developing, it may not be the case that lowering blood glucose in prediabetics to reduce the risk of diabetes developing will do the same. Even if treating a disease prevents

---

[6]The American Diabetes Association defines asymptomatic type 2 diabetes in terms of fasting plasma glucose (higher than 7 mmol/L), or the oral glucose tolerance test (greater than 11.1 mmol/L after 2 hours), or the A1C (glycated hemoglobin) test (greater than 6.5% of haemoglobin glycated) (Viera, 2011). These are all indicators of persistently high blood glucose.

symptoms developing, and treating a predisease prevents the disease developing, it may not be the case that treating the predisease prevents symptoms developing. Even though addressing pathophysiology (e.g. glucose metabolism) may be the best way to treat symptomatic disease, in order to prevent symptomatic disease developing in non-diseased people it may be better to address something else entirely (e.g. obesity).

Given these problems, it may be better to understand preventive medicine as preventing the development of illness, rather than disease. It may be better to understand pathology and pathophysiology as causal factors for disease, rather than as the disease itself. Thus, addressing these causal factors may be seen as preventing disease rather than as treating disease. Understood like this, preventive medicine is no longer solely concerned with risk factors, but also with things traditionally understood as the disease itself. Furthermore, epidemiological investigation of probability of symptoms occurring is relevant to identifying explanations for symptoms, as well as targets for therapeutic and preventive intervention (Giroux 2015a, b). Thus, concepts of pathology, pathophysiology, dysfunction and risk all appear in both therapeutic and preventive medicine.

With these comments in mind, we might ask: what is the value in distinguishing preventive medicine from therapeutic medicine? What role should epidemiology play in defining pathology, pathophysiology and risk factors? Do causal factors have a different epistemic and/or ontological status to risk factors when it comes to defining disease and what are the actual consequences of such definitions for the practice of (preventive) medicine?

2.4.4   Overdiagnosis

Overdiagnosis, the diagnosis of disease that does not benefit the patient, has been recognized as a serious problem in medicine. Overdiagnosis is problematic in itself because it can harm patients who are labelled as 'diseased', and it is problematic because it will often lead to overtreatment which can also be harmful. One cause of overdiagnosis may lie in the interests and needs of large pharmaceutical companies, which profit from expanding the boundaries of disease to include patients previously considered well. But there are also conceptual drivers of overdiagnosis, which we want to focus on here.

One of the main mechanisms identified has been the use of young adult reference classes to define disease in elderly patients, thus pathologizing physiological states that were formally considered part of normal aging. One way to address this problem is to champion concepts of disease that insist on using age-adjusted reference classes, such as Boorse's BST. Some philosophers argue that this is not enough, and that in order to prevent overdiagnosis we must also pay close attention to the risk that a patient will go on to suffer negative consequences, i.e. symptoms, in addition to satisfying naturalistic criteria for disease (Walker and Rogers 2017; Rogers and Walker 2018). Whilst we commend this philosophical work, we suggest a different analysis of the conceptual drivers of overdiagnosis and recommend a different solution, based on historical analysis of some exemplars of overdiagnosis.

Take osteoporosis, a condition defined using bone mineral density measurements. According to the WHO, for a person to be diagnosed with osteoporosis they must have a low bone mineral density compared to the average for a young adult. As bone mineral density tends to fall in all people as they age, many medical researchers have expressed concerns that comparing elderly people to young people will make it seem as though many people have low bone mineral density, when they have normal bone mineral density for their age. This, they claim, inevitably leads to overdiagnosis. Furthermore, medical professionals also argue that low bone mineral density is not associated with an especially high risk of sustaining a fragility fracture and experiencing symptoms. Consequently, the WHO definition is said to promote overdiagnosis by paying insufficient attention to the risk of developing symptoms.

An analysis of the historical development of the definition about osteoporosis, however, reveals that medical professionals have come to define osteoporosis in terms of bone mineral density precisely because they wanted to predict fracture risk. Far from being ignored, risks of developing symptoms were central to this way of defining the disease. Furthermore, the decision to define the disease by comparing all patients to the young adult reference class was also motivated by a desire to assess the risk of fracture (for a full analysis see Binney 2022). It may be the case that current definitions of osteoporosis do not predict the risk of fracture sufficiently well, but adopting age-adjusted reference classes may not resolve the problem. It may lead to underdiagnosis, for example. In the light of the historical development of this medical field, advice that looks sensible at first glance may look much more problematic.

Another key example of overdiagnosis is thyroid cancer. Some of the overdiagnosis of thyroid cancer can be attributed to certain types of "follicular variant of the papillary thyroid carcinoma" (FVPTC), which often have an indolent behavior (Tallini et al. 2017; Xu and Ghossein 2018). Understanding the history of the classification of thyroid tumors can help explain how the conceptualization of FVPTC has driven overdiagnosis (see Binney, Chap. 7, this volume).

In general, we believe historical analysis is often key to understanding how overdiagnosis arose, and to explore whether particular conceptions of disease had a role to play in this development. Other instances of overdiagnosis should be assessed individually, however, and this opens up yet another area for inquiry.

2.4.5   The False Presumption that Patients with the Same Disease Are Homogeneous

We have encountered several problematic situations in which the assumption that patients with the same disease are highly homogeneous whilst being different to patients without that disease appears to be at the heart of the issue.

One situation in which we have encountered this problem is the evaluation of diagnostic tests. Traditional assumptions of diagnostic test evaluation include (1) that sensitivity and specificity (the usual indices used to evaluate tests) are intrinsic properties of the test, which are constants in any clinical context; and (2) that tests need to accurately distinguish disease from non-disease if they are to be useful. These assumptions only make sense if patients with the same disease are highly homogeneous. As they are not, we have found that this leads to incorrect beliefs about the accuracy and utility of medical tests in many areas of medicine.

We have also encountered this problem in debates about how to classify patients into groups with different diseases. Some doctors have expressed concerns that currently accepted diagnostic categories are not acceptable, as the patients it captures are too heterogeneous to allow general rules about treatment or for accurate prognostication. One response to this problem, seen especially in precision oncology, has been to subclassify patients into smaller groups, using additional information provided by genetic, molecular and immunohistochemical biomarkers. Another has been to say that patients need to be reclassified, such that patients previously understood to have the same disease should instead be understood to have different diseases, in the hope that these new, smaller disease categories will produce the necessary homogeneity for optimal medical practice. For example, type II diabetes researchers have argued that this condition is not a true specific disease entity, but is rather a collection of several different disease entities that need separating if progress in treatment and prognostication are to be made (Gale 2013; Philipson 2020). This solution to the problem of heterogeneity maintains the specific disease entity model.

A completely different way of addressing the problem of patient heterogeneity is to classify patients into multiple crosscutting categories, such that two patients will sometimes be seen as the same sort of patient and sometimes be seen as a different sort of patient. An example is provided by polycystic ovary syndrome (PCOS). Following rancorous debate about a single correct classification, researchers in this field have reportedly settled for a pluralistic system, which makes uses of several overlapping classifications. These overlapping classifications are produced by combining different "phenotypes" of PCOS, and researchers pay attention to different combinations of phenotypes depending on their particular clinical interest. So, if doctors and researchers are interested in using certain treatments for PCOS patients with infertility they may pay attention to one set of phenotypes, but if they are interested in patients at high risk of metabolic consequences of PCOS they may pay attention to another set (Azziz 2021; Sachdeva et al. 2019).

The suggestion that making use of subdividing and crosscutting classifications, including phenotypes, endotypes, regiotypes and theratypes, might be a useful way to cope with patient heterogeneity is also made for several other conditions[7] and may represent interesting philosophical innovations. Similar innovations to cope with the problem of patient heterogeneity include the use of "target conditions"

---

[7] For example: Ozdemir et al., 2018; Agache and Akdis, 2019; Battaglia et al., 2019; Mobasheri et al., 2019; Petrelli et al., 2021)

(e.g. infectious patients, current infection, past infection, etc.) tailored to the particular purpose of testing when evaluating medical tests (Bossuyt et al. 2003); and the fitting of patients into a multidimensional matrix, rather than classification into discrete categories or placing them on a continuum with one dimension, to express their disease status (Kanis et al. 2008; Levey et al. 2011). From a pragmatist perspective, such conceptual solutions arising from within medical practice itself are a great starting point for further philosophical inquiry.

2.4.6   Problems with the Notion of Health

Besides defining disease and specific disease categories, there is also much debate about how to understand 'health'. Especially in the domain of health policy, which aims to promote and protect health of individuals and the population at large, the question what exactly 'health' entails frequently pops up. In the Netherlands, the past decade has seen the emergence of a practice and a concept know as 'positive health'. This new concept, defining health as: "the ability to adapt and self-manage, in light of the physical, emotional and social challenges of life" (Huber et al. 2011) was coined to remedy the shortcomings of the well-known 1948 WHO definition of health, which was said to be outdated, contra-productive and possibly harmful. The WHO criterium of complete emotional, physical and social well-being was viewed as too demanding, leading to medicalization and difficult to use in studies on (public) health. Moreover, the increased prevalence of chronic diseases in our ageing societies was said to challenge our definition of health, raising the question whether it is possible to be healthy and have a disease at the same time. Positive health was thus explicitly presented and perceived as a conceptual resolution to problematic situations encountered in health practices. Besides the promise to be a better equipped definition to use for research purposes, positive health is also said to enable better conversation with patients, to empower them and to enhance cross-disciplinary cooperation in the social domain.

Interestingly, at the same time, the new concept also appears to *create* problematic situations. Kingma (2017), for example, has assessed positive health on both internal/conceptual as well as external/pragmatic level, and concludes that the new concept is problematic in both senses. It is unable to distinguish, on a theoretical level, between the normal and the pathological, and also does not solve the societal problems that it is supposed to solve (e.g., medicalization). Other scholars have also criticized the concept for various reasons. Yet the concept appears to be successful and popular in both medical practice and policy.

This makes Positive Health, including its historical genesis, a very suitable case-study for addressing the dynamics of problematic situations, the performativity of health concepts, and their embeddedness in institutional systems and power-relations. More in general, we can say that the ongoing debates on what exactly constitutes 'health', especially when related to public health and health policy, signal the presence of a problematic situation—or perhaps of multiple problematic situations (cf Haverkamp et al. 2018).

2.4.7 Institutional Designation of the Sick Role

Consider patients with conditions such as ME/CFS, chronic pain syndromes, and more recently, long-covid. These patients have—or claim they have—symptoms that invalidate them and make them suffer, but those symptoms cannot be objectified or cannot be (fully) explained by pathology. These and other cases of 'medically unexplained symptoms' (MUS) create problems with regard to diagnosis and treatment (see Sect. 2.4.1.), but also in relation to numerous social and institutional issues. Disease status gives patients access to healthcare and medical interventions and the reimbursement thereof by health insurance; it determines their right to receive particular social benefits and exempts them from certain social duties and moral accountabilities. Conditions whose status as 'real' disease is controversial, or patient's whose symptoms are not clearly caused by a real disease, miss out on the social and institutional benefits that this status provides. The disagreements about such cases between various groups of stakeholders (patients, doctors, insurance companies, employers, welfare organizations) constitute a problematic situation. In those cases, it is unclear whether some patients who actually deserve certain benefits are treated unjustly by the social and bureaucratic system that denies them those rights, or whether some people unjustly benefit from arrangements they do not deserve.

There is a clear conceptual component to this problem. According to medical historian Charles Rosenberg (2002) the concept of disease has social power and utility in medicine and our society. In particular the idea of a 'disease entity' has gained bureaucratic status, as it serves the administrative system of healthcare and related institutions. Classifying a condition as healthy or diseased can be seen as a decision that confers value within a certain institutional context, and we might pragmatically choose to define conditions as diseases in order to attain desired outcomes. The view of disease as a practical concept that has social and institutional value and implications is, however, controversial. Many have argued that these practical aspects should be considered separately from the question of what defines disease. It has even been argued that we do not need a clear definition of disease to make normative decisions on practical issues (Hesslow 1993), and that there is not necessarily a one-on-one relationship between the scientific concept of disease and the sociological concept of sickness. Nevertheless, the fact is that in our current society we do base many normative decisions on our health and disease definitions, and we do use these terms in a value-laden way, as 'thick concepts' (Keil and Stoecker, 2017; Haverkamp et al., 2018).

All of this raises the question to what extent we should allow for handling these kinds of social and normative issues by focusing on the disease status. If so, which disease concept(s) is/are suitable for that purpose, and if not which other concepts might be helpful?

2.5   Final Remarks

We have briefly outlined our theoretical starting points: pluralism, contextualism, the practical nature of concepts, the centrality of human suffering to medicine, and a pragmatist epistemological position between simplistic realism and silly relativism. We have also presented some preliminary explorations of problematic situations encountered in medical practice in relation to conceptualization of health and disease. The following chapters and commentaries will explore some of these problematic situations more in depth or address related issues concerning health and disease concepts. In the final chapter of this volume, we'll take stock of what we have learned from all these contributions, as well as from the discussions during the workshop, for the further development of a pragmatist approach to conceptualizing health and disease.

References

Agache, Ioana, and Cezmi A. Akdis. 2019. Precision medicine and phenotypes, endotypes, genotypes, regiotypes, and theratypes of allergic diseases. *The Journal of Clinical Investigation* 129 (4). American Society for Clinical Investigation: 1493–1503. https://doi.org/10.1172/JCI124611.

Azziz, Ricardo. 2021. How Polycystic Ovary syndrome came into its own. *F&S Science* 2 (1). Elsevier: 2–10. https://doi.org/10.1016/j.xfss.2020.12.007.

Battaglia, Manuela, Simi Ahmed, Mark S. Anderson, Mark A. Atkinson, Dorothy Becker, Polly J. Bingley, Emanuele Bosi, et al. 2019. Introducing the endotype concept to address the challenge of disease heterogeneity in Type 1 diabetes. *Diabetes Care* 43 (1): 5–12. https://doi.org/10.2337/dc19-0880.

Binney, Nicholas. 2022. Osteoporosis and risk of fracture: Reference class problems are Real. *Theoretical Medicine and Bioethics* 43 (5): 375–400. https://doi.org/10.1007/s11017-022-09590-3.

———. 2023. Ludwik Fleck's reasonable relativism about science. *Synthese 201* (2): 40. https://doi.org/10.1007/s11229-022-04018-w.

Bossuyt, Patrick M., Johannes B. Reitsma, David E. Bruns, Constantine A. Gatsonis, Paul P. Glasziou, Les M. Irwig, David Moher, Drummond Rennie, Henrica C.W. de Vet, and Jeroen G. Lijmer. 2003. The STARD statement for reporting studies of diagnostic accuracy: Explanation and elaboration. *Annals of Internal Medicine* 138 (1). American College of Physicians: W1–12. https://doi.org/10.7326/0003-4819-138-1-200301070-00012-w1.

Burke, Tom. 1994. *Dewey's new logic: A reply to Russell*. Chicago: University of Chicago Press. https://press.uchicago.edu/ucp/books/book/chicago/D/bo3618669.html.

Capps, John. 2023. The pragmatic theory of truth. In *The Stanford encyclopedia of philosophy*, Summer 2023, ed. Edward N. Zalta and Uri Nodelman. Metaphysics Research Lab, Stanford University. https://plato.stanford.edu/archives/sum2023/entriesruth-pragmatic/.

Cefalu, William T. 2016. "Prediabetes": Are there problems with this label? No, we need height-ened awareness of this condition! *Diabetes Care* 39 (8): 1472–1477. https://doi.org/10.2337/dc16-1143.

Chang, Hasok. 2022. *Realism for realistic people: A New pragmatist philosophy of science.* Cambridge: Cambridge University Press. https://doi.org/10.1017/9781108635738.

Dewey, J. 1938. *Logic: The theory of inquiry.* Oxford: Holt.

Dewey, John, Addison Webster Moore, Harold Chapman Brown, George Herbert Mead, Boyd Henry Bode, Henry Waldgrave Stuart, James Hayden Tufts, and Horace Meyer Kallen. 1917. The need for a recovery of philosophy. In *Creative intelligence: Essays in the pragmatic attitude*, vol. 1, 3–69. New York: Henry Holt and Company. https://ia801604.us.archive.org/8/items/creativeintellig00dewe/creativeintellig00dewe.pdf

Dubroff, Jacob G., and Ilya M. Nasrallah. 2015. Will PET Amyloid imaging lead to overdiagnosis of Alzheimer Dementia? *Academic Radiology* 22 (8): 988–994. https://doi.org/10.1016/j.acra.2015.02.005.

Fleck, Ludwik. 1979 [1935]. *Genesis and development of a scientific fact.* University of Chicago Press.

Gale, Edwin A.M. 2013. Is Type 2 diabetes a category error? *The Lancet* 381 (9881). Elsevier: 1956–1957. https://doi.org/10.1016/S0140-6736(12)62207-7.

Giroux, Élodie. 2015a. Risk Factor and Causality in Epidemiology. In *Classification, disease and evidence: New essays in the philosophy of medicine*, History, philosophy and theory of the life sciences, ed. Philippe Huneman, Gérard Lambert, and Marc Silberstein, 179–192. Dordrecht: Springer. https://doi.org/10.1007/978-94-017-8887-8_9.

———. 2015b. Epidemiology and the bio-statistical theory of disease: A challenging perspective. *Theoretical Medicine and Bioethics* 36 (3): 175–195. https://doi.org/10.1007/s11017-015-9327-7.

Haalboom, Floor. 2023. Sugar-Sick yet healthy: changing concepts of disease in the Dutch Diabetics Association (1945–1970). *Social History of Medicine*, September, hkac073. https://doi.org/10.1093/shm/hkac073.

Haverkamp, Beatrijs, Bernice Bovenkerk, and Marcel F. Verweij. 2018. A practice-oriented review of health concepts. *The Journal of Medicine and Philosophy: A Forum for Bioethics and Philosophy of Medicine* 43 (4): 381–401. https://doi.org/10.1093/jmp/jhy011.

Hesslow, Germund. 1993. Do we need a concept of disease? *Theoretical Medicine* 14 (1): 1–14. https://doi.org/10.1007/BF00993984.

Huber, Machteld, J. André Knottnerus, Lawrence Green, Henriëtte van der Horst, Alejandro R. Jadad, Daan Kromhout, Brian Leonard, et al. 2011. How should we define health? *BMJ* 343 (July). British Medical Journal Publishing Group: d4163. https://doi.org/10.1136/bmj.d4163.

James, William. 2014. *Pragmatism and the meaning of truth.* CreateSpace Independent Publishing Platform.

Kanis, J.A., E.V. McCloskey, H. Johansson, O. Strom, F. Borgstrom, A. Oden, and National Osteoporosis Guideline Group. 2008. Case finding for the management of osteoporosis with FRAX®—Assessment and intervention thresholds for the UK. *Osteoporosis International* 19 (10): 1395–1408. https://doi.org/10.1007/s00198-008-0712-1.

Keil, Geert, and Ralf Stoecker. 2017. Disease as a vague and thick cluster concept. In *Vagueness in Psychiatry*, ed. Geert Keil, Lara Keuck, and Rico Hauswald, 46–74. Oxford: Oxford University Press.

Khunti, Kamlesh, and Melanie J. Davies. 2018. Clinical Inertia versus overtreatment in glycaemic management. *The Lancet Diabetes & Endocrinology* 6 (4). Elsevier: 266–268. https://doi.org/10.1016/S2213-8587(17)30339-X.

Kingma, Elisabeth. 2017. Kritische Vragen Bij Positieve Gezondheid. *Tijdschrift Voor Gezondheidszorg En Ethiek* 3: 81–83.

Levey, Andrew S., Paul E. de Jong, Josef Coresh, Meguid El Nahas, Brad C. Astor, Kunihiro Matsushita, Ron T. Gansevoort, Bertram L. Kasiske, and Kai-Uwe Eckardt. 2011. The definition, classification, and prognosis of chronic kidney disease: A KDIGO controversies conference report. *Kidney International* 80 (1): 17–28. https://doi.org/10.1038/ki.2010.483.

Linden, Rik van der, Timo Bolt, and Mario Veen. 2022. "If it can't be coded, it doesn't exist". A historical-philosophical analysis of the new ICD-11 classification of chronic pain. *Studies in History and Philosophy of Science* 94 (August): 121–132. https://doi.org/10.1016/j.shpsa.2022.06.003.

Mobasheri, Ali, Simo Saarakkala, Mikko Finnilä, Morten A. Karsdal, Anne-Christine Bay-Jensen, and Willem Evert van Spil. 2019. Recent advances in understanding the phenotypes of Osteoarthritis. *F1000Research* 8 (December): F1000 Faculty Rev-2091. https://doi.org/10.12688/f1000research.20575.1.

Montine, Thomas J., Creighton H. Phelps, Thomas G. Beach, Eileen H. Bigio, Nigel J. Cairns, Dennis W. Dickson, Charles Duyckaerts, et al. 2012. National Institute on Aging-Alzheimers Association Guidelines for the neuropathologic assessment of Alzheimer's disease: A practical approach. *Acta Neuropathologica* 123 (1): 1–11. https://doi.org/10.1007/s00401-011-0910-3.

Nugraha, Boya, Christoph Gutenbrunner, Antonia Barke, Matthias Karst, Jörg Schiller, Peter Schäfer, Silke Falter, et al. 2019. The IASP classification of chronic pain for ICD-11: Functioning properties of chronic pain. *PAIN* 160 (1): 88–94. https://doi.org/10.1097/j.pain.0000000000001433.

O'Leary, Diane. 2020. A concerning display of medical indifference: Reply to "Chronic fatigue syndrome and an illness-focused approach to care: Controversy, morality and paradox". *Medical Humanities* 46 (4). Institute of Medical Ethics: e4–e4. https://doi.org/10.1136/medhum-2019-011743.

Ozdemir, Cevdet, Umut Can Kucuksezer, Mubeccel Akdis, and Cezmi A. Akdis. 2018. The concepts of Asthma endotypes and phenotypes to guide current and novel treatment strategies. *Expert Review of Respiratory Medicine* 12 (9). Taylor & Francis: 733–743. https://doi.org/10.1080/17476348.2018.1505507.

Petersen, Ronald C. 2018. How early can we diagnose Alzheimer disease (and is it sufficient)?: The 2017 Wartenberg lecture. *Neurology* 91 (9). Wolters Kluwer Health, Inc. on behalf of the American Academy of Neurology: 395–402. https://doi.org/10.1212/WNL.0000000000006088.

Petrelli, Alessandra, Anna Giovenzana, Vittoria Insalaco, Brett E. Phillips, Massimo Pietropaolo, and Nick Giannoukakis. 2021. Autoimmune inflammation and insulin resistance: Hallmarks so far and yet so close to explain diabetes endotypes. *Current Diabetes Reports* 21 (12): 54. https://doi.org/10.1007/s11892-021-01430-3.

Philipson, Louis H. 2020. Harnessing heterogeneity in Type 2 diabetes mellitus. *Nature Reviews Endocrinology* 16 (2). Nature Publishing Group: 79–80. https://doi.org/10.1038/s41574-019-0308-1.

Raffaeli, William, and Elisa Arnaudo. 2017. Pain as a disease: An overview. *Journal of Pain Research* 10 (August): 2003–2008. https://doi.org/10.2147/JPR.S138864.

Rogers, Wendy A., and Mary J. Walker. 2018. Précising definitions as a way to Combat overdiagnosis. *Journal of Evaluation in Clinical Practice* 24 (5): 1019–1025. https://doi.org/10.1111/jep.12909.

Rosenberg, Charles E. 2002. The Tyranny of diagnosis: Specific entities and individual experience. *The Milbank Quarterly* 80 (2): 237–260. https://doi.org/10.1111/1468-0009.t01-1-00003.

Russell, Bertrand. 1967. *A history of western philosophy*. New York: Simon & Schuster/Touchstone.

Sachdeva, Garima, Shalini Gainder, Vanita Suri, Naresh Sachdeva, and Seema Chopra. 2019. Comparison of the different PCOS phenotypes based on clinical metabolic, and hormonal profile, and their response to Clomiphene. *Indian Journal of Endocrinology and Metabolism* 23 (3): 326–331. https://doi.org/10.4103/ijem.IJEM_30_19.

Schermer, Maartje H.N. 2023. Preclinical disease or risk factor? Alzheimer's disease as a case study of changing conceptualizations of disease. *The Journal of Medicine and Philosophy* 48 (4): 322–334. https://doi.org/10.1093/jmp/jhad009.

Schermer, Maartje H.N., and Edo Richard. 2019. On the reconceptualization of Alzheimer's disease. *Bioethics* 33 (1): 138–145. https://doi.org/10.1111/bioe.12516.

Sharpe, Michael, and Monica Greco. 2019. Chronic fatigue syndrome and an illness-focused approach to care: Controversy, morality and paradox. *Medical Humanities* 45 (2). Institute of Medical Ethics: 183–187. https://doi.org/10.1136/medhum-2018-011598.

Sleath, Jonathan D. 2015. In pursuit of Normoglycaemia: The overtreatment of Type 2 diabetes in general practice. *British Journal of General Practice* 65 (636): 334–335. https://doi.org/10.3399/bjgp15X685525.

Tallini, Giovanni, R. Michael Tuttle, and Ronald A. Ghossein. 2017. The history of the follicular variant of papillary thyroid carcinoma. *The Journal of Clinical Endocrinology & Metabolism* 102 (1): 15–22. https://doi.org/10.1210/jc.2016-2976.

Tesio, Luigi, and Marco Buzzoni. 2021. The illness-disease dichotomy and the biological-clinical splitting of medicine. *Medical Humanities* 47 (4). Institute of Medical Ethics: 507–512. https://doi.org/10.1136/medhum-2020-011873.

Thayer, Horace Standish. 1970. *Pragmatism: The classic writings*. New York: New American Library.

Treede, Rolf-Detlef, Winfried Rief, Antonia Barke, Qasim Aziz, Michael I. Bennett, Rafael Benoliel, Milton Cohen, et al. 2019. Chronic pain as a symptom or a disease: The IASP classification of chronic pain for the international classification of diseases (ICD-11). *PAIN* 160 (1): 19–27. https://doi.org/10.1097/j.pain.0000000000001384.

van der Linden, Rik and Maartje Schermer. 2022. Health and disease as practical concepts: exploring function in context-specific definitions. *Medicine, Healthcare and Philosophy* 25 (1): 131–140. https://doi.org/10.1007/s11019-021-10058-9.

Viera, Anthony J. 2011. Predisease: When does it make sense? *Epidemiologic Reviews* 33 (1): 122–134. https://doi.org/10.1093/epirev/mxr002.

Walker, Mary Jean, and Wendy Rogers. 2017. Defining disease in the context of overdiagnosis. *Medicine, Health Care and Philosophy* 20 (2): 269–280. https://doi.org/10.1007/s11019-016-9748-8.

Wilshire, Carolyn, and Tony Ward. 2020. Conceptualising illness and disease: Reflections on Sharpe and Greco (2019). *Medical Humanities* 46 (4). Institute of Medical Ethics: 532–536. https://doi.org/10.1136/medhum-2019-011756.

Xu, Bin, and Ronald Ghossein. 2018. Evolution of the histologic classification of thyroid neoplasms and its impact on clinical management. *European Journal of Surgical Oncology: The Journal of the European Society of Surgical Oncology and the British Association of Surgical Oncology* 44 (3): 338–347. https://doi.org/10.1016/j.ejso.2017.05.002.

Yudkin, John S. 2016. "Prediabetes": Are there problems with this label? Yes, the label creates further problems! *Diabetes Care* 39 (8): 1468–1471. https://doi.org/10.2337/dc15-2113.

Chapter 3
Nature and Culture in Health and Disease: Historical Strategies in Medical Diagnostics for Navigating Between Critical Dichotomies

Heiner Fangerau

3.1 Introduction: Diagnostic Essentialism and Nominalism

In 1963, the *Archives of Internal Medicine* published a three-part article by Ralph Engle and B. J. Davis on the present, past and future of medical diagnosis. The impetus for this series was their involvement in an attempt to use "modern electronic computers" to "aid physicians in diagnosis" (Engle and Davis 1963). In their attempt to cluster and logically organise diagnostic data for computer-calculated diagnosis (Greene and Lea 2019), they saw a need to "define terms more precisely" and to review some fundamental historical and philosophical questions about the status and role of the diagnosis in medicine.

This urge to grapple with the concept of diagnosis seems to resurface from time to time, whenever new approaches or previously unknown or unused instruments enter the medical realm (see Boenink and van der Molen, Chap. 11, this volume). In the long nineteenth century, similar discussions were nurtured by newly established nosologies, specialization and the augmentation of the physicians' traditional five senses through an increasing number of technical devices (Barker 1916) as well as by the developments of microbiology, statistics and genetics (Allbutt 1896). Today, the discourse about the use of 'artificial intelligence' in medicine, a kind of sequel to the proposal for computerised diagnosis in the 1960s, has reignited the debate. The diagnostic process is described as "sophisticated", "highly complex" and error-prone. The conceptualisation of the diagnostic process is still a challenge (Mirbabaie et al. 2021: 694 f.).

H. Fangerau (✉)
Department for the History, Philosophy and Ethics of Medicine, Medical Faculty, Heinrich Heine University Duesseldorf, Düsseldorf, Germany
e-mail: heiner.fangerau@hhu.de

© The Author(s) 2024
M. Schermer, N. Binney (eds.), *A Pragmatic Approach to Conceptualization of Health and Disease*, Philosophy and Medicine 151,
https://doi.org/10.1007/978-3-031-62241-0_3

Thus, the definition of diagnosis has remained a topic of ongoing debate. Engle's and Davis's expansive statement that "diagnosis encompasses the entire art and science of medicine" (Engle and Davis 1963: 513) appears to be as accurate as it is vague. Above all, it highlights the absence of a unified concept. Common parlance embraces two understandings of diagnosis. Both are outcomes of medical practice and logic: one refers to the present condition of a particular individual patient; the other refers consistently to a universally regarded medical condition that transcends the immediate moment (Fangerau 2021; Nicolson 1993; Galdston 1941). The link between these two meanings is the process of recognizing, interpreting and classifying the individual patient-centred signs of a general, named disease. In this capacity, the process encompasses social, moral and bureaucratic elements (Rosenberg 2002).

The modern understanding of diagnosis has its origins in nineteenth-century medicine, in the course of which medical semiotics—as a medical activity—was gradually transformed into diagnosis, and the recording of signs became increasingly technical. Phenomena that could be perceived by the senses were translated into objective, comparable numerical values and graphical curves by instruments, and signs of disease that were not accessible to the senses were detected and recorded with the help of technical devices (Martin and Fangerau 2007, 2013). This established a way of thinking in terms of 'clinical pictures' and disease classifications. With these classifications in mind, the physician was faced with the task of reconciling the signs of an individual patient's illness with a generalised ordering of signs. The systematically ordered signs, in turn, had to be incorporated into a nosology that was valid in a given context (Wieland 1975; Eich 1986; Eckart 1998; Rosenberg 2002). The starting point for this approach was the move towards systematic clinical observation of numerous patients in large hospitals and the correlation of diagnostic findings with post-mortem examinations (Foucault 1973; Risse 1987). External symptoms and organ changes were systematically correlated. In the context of pathological anatomy, this approach became the central point of reference for diagnostics which, in the next step, consisted of finding the pathologies of the dead in the signs of the living.[1] A particular constellation of signs had to be clearly and convincingly associated with a clinical picture. Thus, the focus on a theory of signs—semiotics—was increasingly replaced by differential constellations of diagnostic findings, pathophysiological causal chains and numerical approaches (Hess 1993). In practice, this replacement was accompanied by a drastic increase in the number of new technical diagnostic procedures, which gained popularity as 'physical diagnostics' (Eckart 1996). Laennec's stethoscope for indirect auscultation, Piorry's plessimeter for indirect percussion and later the micro- and endoscopes for expanding the visual space were to become paradigmatic icons of this development.

Much thought has been given in medicine, especially in the nineteenth and twentieth centuries, to the relationship between disease and diagnosis. The question

---

[1] Günter B. Risse called this transition "a shift in medical epistemology" (Risse 1987).

was whether its goal was to distinguish one disease from the other or to come to individual diagnoses considering the patient's complex context. Galdston called the later process "the summation of all that deviates from the normal, sick individual of definite psychological and physical endowments living under particular circumstances in a given milieu" (Galdston 1941: 373). Central to this aspect was the interpretation of symptoms either as pathognomonic indications of general disease entities or as specific phenomena that indicated individually represented illnesses.

These signs of illness assume a strange position in diagnostics, understood as the practice of finding a diagnosis, with interpretation oscillating between the biological and the culture-bound. It is precisely this tension that this article addresses. The starting point is the classical opposition between essentialism and nominalism. Since ancient philosophy, so-called nominalist positions, which oppose the idea of universal entities in the world with the idea that only human interpretation and naming allow universals to become perceptible things, have been opposed to materialist, realist or (since Popper) essentialist interpretations of disease, its signs and diagnosis (Scadding 1996).

This article asks how the concept of diagnosis in the nineteenth and early twentieth centuries dealt with biological, naturalistic conditions and constructivist, culture-bound interpretations of signs and symptoms, and how, from a historical perspective, the convergence of 'nature' and 'culture' in diagnosis can be seen as a process.[2] The initial focus will be on the discussions surrounding the production of diagnostic signs, their difference from symptoms and their widespread recognition as an inherent part of diagnosis during the nineteenth century. Symptoms and signs have often been used as synonyms, but for physicians in the nineteenth century they referred to different levels of the diagnostic process. This era witnessed the emergence of novel disease concepts, a heightened mechanization of creating diagnostic indicators, the proliferation of research resources, the expansion of hospital frameworks, and the growth of medical professionalism and specialization, all of which contributed to a fresh comprehension of diagnosis. The paper then presents two approaches from the 1920s that attempted to reconcile nominalism and essentialism. These attempts came at a time when, after the boom of a reductionist and materialist medicine that saw itself as a natural science, voices were being raised that placed more emphasis on holism (including functional and behavioural disease concepts) and reference disciplines beyond the natural sciences (including history) as sources of knowledge for medicine (Warner 2013). The emphasis is on the approaches of physicians such as Richard Koch, who, drawing on Hans Vaihinger's philosophy of As-If, attempted to give the idea of diagnosis an intentional and relational orientation that saw nature and culture in diagnosis not as opposites but as interrelated elements.

---

[2]This paper is a further development of ideas that have previously been published in German (Fangerau 2021, 2023; Fangerau and Martin 2015; Martin and Fangerau 2021). Especially the theoretical elements have been extended.

3.2   Diagnosis and Diagnostics Since the Nineteenth Century

Until the twentieth century, efforts to define diagnosis were primarily concerned with characterising it as an attempt to recognise nosological entities that appeared in a patient (Martin and Fangerau 2021). Diseases were understood as clearly classifiable, ontologically graspable deviations from normal physiology that universally took possession of the body as natural entities. For Karl Friedrich Burdach (1776–1847) and Johann C. F. Leune (1757–1825) diagnostics at the beginning of the nineteenth century was a "special semiotics" that was to be understood as the "comparison and compilation of various symptoms into a whole" in order to recognise the "genus and nature of the disease present" (Burdach and Leune 1803: 93 f.). A little later, Jacob Friedrich Sebastian (1771–1840) defined diagnosis as the "knowledge of the present disease" and described diagnostics as "the art (...) and science of recognising the present disease" together with seeing its "peculiarity and variety" (Sebastian 1819: 9). In 1883, Hermann Baas (1838–1909) summarised this view, which was still held some decades later, by stating that "Diagnostics is that branch of medical science and art which teaches to recognise and separate the disease individuals or disease pictures established in pathology according to the respective state of science. These are always considered as a whole, consisting of a sequence of phenomena which, although always the same in general, may vary in detail from case to case" (Baas 1877: 1).

The protagonists were thus well aware of the precariousness of a fluid, ever-changing classification of diseases, which made it difficult to name a defined disease that transcended time and individuals. For this reason, in the first third of the nineteenth century, efforts were made to differentiate symptoms and signs from each other in order to distinguish pathophysiological phenomena and physical symptoms from their interpretation as signs of a specific disease (Eckart 1998; King 1982). At the end of the century the English physician Thomas Clifford Allbutt called all these attempts "otiose". For him, they only added to the confusion, but even he saw the need to make a distinction by stating: "everything that befalls a patient is a 'symptom,' and his symptoms are the sign of his malady" (Allbutt 1896: xxxii). This definition was close to those given 60 years earlier. For example, according to Friedrich Ludwig Meissner's (1796–1860) *Encyclopädie der medicinischen Wissenschaften* of 1833,[3] symptoms were understood as general manifestations of disease, as "a change perceptible to the senses, which occurs in the physical state of an organ or its activity, and which is associated with the presence of a disease" (Meissner 1830–1834, vol. 11: 435). The senses referred to include the patient's senses as well as the examining physician's senses. The sign went beyond this appearance: a symptom charged with meaning was considered a sign. According to Meissner's *Encyclopädie*, the symptom based on (pathological) bodily functions or structures was a "simple sensation which only becomes a sign

---

[3] This encyclopaedia was in turn modelled on the French *Dictionnaire de médecine* by Adelon et al. (1821–1828).

through a special operation of the mind. . ." (Meissner 1830–1834, vol. 13: 199). Here, meaningful transformations of symptoms to signs relied on medical knowledge. In other words, the individual symptom only became a sign when it had been judged by the medical observer in terms of its significance for a disease and in terms of its value for distinguishing diseases. The symptom received its 'meaning' and its evidence for something only through its association with a nosologically predefined disease.[4]

By the mid-nineteenth century diagnosticians had developed a whole categorical scheme of sign types, attempting to distinguish natural signs from the artificial. Natural signs were those that could be detected by the five senses. Artificial signs were those that required technical or chemical intervention to produce (Sprengel 1801; see also Sebastian 1819). At the beginning of the century, for example, the physician, botanist and medical historian Kurt Sprengel (1766–1833) placed higher value on "natural signs" than on "artificial signs". Sprengel also distinguished between signs that could be perceived by everyone, signs that required "artful" examination, and signs that could be perceived "only by the sick person". He considered the latter to be the least reliable (Sprengel 1801: 4ff., 19ff.).[5] The reliability of a sign should be increased by intellectual connections and logical argumentation, for example, if the reason for the meaning of a sign could be stated or deduced from experience (Sprengel 1801). However, the basis of the sign should be the observation of nature. Consequently, he argued, if a physician was "a follower of this or that school, who distinguished himself by pointed theories and philosophical theses, he deserves far less credibility than if he did not belong to any school at all, but (. . .) observed nature itself, uninfluenced by hypotheses" (Sprengel 1801: 10).

At the beginning of the nineteenth century, the sense of sight was considered superior to the other senses for detecting natural signs (Wichmann 1794). In contrast to the purely theoretical observations in the "nosological tables" of "philosophical" physicians (Wichmann 1794: 74), the German physician and author Johann Ernst Wichmann (1740–1802), for example, praised the natural authenticity of sight in the examination of corpses as a shortcut in investigating the cause of a disease (Wichmann 1794). Overall, he emphasised the value of observational authenticity in the diagnostic assessment of signs. The descriptions of others could not be relied upon, and indirect views through illustrations could at best be regarded as

---

[4] This differentiation, which had already become established around 1830, seems to have become blurred again around 1900. In Eulenburg's *Realencyclopädie der gesammten Heilkunde*, for example, symptoms and signs of disease are used synonymously. At the same time, however, in the process of diagnostics, the attribution of meaning is regarded as inherent to the concept of a symptom, for example when the author of the lemma 'symptom' notes: "The diagnosis is therefore always a conclusion based on the consideration of all the individual symptoms", (Eulenburg 1900: 623).

[5] With regard to the difference between pathology and semiotics Sprengel stated that pathology was proceeding from the causes to the effects whereas semiotics started with sensory consequences from which the causes are inferred. The signs were to be derived from these sensory consequences.

substitutes, although he admitted that engravings were "perhaps the only and best way" of representing symptoms and signs (Wichmann 1794: 35, 37).

The Heidelberg professor Jacob Friedrich Christian Sebastian (1771–1840), for his part, warned of the danger of deceptive, arbitrary signs that were not always necessarily related to the disease being diagnosed (Sebastian 1819). In such cases, artificial signs could be used to overcome deceptive signs that might occur, for example, in simulants.[6] Thus, in his *Grundriss der allgemeinen pathologischen Zeichenlehre* of 1819 (the same year as Laennec's *Traité de l'auscultation mediate* was published), artificial "indirectly sensual" signs were given greater importance than in the overviews of Wichmann and Sprengel. In addition to the chemical examination of urine, already discussed by Sprengel, as an example of useful artificial signs, he also included those that went beyond the limits of the natural senses. He included "skilful hand movements"[7] and signs obtained with the aid of instruments such as probes, catheters or mirrors (Sebastian 1819: 11 f.). In his opinion, the repeated reproducibility of a sign, its production or perception by "several senses, sometimes near, sometimes at a distance", and the comparison as well as the experience of the physician increased the significance of diagnostic features (Sebastian 1819: 21).

Multimodality then gained in importance. Shortly after their initiation and 'staging', percussion and auscultation as artificial signs had already found their way into the presentation of diagnostics in a wide variety of forms, where they remained important producers of evidence well into the twentieth century (Martin and Fangerau 2011). In the middle of the century, the physician Adolf Moser (1810–1856) used this as a basis for developing a wide variety of signs. He distinguished between subjective signs, which could only be perceived by the patient, and objective signs, which could be determined by the doctor, as well as vital and physical signs, natural, artificial, arbitrary, material, functional, visible, rational, absolute, unchangeable, relative, accidental, general, local, true, sufficient, certain, false, insufficient, uncertain and deceptive signs. Moser also ascribed greater significance and reliability to natural signs than to those produced by reflection. At the same time, however, he considered the most reliable signs to be those that could be measured and counted, or that could be obtained by technical means that extended the senses (Moser 1845). For him, measuring and counting were not so much mental abstractions from nature as the highest expression of grasping natural phenomena and their laws.

Moser also explicitly cited the five senses as criteria for distinguishing symptoms of disease. Unlike Sprengel, however, Moser did not see artificiality and the sensual claim to authenticity and truth as contradictory. On the contrary, he understood the technical collection of signs on the one hand as a legitimate extension of the senses

---

[6] In many subsequent works, simulation, i.e. evidence feigned by the patient and its recognition, is a constantly recurring theme (see e.g. Schmalz 1825, XIV).

[7] This could mean manipulation (e.g. to test ranges of motion) and provocative hand movements such as are used to test ocular response or palpation, as used to check the position, hardness, shape and size of abdominal organs.

and on the other hand as an inherently evidential and objective method of sign production, partially detached from the examiner. Objectivity became a scientific value when not only the individual-specific, but also the regular pathological signs of a disease were to be collected independently of the examiner (Moser 1845; see also Piorry 1846). He commented: "While the results of more recent research have gained in definiteness, certainty and accuracy (...) the ancients were more experienced in simple sensory perceptions, and understood how to use for practical medicine the finer, as it were lively nuances in the observations made by means of the sense organs. (...) Newer medicine wants first to present these perceptions objectively, then to enliven them by thinking according to theoretical laws, in order to arrive at a secure basis for action" (Moser 1845: V).

Moser thus praised the use of new tools to technically enhance the perceptible (Moser 1845) and immediately presented a three-page catalogue of equipment ranging from stethoscopes and thermometers to scales and chemicals that, in his opinion, should be part of the diagnostic inventory of, say, a hospital (see similarly Piorry 1846). In the final analysis, however, he still saw the experienced physician himself as the producer of objective signs, if he had sharpened his senses "through practice, through proper training" as opposed to the subjective signs reported by the patient (Moser 1845: 5). It was therefore up to the physician to know the value of different signs, which Moser divided into "true, sufficient, certain" signs on one hand, and "false, insufficient, uncertain, deceptive signs" on the other (Moser 1845: 9). Even the physical signs which, in Moser's eyes, were more certain and definite because they could be objectively recorded, could deceive so they, too, had to be viewed critically and interpreted in context by the expert (Moser 1845).

A little later, the pathological anatomist Carl Ernst Bock (1809–1874) stated rather apodictically in his 1853 textbook on diagnostics: "Only the objective symptoms, which can be perceived mainly by the so-called physical diagnostics, by inspection, palpation, measurement, percussion and auscultation, by chemical and microscopic examinations, have a diagnostic value for the physician" (Bock 1853: 6). In line with Bock's view, quantitative methods of measurement came increasingly to be introduced into diagnostics, in addition to inspection, in order to meet the demand for indirect but reliable signs. The established 'clinical chemistry' had at its disposal a special arsenal of numerous apparatuses and methods (polarimeters, fermentation and detection tests, etc.) by which the individual components of blood and urine in particular could be converted into numerical signs, i.e., quantified, for diagnostic purposes (Martin and Fangerau 2009). However, these quantifications were by no means diagnostic in themselves. A diversion via the definition of 'normal values' was required to provide diagnostic evidence (Büttner 1997). The fundamental requirement was to replace the traditional dichotomy between disease and health. Disease could no longer be understood as a state or entity, but as a constantly changing process that deviated from a norm. This processuality was located between the poles of normal and pathological, which in turn led to the dual problem of having to accept continuity between these two states, while at the same time requiring fixed normative values in order to be able to classify measurements in any meaningful way (cf. Canguilhem 1977; Link 1997; Grmek 1964).

The principle of continuity between the normal and the abnormal was based on the work of Broussais and Comte.[8] François Joseph Victor Broussais (1772–1838) denied that disease was a special ontic quality, rejecting an ontological opposition between disease and health. There was no fundamental difference between the two conditions, only a quantitative one. Jakob Henle (1809–1885), on the other hand, defined disease in his *Rationelle Pathologie* as a "deviation from a normal, typical, i.e. healthy, process of life" (Henle 1846: 90).

After the turn of the century, this meant, for example, for the practitioner Warren T. Vaughan in 1922, that diagnostic work could be compared to detective work. He was convinced that the amount of data collected and the experience based on it were crucial to the outcome. The concept of experience meant for him, as it did for Ludwig Krehl (Krehl 1903), to personally observe, undergo and comprehensively analyse as many clinical cases as possible (in one's career). He saw the cognitive interest of diagnostics in the analysis of function rather than structure. The aim was to have functional tests that would warn of the earliest deviation from normal (Vaughan 1922).

The purpose behind all this contemplation was practical application. Such a focus on practice is related to the legitimacy of medical action (Rothschuh 1978). From the patient's point of view, almost all medical interventions are unpleasant: everything scratches, bites, burns or stings, so the goal of diagnosis first opens up an interpersonal space for intrusive actions that are otherwise forbidden, in order to justify therapeutic action on the basis of a diagnosis in the next step (Fangerau 2017). By the late eighteenth century, Wichmann was convinced that only through accurately distinguishing the disease afflicting a patient could the appropriate treatment be administered (Wichmann 1794). While this emphasis aligns with the notion that distinct disease entities could be differentiated, a century later, the American physician John Herr Musser (1856–1912) adopted a more meticulous approach, concentrating on the unique manifestation of diseases in individual patients. In his textbook on medical diagnosis, he conveyed that the objective of diagnosis was not to give a name to a disease, but to treat it (Musser 1894). As Ludolph Krehl put it, the ultimate goal of diagnosis is the recognition of how a disease condition affects the patient in his or her unique personal circumstances (Krehl 1903). Moreover, he later asserted that a diagnosis is essential for fostering mutual understanding and for the discipline of thought (Krehl 1931).

3.3 Theories of Diagnosis in the 1920s: Crookshank, Koch and Vaihinger's 'As-If'

Such a pragmatist approach to diagnosis became the basso continuo in debates about the philosophy of diagnosis in the 1920s. An example is offered by the British epidemiologist and psychologist Francis Graham Crookshank (1873–1933). At the

---

[8]For detailed information on the Broussais–Comte principle, see Link (1997).

prestigious Bradshaw Lecture, which had been organised by the Royal College of Physicians since 1881, he gave a lecture dedicated to the "Theory of Diagnosis" (Crookshank 1926a, b). In this lecture, he described diagnosis as the process of forming a judgement about the condition of the sick person, which, on the one hand, guides the task of healing and, on the other, involves observation, interpretation of what is observed, and epitomisation of the interpretation (Crookshank 1926a).

However, his primary argument was to critique the prevailing physical realism that had dominated medicine since the mid-nineteenth century. He aimed to challenge this perspective by proposing a philosophy of medicine structured as a conceptual system rather than a perceptual experience. In his historically inspired approach, he contrasted a naturalistic, realist-oriented concept of disease, which included the idea of diagnosis as a practice of differentiating ontological entities of disease, with the nominalist view that disease and diagnosis were interpretative constructs justified by their utility and acceptability rather than by a universal, naturalistic perception of reality. Therefore, his motivation to discuss diagnosis was also rooted in the desire to critique an excessively reductionist approach to medicine. Simultaneously, he was concerned with the flaws of an excessively strong nominalism. He regarded a nominalist "tyranny of names" as "no less pernicious than (. . .) the modern form of scholastic realism. Diagnosis (. . .) too often means in practice the formal and unctuous pronunciation of a name that is deemed appropriate and absolves from the necessity of further investigation" (Crookshank 1923a).

At the end of his Bradshaw lecture, he suggested that the contradiction between the realist and nominalist standpoints should be overcome by what he called conceptualism. By conceptualism, Crookshank meant the idea "(. . .) that the universal is not existent otherwise than mentally and as a terminus, or predicable, a mental concept signifying univocally several singulars; and that the concept is not so much a thing as an act, having no reality besides the act and the singulars of which it is composed, while the act of abstraction does not presuppose any activity of the understanding or will, but is a spontaneous secondary process by which perceptions are, as it were, stored as soon as several similar representations are present, though in a fading or evanescent state" (Crookshank 1926b: 998). This conceptualism, Crookshank continues, has always shown an impulse towards direct observation, a distrust of abstraction and a reluctance to hypostatize abstractions. It thus occupies an intermediate position between "nominalism and realism". Ultimately, from the perspective of conceptualism, the "best diagnosis" is the one that is most likely to satisfy intellectual and affective needs and enables the doctor to do the right thing, subjecting not only the patient but also his fellow men and circumstances to his will.

Shortly before Crookshank's Bradshaw lecture, in 1917, the German Richard Koch had published a book-length account on medical diagnosis, which was reissued in a second, expanded, edition in 1920. Basically, Koch interpreted illness as a "thought construct" ("Denkgebilde") (Koch 1924: 51) and explicitly dissociated diagnosis from ontological or natural concepts of disease. Koch stated that while he believed that real causes of disease as well as real "pathological processes in the body and pathological manifestations on the body" existed, real kinds of disease did not, in fact, exist (Koch 1920: 42). At the same time, Koch shifted the focus from the concept of disease to the suffering patient's individual condition (Töpfer and Wiesing 2005a, b): "What is diagnosed is not a concept of disease, but the condition of the individual" (Koch 1920: 70). In normative terms, he stated that the diagnosis must be "completely geared to therapy", otherwise it would be "something outside medicine" (Koch 1920: 149). Unlike natural science, which seeks "infinite knowledge", medicine needs only "finite knowledge" to be able to act. In this context, the medical diagnosis was, for him, "an expression of the sum of knowledge" that "prompts the doctor to act and behave" (Koch 1920: 17).

From today's perspective, Crookshank seems like a strange alter ego of the German Richard Koch, both thematically and in terms of content. Crookshank started out as a psychiatrist and paediatrician, and, after the First World War, turned increasingly to psychological and philosophical issues in medicine. He became famous for his controversial book, *The Mongol in Our Midst: A Study of Man and His Three Faces*, which was published in 1923. In this book, which was partly the result of a long study of people diagnosed with Down syndrome (Zihini 1989), he postulated that the syndrome's occurrence resulted from the emergence of genes from "Mongolian" ancestors who had intermarried with Europeans in earlier times. This was reflected in varying degrees of social deviance and physical stigmata (Crookshank 1923b). The book certainly struck a racist and eugenic chord with generally educated readers and went through three editions by 1931. Its academic reception was not positive. The American linguist and ethnologist Edward Sapir (1884–1939) dismissed it outright as a joke, since the thesis it put forward could only be rejected with a laugh (Keevak 2011). An obituary of Crookshank in the *British Medical Journal,* following his suicide, stated ambiguously: "Whatever he talked about, he could hold his audience enthralled, even if afterwards his hearer began to wonder whether the well-told tale were not almost too good to be true" (Crookshank 1933).

Richard Koch, for his part, first worked as an internist before turning to the history and theory of medicine.[9] In the 1920s in particular, he published several works on the philosophical foundations of medicine (Preiser 1988),[10] including extensive reflections on the role and function of history in medicine (Töpfer and Wiesing 2005b; Winau 1988; Wiesing 1997). In 1933 he was stripped of his teaching license by the National Socialists. He was an observant Jew and, according to the Nazi interpretation, not Aryan. As early as 1932, he had distanced himself from the Nazi image of the doctor and racial hygiene, rejecting a collectivist understanding of the patient and emphasizing the individuality of the sick person (Töpfer and Wiesing 2005a). In 1936, he emigrated via Brussels to the Soviet Union (Preiser 1988).

More distinctive than Crookshank's conceptualism was Koch's specific reference to the philosopher Hans Vaihinger (1852–1933) in his endeavor to elucidate diagnosis from a performative standpoint. In the preface to the second edition of his

---

[9]On Koch's biography see Rothschuh (1980a, b).

[10]A bibliography of Koch's works has been compiled by Rothschuh (1980a, 240–243).

Diagnosis, Koch explicitly referred to Vaihinger's philosophy of 'As If' ('As-Ob') as a fundamental inspiration. Vaihinger basically argued that the production of knowledge consciously and deliberately leads to fictions, and these are characterised by being either necessary for the further development of knowledge about a subject or practically useful for action (Müller and Fangerau 2012a, b).

It appears reasonable to say that Koch, in emphasising diagnosis as a practice-enabling factor, drew from pragmatic aspects within Vaihinger's reasoning. However, upon closer examination, it becomes evident that it was Vaihinger's fictionalism that actually captured his focus. Crookshank also mentions Vaihinger's fictionalism in connection with his own conceptualism when referring to the "fictional nature of general ideas or universals" (Crookshank 1926b: 998).[11] Koch, for his part, not only included an entire chapter on Vaihinger in the second edition of his book on diagnosis but also contributed a 100 page-long essay titled, 'The As-If in Medical Thought' in 1924 (Koch 1924).

Vaihinger had written his *Philosophie des Als-Ob* in 1876–1878, but it was not published until 1911 (Vaihinger 1911). The first version had served for his Habilitation in 1877. Over the subsequent years, he gained recognition as an expert on Immanuel Kant and, in 1883, he was offered a professorship in Halle. Vaihinger established the journal *Kant-Studien* in 1896 and founded the 'Kant-Gesellschaft' scholarly society in 1904. Additionally, in 1919, he initiated the foundation of the journal *Annalen der Philosophie* with the somewhat peculiar subtitle, *mit besonderer Rücksicht auf die Probleme der Als-Ob-Betrachtung* (*with special consideration of the problems of the As-If perspective*) (Simon 2014). Renamed as *Erkenntnis* in 1930, this journal would become the organ of the Vienna circle (Hegselmann and Siegwart 1991; Dätwyler 2021).

With his As-If philosophy, Vaihinger offered a theory of fictionalism which he linked to Immanuel Kant, Friedrich Albert Lange, Arthur Schopenhauer and Friedrich Nietzsche (Vaihinger 1921). He attempted to define a distinct intermediate position between materialism and idealism bridging a worldview based on reason, a worldview based on the senses, and a worldview based on the intellect (Heidelberger 2014). For him, only empirical perceptions formed the basis of facts. Any intellectual processing, even of empirical sensations was to be regarded as fiction. He called this standpoint a "positivistic idealism" or "idealistic positivism". Fictions without contradictions he called "semifictions". Fictions that could contain contradictions were called "real fictions". These fictions enabled action and survival by allowing one to cope with the external world experienced through the senses. Thus, for Vaihinger (scientific) theories were not a representation of a given reality, but "instruments for practical engagement with the world" (Heidelberger 2014: 51). Beyond his sensational empiricism, Vaihinger gave a further, biological, explanation for the formulation of such fictions by thinking. He linked thinking to the Darwinian

---

[11] Crookshank was also indirectly linked to Vaihinger by Charles Ogden, the editor of his paper on 'the importance of signs' quoted above. Ogden was the translator of Vahinger's book on 'As if' into English.

theory by emphasising its biological function in the struggle for survival (Heidelberger 2014). Fictions that permit action are useful in this struggle. As Kurt Sternberg summarised in his review of Vaihinger's opus magnum: "Between the two poles of sensation and action lies the whole realm of concepts, as constructed by thought to bridge these two poles, so as to enable action within the realm of sensory experience" (Sternberg 1911: 33).

This practical orientation of Vaihinger's fictionalism has been compared to 'American pragmatism' since its publication (Jacoby 1912; Bouriau 2016; Ceynowa 1993; Stoll 2020). However, Vaihinger himself in his preface to the German edition, and even more explicitly in the English edition, separated his standpoint from pragmatism (Vaihinger 1924c). For him, a practically useful idea is not necessarily true, but a fiction can have practical value despite being admittedly untrue. In his words: "Fictionalism does not admit the principle of Pragmatism which runs: 'An idea which is found to be useful in practice proves thereby that it is also true in theory, and the fruitful is thus always true.' The principle of Fictionalism, on the other hand, or rather the outcome of Fictionalism, is as follows: 'An idea whose theoretical untruth or incorrectness, and therewith its falsity, is admitted, is not for that reason practically valueless and useless; for such an idea, in spite of its theoretical nullity may have great practical importance.' But though Fictionalism and Pragmatism are diametrically opposed in principle, in practice they may find much in common. Thus both acknowledge the value of metaphysical ideas, though for very different reasons and with very different consequences." (Vaihinger 1924c: viii). At the same time, the practical instrumentalism inherent in Vaihinger's fictions, together with his reference to the empiriocriticism he shared with Ernst Mach, provides a link to the logical empirism of Rudolf Carnap and—beyond the institutional link of the journal *Erkenntnis*—to the epistemology of the Vienna Circle. However, unlike Vaihinger, who believed that human perception constituted a universal factual starting point for the establishment of fictive thought, Carnap saw perception itself as the result of a process of abstraction (Gabriel 2014).

Vaihinger's philosophy became very popular in the 1920s. By 1927, his book had gone through ten editions, including a popularised German edition ("Volksausgabe") in 1923 and 1924, and an English edition in 1924 (Vaihinger 1924a, b). However, his fictionalism and its popularization also encountered substantial criticism. In 1920 Vaihinger and some of his adherents organized a first As-If Conference in Halle to discuss Einstein's theory of relativity in the light of the As-If. They invited Albert Einstein, but Einstein was dissuaded from attending by colleagues like Paul Ehrenfeld. Ehrenfeld scoffingly labelled the conference a "Witches' Sabbat of As-If-ology" and referred to Vaihinger's followers as "As-If-ists" in order to underscore his impression that the fervour surrounding As-If had reached a sectarian dimension (Hentschel 2014).

The response from medical professionals was notably more favourable. Richard Koch, who spoke at the second As-If Conference in 1922, highlighted the practical significance of Vaihinger's approach for medicine, likening it to the roles of chemistry, physics, and biology in the medical field (Schmidt 1923). He was not alone: several physicians took up these ideas with enthusiasm in the 1920s.[12] They especially discussed Vaihinger's idea of a "useful fiction" and its value for medical practice, for research, for basic theoretical assumptions (disease and diagnosis as fictions) and as a "classification system" (Kulenkampff 1925; Rietti 1924/1925; Coerper 1919).

The As-If was also linked to diagnosis and diagnostics. The Italian Fernando Rietti, for instance, identified several fictions in medicine that were important for diagnostics. He classified representational formats in diagnostics as useful fictions: schemata, for instance, he called fictions "which contain the essence of reality, but in a much simpler and purer form" (Rietti 1924/1925: 398). He also identified symbolic and analogical fictions in the description of symptoms, or illustrative fictions. The diagnosis itself was according to him a "summatory fiction" that constituted a general concept about many individual diagnostic phenomena. "The clinical pictures whose names form the chapter headings of our textbooks" were for him "practically useful and necessary mental constructs without any real basis (...). Abdominal typhus, pneumonia, etc., are fictitious stocks, the generalisation – and also the abstraction – of the cases that have come into the physician's sphere of observation; but in reality there are only sick people, i.e. people in whom the processes of life take place in a way that deviates from the norm. Therefore, the creation of these fictitious beings allows us to order our knowledge and regulate our behaviour in the treatment of the sick: like all other fictions, they have no claim to reality, but they do have a claim to usefulness" (Rietti 1924/1925: 402).[13]

The surgeon Diedrich Kulenkampff (1880–1967), for his part, emphasized the importance of heuristic, "pathfinding" fictions in guiding subsequent action. For medicine, he considered among the most important fictions suggested by Vaihinger (Vaihinger 1922) to be the "artful, ingenious procedures", artificial methods, schemes, auxiliary methods, auxiliary hypotheses, provisional assumptions, and means of orientation (Kulenkampff 1925: 337 f.). In doing so, he also drew a line in the sand towards diagnostics and examination. What was important to him here was Vaihinger's statement that "fiction" is not a fallacy, but a "point of passage of thought". With regard to diagnostic examination, Kulenkampff made it clear that sensory examination and findings could ultimately only be combined in the sense of a fiction in order to produce a diagnostic sign. He was explicit about this: "We must not pretend that abdominal palpation is essentially a mechanical problem, but must remain aware that we only feel differences in tension and density, which we feel with the head, and if everything else is right, we can use the head to prove that the palpating finger is right. In other words, the mechanical notion of abdominal palpation is a fictional one that seems to have been forgotten in the immediate sensuality of the event" (Kulenkampff 1925: 335).

---

[12] In his consideration of the philosophy of As If in medicine, Richard Koch offered a literature review of its reception in medical literature (Koch 1924). See also Büttner (1997: 28).

[13] As early as 1836, Piorry had also already stated that disease was "an abstraction" composed of "organic, primary or secondary conditions" (Piorry 1846: 43).

Finally, Richard Koch, in his essay on "Das Als-Ob im ärztlichen Denken" ("As-If in medical thought") described the importance of maintaining fictions of recognisability and certainty in medical diagnosis. He confessed his long ambivalence about whether the diagnosis recognised something that, "strictly speaking", did not exist, namely the disease, while at the same time it was clear to him that the disease was "often, but not always, an abstraction of something that exists". Vaihinger's idea had helped him out of this dichotomy. Using the example of a differential diagnostic complex in which there is ambiguity about the diagnosis, Koch explained his idea that the physician must behave "as if one of these possibilities were reality. (. . .) We make an inventory of what could be present and arrange our measures in such a way that they do justice to all the possibilities. We behave as if the only possible state *A*, *B* or *C* were really present." He calls such fictions existential fictions, necessary to justify action (Koch 1924: 40). Furthermore, a diagnosis went hand in hand with a fiction of recognisability, because a diagnosis was a momentary description of a longer course of events, which by no means allowed a disease to develop or become a disease only at the time of the examination (Koch 1924).

Finally—and here he returns to the means of diagnostics and the question of what is normal—there is a norm fiction in medicine. According to Koch, what is healthy and what is ill "also depends on the state of the diagnostic tools. Before the introduction of Wassermann's reaction and X-ray diagnostics, many people were healthy who are now considered ill. Some will be healthy for the doctor working without diagnostic aids, while they will be considered sick after a specialist examination, after a stay in hospital" (Koch 1924: 57).

For Koch, the first means of gaining diagnostic knowledge was—in full accordance with Vaihinger's emphasis on the senses—perception (which includes senses beyond vision). The second means was examination. The third knowledge, intuition and mental processes. This last means linked the idea of 'as-if' with the question of whether a diagnosis was persuasive. Here, he argues with intuition, which he tries to characterize as subliminal cognition through practice.[14] Diagnosis by "perfect intuition" does not proceed by cognitive rationale, "but only freedom from error" (Koch 1920: 122). The experienced physician developed a feeling of security, which Koch equates with the concept of "self-evidence", because judgements made on the basis of this feeling of security were plausible. Experience, thought and belief lead to this feeling of security, which could also be deceptive and illusory. Therefore, in the end, the feeling should be subjected to reason again in an iterative process.

After the Second World War, for various reasons, Koch and Crookshank's diagnostic theory and Vaihinger's philosophy of as-if were almost forgotten. Crookshank died in 1933. His unfortunate *Mongol in our midst* seems to have

---

[14]Wolfgang Wieland clarified this 60 years later when he wrote that the statement of a diagnosis on the basis of direct observation only had be derived and inferred, and that even a diagnostic intuition, on which the experienced person relied, was in reality characterized by chains of reasoning that only ran "subliminally" on the basis of experience (Wieland 1983: 22).

eclipsed his epistemological writings on diagnosis. Richard Koch was persecuted by the Nazi regime in Germany, his books and articles were no longer cited and erased from the collective memory. Something similar happened to Vaihinger, who also died in 1933. Although some of his co-workers, such as Raymund Schmidt, or early adopters, such as the physician Carl Coerper, became National Socialists, Vaihinger's legacy seems to have suffered from anti-Semitic accusations against him and his philosophy. At the least, he was negated by the regime, and after 1945 he and his philosophy were not reinstated into the philosophical canon. As Gerd Simon put it, he no longer fitted into any fashion (Simon 2014).

3.4 Conclusion

The examples of Koch, Crookshank, Rietti and Kulenkampff show that, by the 1920s, the dominant image of diagnosis as the differentiation of universal diseases from individual patients which had developed in the 19th century had become nuanced in terms of medical theory. Koch had moved away from a naturalistic or realist approach to disease and diagnosis, in favour of a fictionalist approach without negating the condition of being sick. Crookshank followed a similar path, attempting to reconcile realism and nominalism in relation to illness and diagnosis through perceptual conceptualism. In addition, both authors emphasized the processual nature of diagnosis and the therapeutic effect of action that they associated with medical diagnosis.

In their reflections on diagnosis, Koch and Crookshank (like many others before and after them) were driven by the problem of accepting a nature-bound reality of symptoms and diseases while still holding on to the perception of their historical contingency as well as the realisation that symptoms, diagnoses and diseases can only become realities through denotation and conceptualisation. Only the fiction of this reality made it possible to talk about diagnoses and to act upon them.

In today's reading, both authors attempted to imagine naturalistic and social constructivist perspectives on diagnosis and illness as compatible. In the 1920s, they laid a foundation, for example, for work arguing for a "naturalistic social constructivism" by focusing on dysfunction and social norm (cf. Conley and Glackin 2021). Both referred to Hans Vaihinger, although Koch, in his 'Als Ob im ärztlichen Denken', dealt with Vaihinger's theory in a more differentiated and detailed way than Crookshank, who referred to him only superficially—but also explicitly when stating, for example: "Accepting the purely conceptual values of our universals, laws, and generalizations as convenient interpretative expressions of what we perceive, and so AS IF true, we recognize the relative and related values of the interpretations of the past as well as of those of today" (Crookshank 1926c: xxiv).

What made Vaihinger's philosophy so appealing to Koch and his contemporaries? Some theses may be presented as a potential explanation.

First of all, Vaihinger's criticism of materialism linked well with Koch's and Crookshank's criticism of mechanistic materialism in medical diagnosis. Both argued (like also, for example, the later social and racial hygienist Carl Coerper (1919)) for a reintegration of history and philosophy in medical epistemology. For Crookshank, the history of medicine should be written from a philosophical point of view, because medicine as a discipline was ultimately a natural philosophy (Crookshank 1926c). For Koch, however, history and nature were intertwined. He saw medicine as a "basic function of life", which was at the same time a "basic human ability and above all an art" (Koch 1930: 25). The physician, for his part, was a practitioner of his art, taking advantage of the beneficial reactions of the organism and counteracting the harmful ones (Wiesing 1997). The historical nature of this operation should be taken seriously (Koch 1930). Viewing the concepts offered by the natural sciences as useful fictions provided an avenue for alternative interpretations of medical phenomena through philosophical and historical lenses.

Secondly, fictionalism offered a bridge to explain the diagnostic step from sensually perceivable symptoms to their interpretation as signs as well as the special status of technically produced symptoms and signs in diagnostics. Their intermediate position between direct physical perceptions and indirect representations of physical phenomena made it difficult to define their epistemological status. They were, in an ideal way, objective, neutral, material, and simultaneously remote from natural perceptions. From this perspective, their status as diagnostic indicators could more effectively be portrayed in 'As-If' terms than as strictly naturalistic entities. Simultaneously, Koch and Crookshank were cognisant of the interplay between what may be perceived as cultural interpretation and natural occurrences.

Thirdly, Vaihinger's fictionalism offered them a more useful avenue for medicine than pragmatism. Since medicine requires justification for its actions and the ability to convince patients of the necessity of intervention, useful fictions that could potentially be true appeared more compelling to Koch and his contemporaries than what they understood as pragmatic approaches devoid of any claim to theoretical truth.[15]

Finally, Vaihinger's naturalization of epistemology through his reference to Darwin's theory of evolution (Heidelberger 2014), along with the accompanying biological and psychological foundation of his philosophy, resonated with biology-trained physicians during the 1920s. They could readily identify direct links between Vaihinger's arguments and their own biological and psychological orientation.

In conclusion, Koch and Crookshank wanted to overcome an artificial separation of nature and culture, or a reduction of diagnosis (and medicine itself) to the natural sciences, by looking at action, focusing on useful fictions, and including the historical dimension in medical thinking. As-If served as a perfectly fitted, functional and effective reference point for this ideal. Although Koch's theory of diagnosis is over a century old and he, along with Vaihinger, no longer occupies the central stage in discussions about diagnosis, health, and disease, Koch's line of thinking presents valuable perspectives for contemporary deliberations concerning the nature of medical diagnosis and the significance of essentialism and nominalism in medical

---

[15]This is not the only way to understand pragmatism (see the Prologue, Chap. 2, this volume).

practice. For example, since the 1960s it has been proposed that computer-based diagnostic systems that rely on statistically informed decision trees, probabilistic models or pattern recognitions offer potential assistance to physicians. In the 1980s, great expectations were pinned on programmes simulating expert reasoning, then called artificial intelligence (Szolovits et al. 1988). However, the epistemological nature of their information and the extent to which they can be integrated into medical practice are not yet fully understood. In particular, it seems useful not to see nature and culture as opposites, nor to see essentialism or nominalism as useful approaches for diagnostic action. Rather, nature and culture can be seen, in Koch's spirit, as part of a network of observation, thought and action grouped around at least four points: reality is disease, its basis is physical nature, the practice and result of its interpretation is culture, and its naming is a mental orientation.

References

Adelon, Nicolas Philibert, Gabriel Andral, and Pierre A. Béclard, eds. 1821–1828. *Dictionnaire de médecine*. Paris: Béchet.

Allbutt, Thomas Clifford. 1896. Introduction. In *A system of medicine*, ed. Thomas Clifford Allbutt, vol. 1, xix–xxxix. New York: Macmillan.

Baas, Johann Hermann. 1877. *Medicinische Diagnostik mit besonderer Berücksichtigung der Differentialdiagnostik*. Stuttgart: Encke.

Barker, Lewellys F. 1916. The development of the science of diagnosis. *The Journal of the South Carolina Medical Association* 12 (9): 278–284.

Bock, Carl Ernst. 1853. *Lehrbuch der pathologischen Anatomie und Diagnostik; 2: Lehrbuch der Diagnostik mit Rücksicht auf Pathologie und Therapie*. Leipzig: Wigand.

Bouriau, Christophe. 2016. Hans Vaihingers Die Philosophie des Als-Ob: Pragmatismus oder Fiktionalismus? *Philosophia Scientiae* 20: 77–93.

Burdach, Karl Friedrich, and Johann Carl Friedrich Leune. 1803. *Realbibliothek der Heilkunst*. Leipzig: Friedrich Gotthold Jakobäer.

Büttner, Johannes. 1997. Die Herausbildung des Normalwert-Konzeptes im Zusammenhang mit quantitativen diagnostischen Untersuchungen in der Medizin. In *Normierung der Gesundheit. Messende Verfahren der Medizin als kulturelle Praktik um 1900*, ed. Volker Hess, 17–32. Husum: Matthiesen.

Canguilhem, Georges. 1977. *Das Normale und das Pathologische*. Frankfurt/Main/Berlin/Wien: Ullstein.

Ceynowa, Klaus. 1993. *Zwischen Pragmatismus und Fiktionalismus: Hans Vaihingers Philosophie des 'Als Ob'*. Würzburg: Köngishausen & Neumann.

Coerper, Carl. 1919. Die Bedeutung des fiktionalen Denkens für die medizinische Wissenschaft. *Annalen der Philosophie* 1: 191–202.

Conley, Brandon A., and Shane N. Glackin. 2021. How to be a naturalist and a social constructivist about diseases. *Philosophy of Medicine* 2 (1): 1–21. https://doi.org/10.5195/pom.2021.18.

Crookshank, Francis Graham. 1923a. The importance of a theory of signs and a critique of language in the study of medicine. In *The meaning of meanings. A study of the influence of language upon thought and of the science of symbolism*, ed. Charles K. Ogden and Ivor A. Richards, 337–355. New York: Harcourt.

———. 1923b. *The Mongol in our midst: A study of man and his three faces*. London: Kegan Paul, Trench, Trubner.

———. 1926a. Bradshaw lecture on the theory of diagnosis. Part 1. *The Lancet* 208 (5384): 939–942.

———. 1926b. Bradshaw lecture on the theory of diagnosis. Part 2. *The Lancet* 208 (5385): 995–999.

———. 1926c. The relation of history and philosophy to medicine. Introductory essay. In *An introduction to the history of medicine. From the time of the Pharaos to the end of the XVIIIth century*, ed. Charles Greene Cumston, xiii–xxxii. London: Dawsons of Pall Mall.

———. 1933. *British Medical Journal* 2 (3800): 848.

Dätwyler, Maria. 2021. *Im Auftrag der Wahrheit. Selbstpositionierungsstrategien der Philosophie im 20. Jahrhundert*. Paderborn/Leiden: Brill Mentis.

Eckart, Wolfgang U. 1996. 'Und setzt eure Worte nicht auf Schrauben'. Medizinische Semiotik vom Ende des 18. bis zum Beginn des 20. Jahrhunderts – Gegenstand und Forschung. *Berichte zur Wissenschaftsgeschichte* 19: 1–18.

———. 1998. Zeichenkonzeptionen in der Medizin vom 19. Jahrhundert bis zur Gegenwart. In *Semiotik. Ein Handbuch zu den zeichentheoretischen Grundlagen von Natur und Kultur. 2. Teilband*, ed. Roland Posner, Klaus Robering, and Thomas A. Sebeok, 1694–1712. Berlin: Mouton.

Eich, Wolfgang. 1986. *Medizinische Semiotik (1750–1850)*. Freiburg: Hans Ferdinand Schulz Verlag.

Engle, Ralph L., and B.J. Davis. 1963. Medical diagnosis: Past, present and future. *Archives of internal medicine* 112: 512–543.

Eulenburg, Albert, ed. 1900. *Real-Encyclopädie der gesammten Heilkunde. Medicinisch-chirurgisches Handwörterbuch für praktische Aerzte*. Dritte, gänzlich umgearbeitete Auflage ed. Vol. 23. Berlin/Wien: Urban & Schwarzenberg.

Fangerau, Heiner. 2017. Bilder, Zahlen und die Medizin der Zukunft: Das 'Als ob' als Philosophie der Prävention. In *Präventionsentscheidungen: Zur Geschichte und Ethik der Gesundheitsvorsorge im 21. Jahrhundert*, ed. Sebastian Kessler, Heiner Fangerau, and Urban Wiesing, 59–94. Stuttgart: Frommann-Holzboog.

———. 2021. Krankheitszeichen und Stigma: Die Differenzierung der ärztlichen Sinne in der Diagnostik des 19. Jahrhunderts und ihre sozialen Folgen. *Werkstatt Geschichte* 83: 37–48.

———. 2023. *Zwischen Natur und Kultur: Der Diagnosebegriff in den 1920er Jahren bei Richard Koch und Francis Crookshank*. In: *Vita brevis, ars longa. Aktuelle Perspektiven zu Geschichte, Theorie und Ethik der Medizin*, ed. Hans-Jörg Ehni, Georg Marckmann, Robert Ranisch, and Henning Tümmers (45–53). Stuttgart: Kohlhammer.

Fangerau, Heiner, and Michael Martin. 2015. Medizinische Diagnostik und das Problem der Darstellung: Methoden der Evidenzerzeugung. *Angewandte Philosophie: Themenheft Medizinische Erkenntnistheorie* 1: 38–68.

Foucault, Michel. 1973. *Die Geburt der Klinik: Eine Archäologie des ärztlichen Blicks*. München: Hanser.

Gabriel, Gottfried. 2014. Fiktion und Fiktionalismus. Zur Problemgeschichte des 'Als Ob'. In *Fiktion und Fiktionalismus. Beiträge zu Hans Vaihingers Philosophie des Als Ob*, ed. Matthias Neuber, 65–87. Würzburg: Königshausen und Neumann.

Galdston, Iago. 1941. Diagnosis in historical perspective. *Bulletin of the History of Medicine* 9 (4): 367–384.

Greene, Jeremy A., and Andrew S. Lea. 2019. Digital futures past – The long arc of big data in medicine. *New England Journal of Medicine* 381 (5): 480–485.

Grmek, Mirko. 1964. La conception de la maladie et de la santé chez Claude Bernard. In *L'aventure de la science, tome 1*, ed. Alexandre Koyré, 208–227. Paris.

Hegselmann, Rainer, and Geo Siegwart. 1991. Zur Geschichte der 'Erkenntnis'. *Erkenntnis* 35 (1/3): 461–471.

Heidelberger, Michael. 2014. Hans Vaihinger und Friedrich Albert Lange mit einem Ausblick auf Ludwig Wittgenstein. In *Fiktion und Fiktionalismus. Beiträge zu Hans Vaihingers Philosophie des Als Ob*, ed. Matthias Neuber, 43–63. Würzburg: Königshausen und Neumann.

Henle, Jakob. 1846. *Handbuch der rationellen Pathologie.* Vol. 1. Braunschweig.

Hentschel, Klaus. 2014. Zur Rezeption von Vaihingers Philosophie des Als Ob in der Physik. In *Fiktion und Fiktionalismus. Beiträge zu Hans Vaihingers Philosophie des Als Ob,* ed. Matthias Neuber, 161–184. Würzburg: Königshausen und Neumann.

Hess, Volker. 1993. *Von der semiotischen zur diagnostischen Medizin: Die Entstehung der klinischen Methode zwischen 1750 und 1850. Vol. 66. Abhandlungen zur Geschichte der Medizin und der Naturwissenschaften.* Husum.

Jacoby, Günther. 1912. Der amerikanische Pragmatismus und die Philosophie des Als Ob. *Zeitschrift für Philosophie und philosophische Kritik* 147: 172–184.

Keevak, Michael. 2011. *Becoming yellow: A short history of racial thinking.* Princeton: Princeton University Press.

King, Lester S. 1982. *Medical thinking. A historical preface.* Princeton: Princeton University Press.

Koch, Richard. 1920. *Die ärztliche Diagnose. Beitrag zur Kenntnis des ärztlichen Denkens.* Zweite umgearbeitete Auflage ed. Wiesbaden: J. F. Bergmann.

———. 1924. *Das Als-Ob im ärztlichen Denken.* München: Rösl & Cie.

———. 1930. *Philosophische Grenzfragen der Medizin. Der Begriff der Medizin,* Vorträge des Instituts für Geschichte der Medizin an der Universität Leipzig. Vol. 3. Leipzig: Thieme.

Krehl, Ludwig. 1903. *Über die Entstehung der Diagnose (Einladung zur akademischen Feier des Geburtsfestes Seiner Majestät des Königs Wilhelm II von Württemberg auf den 26. Februar 1903 im Namen des Rectors und akademischen Senats der Königlichen Eberhard-Karls-Universität Tübingen).* Tübingen: G. Schnürlen.

———. 1931. Arzt, Diagnose, Beurteilung. In *Die Erkennung innerer Krankheiten. Band 2,* ed. Ludwig Krehl, 1–25. Berlin: Vogel.

Kulenkampff, Diedrich. 1925. Über den Wert und die Bedeutung der Als-Ob-Betrachtung im medizinischen Denken. Ein Versuch. *Virchows Archiv* 255: 332–359.

Link, Jürgen. 1997. *Versuch über den Normalismus. Wie Normalität produziert wird.* 2. erw. ed. Opladen: Westdeutscher Verlag.

Martin, Michael, and Heiner Fangerau. 2007. Listening to the heart's power: Designing blood pressure measurement. *ICON* 13: 86–104.

———. 2009. Technisierung der Sinne – von der Harnschau zur Urinanalyse. Das Beispiel der Harnzuckerbestimmung. *Der Urologe A* 48 (5): 535–541. https://doi.org/10.1007/s00120-009-1956-x.

———. 2011. Töne sehen? Zur Visualisierung akustischer Phänomene in der Herzdiagnostik. *NTM. Zeitschrift für Geschichte der Wissenschaften, Technik und Medizin* 19 (3): 299–327. https://doi.org/10.1007/s00048-011-0055-4.

———. 2013. Seeing sounds? Styling vision? The mechanical visualisation of acoustic phenomena in cardiac diagnostics around 1900. *ICON* 18: 45–62.

———. 2021. *Evidenzen der Bilder. Visualisierungsstrategien in der medizinischen Diagnostik um 1900.* Stuttgart: Steiner.

Meissner, Friedrich Ludwig, ed. 1830–1834. *Encyclopädie der medicinischen Wissenschaften. Nach dem Dictionnaire de Médecine frei bearb. und mit nöthigen Zusätzen versehen.* Leipzig: Fest.

Mirbabaie, Milad, Stefab Stieglitz, and Nicholas R. Frick. 2021. Artificial intelligence in disease diagnostics: A critical review and classification on the current state of research guiding future direction. *Health Technology* 11: 693–731. https://doi.org/10.1007/s12553-021-00555-5.

Moser, Adolf. 1845. *Die medicinische Diagnostik und Semiotik, oder die Lehre v. der Erforschung u. der Bedeutung d. Krankheitserscheinungen bei den innern Krankheiten des Menschen.* Leipzig: Brockhaus.

Müller, Irmgard, and Heiner Fangerau. 2012a. Die Repräsentation des Unsichtbaren: Darstellung als Problem und Promotor in der Entstehung von Wissen. In *Faszinosum des Verborgenen. Der Harnstein und die (Re-) Präsentation des Unsichtbaren in der Urologie,* ed. Heiner Fangerau and Irmgard Müller, 11–29. Stuttgart: Steiner.

————. 2012b. Medical imaging: Pictures 'as if' and the power of evidence. In *Medical imaging and philosophy. Challenges, reflections and actions*, ed. Heiner Fangerau, Rethy K. Chhem, Irmgard Müller, and Shih-chang Wang, 45–60. Stuttgart: Franz Steiner Verlag.

Musser, John Herr. 1894. *A practical treatise on medical diagnosis: For students and physicians*. Philadelphia: Lea Brothers & Co.

Nicolson, Malcolm. 1993. The art of diagnosis. Medicine and the five senses. In *Companion Encyclopedia of the history of medicine*, ed. William F. Bynum and Bd Roy Porter, vol. 2, 801–825. London/New York: Routledge.

Piorry, Pierre A. 1846. *Diagnostik und Semiotik mit vorzüglicher Berücksichtiung der neuesten mechanisch-nosognostischen Hülfsmittel. Aus dem Französischen von Gustav Krupp. Neue Ausg*. Leipzig: Fischer.

Preiser, Gert. 1988. Richard Koch. Zu Leben und Werk eines Frankfurter Arztes. In *Richard Koch und die ärztliche Diagnose*, ed. Gert Preiser, 48–60. Hildesheim: Olms.

Rietti, Fernando. 1924/1925. Das Als Ob in der Medizin. *Annalen der Philosophie und philosophischen Kritik* 4 (8): 385-416.

Risse, Guenter B. 1987. A shift in medical epistemology: Clinical diagnosis, 1770–1828. In *History of diagnostics. Proceedings of the 9th international symposium on the comparative history of medicine – East and West*, ed. Yosio Kawakita, 115–147. Taniguchi Foundation: Osaka.

Rosenberg, Charles E. 2002. The Tyranny of diagnosis: Specific entities and individual experience. *Milbank Quarterly* 80 (2): 237–260.

Rothschuh, Karl Eduard. 1978. *Konzepte der Medizin in Vergangenheit und Gegenwart*. Stuttgart: Hippokrates.

————. 1980a. Richard Hermann Koch (1882–1949): 2. Teil: Werk Und Würdigung. *Medizinhistorisches Journal* 15 (3): 223–243.

————. 1980b. Richard Hermann Koch (1882–1949): Arzt, Medizinhistoriker, Medizinphilosoph (Biographisches, Ergographisches) 1. Teil. Zur Biographie. *Medizinhistorisches Journal* 15 (1/2): 16–43.

Scadding, John Guyett. 1996. Essentialism and nominalism in medicine: Logic of diagnosis in disease terminology. *The Lancet* 348 (9027): 594–596.

Schmalz, Carl Gustav. 1825. *Versuch einer medizinisch-chirurgischen Diagnostik in Tabellen oder Erkenntniß und Unterscheidung der innern und äußern Krankheiten mittels Nebeneinanderstellung der ähnlichen Formen*. Leipzig: Arnold.

Schmidt, Raymund. 1923. Die zweite 'Als-Ob'-Konferenz in Halle am 7. Juni 1922. Ein Bericht. *Annalen der Philosophie* 3 (4): 568–578.

Sebastian, Jacob Friedrich Christian. 1819. *Grundriss der allgemeinen pathologischen Zeichenlehre für angehende Ärzte und Wundärzte. Zum Gebrauch bey seinen Vorlesungen entworfen von F. J. Chr. Sebastian*. Darmstadt: Heyer & Leske.

Simon, Gerd. 2014. Leben und Wirken Hans Vaihingers. In *Fiktion und Fiktionalismus. Beiträge zu Hans Vaihingers Philosophie des Als Ob*, ed. Matthias Neuber, 21–41. Würzburg: Königshausen und Neumann.

Sprengel, Kurt. 1801. *Kurt Sprengels Handbuch der Semiotik*. Halle: Gebauer.

Sternberg, Kurt. 1911. Rezension: Vaihinger, Hans, Die Philosophie des Als Ob. System der theoretischen, praktischen und religiösen Fiktionen der Menschheit auf Grund eines idealistischen Positivismus. *Kant Studien* 16 (1–3): 328–338.

Stoll, Timothy. 2020. Hans Vaihinger. In *The Stanford Encyclopedia of philosophy*, ed. Edward N. Zalta. https://plato.stanford.edu/archives/spr2020/entries/vaihinger/.

Szolovits, Peter, Ramesh Patil, and William B. Schwartz. 1988. Artificial Intelligence in medical diagnosis. *Annals of Internal Medicine* 108 (1): 80–87.

Töpfer, Frank, and Urban Wiesing. 2005a. The medical theory of Richard Koch I: Theory of science and ethics. *Medicine, Health Care and Philosophy* 8 (2): 207–219. https://doi.org/10.1007/s11019-004-7445-5.

————. 2005b. The medical theory of Richard Koch II: Natural philosophy and history. *Medicine, Health Care and Philosophy* 8 (3): 323–334. https://doi.org/10.1007/s11019-004-7446-4.

Vaihinger, Hans. 1911. *Die Philosophie des Als-Ob. System der theoretischen, praktischen und religiösen Fiktionen der Menschheit auf Grund eines idealistischen Positivismus. Mit einem Anhang über Kant und Nietzsche.* Berlin: Reuther & Reichard.

———. 1921. Wie die Philosophie des Als Ob entstand. In *Die Philosophie der Gegenwart in Selbstdarstellungen, Bd. 2,* ed. Raymund Schmidt, 174–203. Leipzig: Felix Meiner.

———. 1922. *Die Philosophie des Als Ob. Siebente und achte Auflage.* Leipzig: Felix Meiner.

———. 1924a. *Die Philosophie des Als Ob. Volksausgabe, Zweite Auflage.* Leipzig: Felix Meiner.

———. 1924b. *The philosophy of 'as if', a system of the theoretical, practical and religious fictions of mankind. International library of psychology, philosophy, and scientific method.* London/ New York: K. Paul, Trench, Trubner & Co./Harcourt, Brace & Company, Inc.

———. 1924c. *The philosophy of 'as if': A system of the theoretical, practical and religious fictions of mankind.* London: Paul, Trench, Trubner.

Vaughan, Warren T. 1922. The philosophy of medical diagnosis. *Clinics and Collected Papers of Saint Elizabeth's Hospital, Richmond, Virginia* 23: 440–446.

Warner, John Harley. 2013. The humanizing power of medical history: Responses to biomedicine in the 20th-century United States. *Procedia – Social and Behavioral Sciences* 77: 322–329.

Wichmann, Johann Ernst. 1794. *Ideen zur Diagnostik: beobachtenden Aerzten mitgetheilet.* Hannover: Helwing.

Wieland, Wolfgang. 1975. *Diagnose. Überlegungen zur Medizintheorie.* Berlin/New York: Walter de Gruyter.

———. 1983. Systematische Bemerkungen zum Diagnosebegriff. In *Anamnese, Diagnose, Therapie,* ed. Richard Toellner and Kazem Sadegh-zadeh. Tecklenburg: Burgverlag.

Wiesing, Urban. 1997. Die Einsamkeit des Arztes und der 'lebendige Drang nach Geschichte'. Zum historischen Selbstverständnis der Medizin bei Richard Koch. *Gesnerus* 54 (3–4): 219–241.

Winau, Rolf. 1988. Die Funktion der Medizingeschichte in der Sicht Richard Kochs. In *Richard Koch und die ärztliche Diagnose,* ed. Gert Preiser, 142–161. Hildesheim: Olms.

Zihini, Lilian Serife. 1989. *The history of the relationship between the concept and treatment of people with Down's syndrome in Britain and America from 1866 to 1967.* Doctoral thesis, University of London.

Chapter 4
Epistemic Inclusion and the Silence of the Patients

Hub Zwart

4.1 Introduction

In *The Birth of the Clinic*, Michel Foucault (1963) practiced historical epistemology, combining two scholarly crafts, history of medicine and philosophy of science. With his philosophical hammer and stethoscope, he assessed epistemic configurations: landscapes of discourse and practice. Philosophical readers are intrigued by the epistemological ruptures he discerned, the sudden upheavals in the ways in which physicians observed, treated, looked at, listened to, and wrote about their patients. Foucault was an "archaeologist", signalling the sudden emergence and disappearance of "discursive formations", pointing to abrupt and remarkable changes in the ways in which phenomena are described, analysed, and categorised. (Foucault 1969). For an archaeologist, all (anonymous) potsherds and coins stemming from a particular formation (or discursive layer) share a number of basic similarities: they were all produced in a similar vein. Likewise, within a particular discursive formation, all texts are basically similar. And although author names are mentioned by Foucault, the author's identity does not really count. Foucault is interested in the basic logic at work, anonymously almost. Notwithstanding the countless discussions and debates between authors, a spontaneous unanimity can be discerned, concerning basic presuppositions that remain unquestioned. At the same time, there are striking differences *between* discursive formations (between layers within libraries). Historians tend to be critical about this approach. Add more detail, and the epistemic fault lines become increasingly diffuse,—the theorem of epistemic discontinuity becomes difficult to uphold.

H. Zwart (✉)
Erasmus School of Philosophy, Rotterdam, The Netherlands
e-mail: zwart@esphil.eur.nl

© The Author(s) 2024
M. Schermer, N. Binney (eds.), *A Pragmatic Approach to Conceptualization of Health and Disease*, Philosophy and Medicine 151,
https://doi.org/10.1007/978-3-031-62241-0_4

Heiner Fangerau's chapter (Chap. 3, this volume) indicates, however, that the perspectives of historians and philosophers are not incompatibly different. Even in detailed accounts that address multiple transitions, lines of influence and discussions, we notice drastic changes of perspective. Both historians and philosophers are interested in assessing and problematising *dichotomies*, moreover. I will focus on one particular dichotomy, a rather obvious one when it comes to the history of medicine, namely between physician and patient. For philosophers, to focus on the dialectical relationship between physician and patient is something inevitable, because it resonates with the polarity between subject and object around which philosophy of science tends to revolve. Patients play a dual role. For themselves (from a first-person perspective) they are subjects of experience, dwelling in a lifeworld, informing others (physicians, family members, fellow sufferers) about disease as a subjective and physical experience. For a general practitioner, a patient may still be a subject, someone to carefully and respectfully listen to, but in evidence-based medicine, patients become *research subjects*, that is: objects of knowledge. In medicine as a scientific field, the physician is the epistemological subject equipped with tools and technological contrivances to measure and assess the condition of the anonymised patient, as a provider of data,—although philosophy of science (or more specifically, philosophy of medicine) shifts the perspective once again, by studying epistemic interactions, so that now the physicians themselves suddenly become *objects* of research.

4.2   The Silence of the Patients

One of the interesting aspects of Fangerau's chapter, initially focussing on nineteenth-century medicine, is that the voice of the patient is almost completely absent. The focus is on the discourse of physicians acting as authors, a seemingly self-evident point of departure. Patients are *treated* by physicians. In principle, all treatments are problematic from a normative point of view, as Fangerau explains. They are experienced by patients as unpleasant, perhaps even harmful, involving series of physical intrusions that, in normal life, would be considered objectionable or even unacceptable. Therefore, treatments need a *justification*. As Fangerau argues, the justification of questionable interventions is provided by a diagnosis which legitimises further medical treatments and interventions.

Hospital rooms and treatment rooms are areas of normativity, structured by a symbolic order of prohibitions, whose violation requires an explicit legitimation, provided by the diagnosis. But the concept of a diagnosis also reveals that treatment rooms are spaces of power. The patient's subjective experience (more or less diffuse sensations such as pain, malaise, fatigue, etc.) is obfuscated by technoscientific vocabularies. Diagnosis and treatment introduce a tension as the technoscientific terminology will not completely do justice to subjective experience. As Fangerau explains, patients display particular symptoms and the objective of physicians is to recognise these symptoms as signs, indicating an underlying condition which is not immediately visible. The physician has to "reconcile" the individual signs of the patient with the "generalised ordering of signs", i.e., the medical vocabulary.

In *The Birth of the Clinic*, Foucault describes the emergence of the general practitioner, someone who visits patients in their homes. In nineteenth-century novels, narrators are not infrequently doctors who have access to the private sphere of families, observing and eavesdropping on the daily lives of characters. Evidence-based medicine entails a dramatic shift, in the sense that the logic of the laboratory world intrudes into the lifeworld, so that the subjective experience of the lifeworld is obfuscated by the technical language of the laboratory world, where tests are conducted, biomarkers are measured, samples are assessed. Thus, the recording of signs becomes "increasingly technical", as Fangerau explains, translated into numerical values and curves with the help of technical equipment. In a quantified format, a particular configuration of measurable signs can be related to a disease, and subsequently to a treatment. Medicine becomes a technoscience, a practice that is inherently technical.

Fangerau's erudite chapter studies the vicissitudes of medical discourse during a particular stage of history. The focus is on the perspective of the physician, more specifically: on the difficult task of making the connection between signs and the underlying condition (i.e., diagnostics). While remarkable transitions in the practice of diagnostics occur, one element remains unquestioned,—we never hear the patient's voice. Physicians take the floor and enter the discussion. Insofar as patients are present, they provide the input, which means that they either physically *produce* the signs, or verbally *report* the signs. Interestingly, the most active role played by patients comes to the fore through the possibility that patients may *simulate* or *feign* the evidence. In other words, the patient is basically an epistemological obstacle, standing in the way between biomedical science and the disease, that has to be circumvented by carefully measuring and interpreting the signs, even with a certain amount of suspicion, because patients, notably when it comes to *reporting* evidence, may be misinterpreting, exaggerating, or even simulating the facts. In principle, the information provided by the patient is invalid and has to be validated.

Thus, while Fangerau studies transitions occurring in discussions *between* physicians-authors, both influencing and challenging one another, as a philosophical reader I am struck by the basic logic or epistemic profile of the discourse produced by these nineteenth-century physician-authors,—by the contrast between the prolific discursive productivity of the physicians and the remarkable silence of the patients. The physician, equipped with contrivances, transforms the patient's diffuse sensations into discrete, measurable signs. While patients report their sensations, the medical observer explains in writing how to determine the significance of these utterances. Thus, the basic configuration highlighted by Fangerau, in his diagnostics of medical discourse, is that the recording and interpretation of signs becomes "increasingly technical". Sensations are quantified with the help of technical devices. Signs that can only be perceived by the sick persons themselves are discarded as unreliable. Yet, also phenomena directly visible to others are gradually replaced by observations that are made 'indirectly', e.g., obtained through measurements or chemical analysis. Such observations are considered as more precise and, above all, reproducible,—independently of the subjectivity of the observer. In other words, subjective signs (reported by patients) are replaced by objective signs determined by physicians, either in the treatment room or, even more indirectly, with the help of laboratory contrivances producing quantified results. In this manner, deceptive or even simulated sensations give way to validated, tested knowledge. Physicians listen to their patients, but there is an attitude of mistrust.

All the power is now in the hands of the physicians. The availability of validated diagnostics legitimises their intrusive behaviour. Technology enhances the distance between physician and patient. In *Birth of the Clinic*, Foucault describes how the stethoscope was invented, making observations more precise, but also relieving male physicians of the requirement to put their ear against the breasts of female patients. The stethoscope allowed physicians to disregard disturbing sensations and focus on the key signal. Technical contrivances literally created distance between sensations experienced by patients and objective observations by physicians. Diagnostics became increasingly quantified, while patients were silenced.

The essence of diagnostics is not the measurement as such. A normative dimension is involved here: the reference to a 'normal value', a standard of normalcy. The measurement is compared with what is expected (given the gender, age, etc. of the patient). Patient are expected to live up to a normative mean. A symptom is basically a quantifiable deviation. Illness is measurable deviance, a deviance from the norm. There is no strict dividing line between health and disease. Rather, the level of deviance is decisive.

4.3   From Continuity to Discontinuity

In the twentieth century, a change has taken place, a shift from continuity thinking towards discontinuity thinking (Zwart 2020, 2022). Whereas Charles Darwin thought about nature in terms of slow and gradual change, the year 1900 was a turning point when Gregor Mendel's work was rediscovered, and the concept of mutation was introduced. Genetics addressed phenomena of life in terms of *presence or absence* of key components (e.g., genes). In Fangerau's chapter, the epistemic profile of the new era is exemplified by an influential and controversial book, *The Mongol in Our Midst*, written by Francis Graham Crookshank and published in 1923. People may be carriers of atavistic genetic traits, the author argues, inherited from Mongoloid ancestors (e.g., invading Huns). Down's syndrome was considered an atavistic genetic throwback, negating the evolutionary progress made by Caucasians, but also allegedly healthy people could display disconcerting "Mongolian stigmata". This reverberated with National-Socialist concepts of racial pollution and hygiene.

These ideas are juxtaposed by a contrasting perspective, which sees concepts of disease as useful constructs, or fictions even, that actually work in practice and allow physicians to support their patients. One of the results of this perspective was that, depending on the availability of novel diagnostic tools, people who were considered healthy may be suddenly perceived as ill. A new test may point to a health problem they themselves were not aware of. As incompatible as both views may seem, they still have something in common. Power and knowledge remain the privilege of the physician, while patients maintain their silence. Whether diagnosis refers to an atavistic trait or a workable fiction, the voice of the patient remains inaudible.

4.4 Epistemic Inclusion

The patient's silence is not an inevitable epistemological given. Rather, it entails a normative question for the present. Could we strive for epistemic justice or epistemic inclusion by developing a more holistic and comprehensive view on health and disease, involving the knowledge and experience of patients, not as mere reporters of (questionable) sensations, but as subject of knowledge? Epistemic inclusion is the aspiration to bridge the gap between laboratory world and life world, between validated knowledge and existential experiences, by staging a dialogue.

Recently, I was involved in an interesting research project devoted to studying metastatic cancer cells. A lab visit by a group of philosophy students resulted in an interesting confrontation. The researchers confessed that they hardly ever entered into conversations with cancer patients. They studied metastatic cells with a help of sophisticated equipment, fascinated by the intriguing behaviour and beauty of their cancer cells, even though realizing the anxiety which the detection of such cells may cause to patients. Could these two worlds of experience, existing side by side, be brought together, thus developing a more comprehensive view on metastatic cancer? Fangerau's chapter point to a basic dimension of medical discourse which all contributors, notwithstanding their differences, failed to question, namely the silence of the patients. The current epidemic of distrust in science may be addressed by a more participatory and inclusive approach, based on dialogue rather than objectification (Oreskes 2019). As historical research shows, medical knowledge is social knowledge, and rather than silencing patients, knowledge may gain in trustworthiness when the experience and concerns of patients become an important source of insight in a less hierarchical epistemic constellation.

References

Crookshank, Francis G. 1923. *The Mongol in our midst: A study of man and his three faces*. London: Kegan Paul.
Foucault, Michel. 1963. *Naissance de la clinique. Une archéologie du regard médical*. Paris: PUF.
———. 1969. *Archéologie du savoir*. Paris: Gallimard.

Oreskes, Naomi. 2019. *Why trust science?* Princeton: Princeton University Press.
Zwart, Hub. 2020. *Styles of thinking*, Philosophy and Psychology in Dialogue 2. Berlin/Münster/
  Zürich: LIT Verlag.
———. 2022. *Continental philosophy of technoscience*, Philosophy of Engineering and Technology. Dordrecht/London: Springer Nature.

Chapter 5
The Variety of Historiographical Medical Relativism

Martin Kusch

5.1 Introduction

Relativism has been a central theme in the history, philosophy, and social studies of science and technology (HPS-ST) for at least a century. Relativism has not been equally prominent in the history, philosophy and social studies *of medicine* (HPS-M). Although several prominent scholars in HPS-M are leaning towards relativistic positions, there has been little attempt to map these positions and or to relate them to relativistic ideas in HPS-ST more broadly. This chapter is a modest attempt to fill this gap. I seek to identify and analyse what I regard as the three main forms of relativism in the historiography of medicine. In the case of each such form of relativism, I shall pick one paradigmatic author; situate them relative to a 'spectrum of relativistic views'; and identify a counterpart position in HPS-ST. I shall also make some critical comments on either previous interpretations of my paradigmatic authors, or on their HPS-ST counterparts. I shall not systematically address the question which form of historiographical relativism is refutable or defensible. My goal here is to understand the 'variety' of relativistic commitments, not to praise or condemn individual versions. Still, it is natural to read all three versions of historiographical relativism discussed here as prima facie plausible forms of relativism: all three avoid both implausible forms of absolutism and extreme forms of 'anything goes' relativism.

M. Kusch (✉)
Department of Philosophy, University of Vienna, Vienna, Austria
e-mail: martin.kusch@univie.ac.at

© The Author(s) 2024
M. Schermer, N. Binney (eds.), *A Pragmatic Approach to Conceptualization of Health and Disease*, Philosophy and Medicine 151,
https://doi.org/10.1007/978-3-031-62241-0_5

5.2   Cunningham on Incommensurability and Retrospective Diagnosis

I begin with the medical historian Andrew Cunningham. Put in a nutshell, Cunningham's relativism surfaces in his Kuhn-inspired discussions of conceptual incommensurability and in his criticism of retrospective diagnosis, that is, the practice of applying modern diagnostic categories in the past.

Cunningham presents his position on incommensurability in different places. The two most convincing presentations are his radio series "The Making of Modern Medicine" (2007) and a paper from 1992 (1992/2016). The former focuses on the case of tuberculosis, the latter on the plague. The following thumbnail sketches must suffice for our purposes.

The guiding thought of both histories is that a disease can have its "identity changed" over time (2007: Ch. 14). The criteria for identifying a disease can radically shift; and such radical shift does not allow for the criteria to be "measured against each other". It is as if one were to measure an item first with a ruler, and later with a colour chart. The length of an item tells us nothing about its colour, and its colour tells us nothing about its length. In other words, the criteria for length and the criteria for colour are "incommensurable" (2007: Ch. 14). We cannot say whether the two disease identities before and after a radical shift are identical or not, "since the criteria of 'sameness' have been changed" (1992/2016: 242).

Early in the nineteenth century, "consumption" was one of the deadliest diseases, alongside cholera and typhus. In Britain, consumption was thought of by many doctors as a hereditary disease and as affecting talented young people in particular. The criteria of consumption were exhaustion, coughing (blood), breathing pains, fits of fever, loss of weight and night sweats. All of these criteria were *symptoms*, not *causes* of consumption. Cunningham calls this the "first identity" of the disease. The second identity was made possible by a new form of medicine emerging in early-nineteenth-century France: clinical medicine. Seriously-ill patients were increasingly cared for in hospitals. Doctors recorded their symptoms and the development of their diseases. And when patients died, doctors performed autopsies. Working in this setting, made it possible for René-Théophile-Hyacinthe Laennec to notice a correlation between consumption symptoms and the presence of "tubercules" in the patients' lungs. The presence of the tubercules was seen as the central "pathology" underlying consumption. Consumption was thereby transformed: it was now the pathologist performing an autopsy who determined whether a patient had been suffering from consumption (2007: Ch. 14).

The third identity of the disease—now increasingly called "tuberculosis"—was due to the French medical doctor Jean-Antoine Villemin. In 1865 he showed that the tubercules contained a poison. As a result, it was no longer the presence of tubercules that defined the disease; it was the poison within them. This allowed for a more precise pathological account. Finally, the fourth identity was introduced of Robert Koch who identified the bacillus causing the pathology, the symptoms and ultimately death. Moreover, Koch's efforts established that tuberculosis was an infection. After Koch, the disease could no longer be defected by a layperson (as was the case with the first identity), or a pathologist; it could only be identified in a laboratory (2007: Ch. 14).

It is a central element of Cunningham's argument that the *extensions* of the four disease identities were not identical. It is not that we have learnt ever more about one and the same disease as, say, identified with the symptoms of the first identity. That is to say, Cunningham's case study is not akin to our discovering the "essence" of gold; in the latter case, we first "baptized" a substance based on "superficial", "observable" properties, before later learning that the essence of this substance is the atomic number 79 (Kripke 1980). The move from one of the four disease identities to the next invariably changed the extension. Roughly put: one could have the symptoms without the tubercules, and the bacillus without the symptoms.

Cunningham's account of the transformation of the plague has the same argumentative structure (1992/2016). Before and after the 1894-discovery of the micro-organism causing plague, the bacillus 'Pasteurella pestis', there existed two different disease identities. Before the discovery, plague was identified on the basis of symptoms (e.g. the buboes) and on the basis of its characteristic course. After the discovery, it is only the "pathogen" that counts: "We are all bacteriologists now, and none of us would attempt to identify plague today without a laboratory. To oppose the claims of bacteriology is now not a rival view, nor an alternative view, nor even a dissident view. It is now a lunatic view" (1992/2016: 239).

In the 1992-paper Cunningham asserts outright that diseases are no "natural kinds".[1] His argument is less to do with changing extensions, and more to do with the complexity and plurality of components in a disease identity:

> ... disease does not seem to be a "natural kind". Rather, a "disease-entity" is a mental construct made up of experiences of pain, distress and debilitation, the outward visible appearances that accompany these experiences, the succession of all these over time together with the outcome (recovery, disablement, death), the changes that the pathologist can find in the parts of the body, together with peoples' thoughts about the origin and reasons for what is happening and why it turns out as it does. (1992/2016: 212)

Cunningham's scepticism about retrospective diagnosis follows naturally from the above considerations. One of his targets is the historian Carlo Cipolla who writes that during the plague in Prato in 1629–30, medical officers "fought against an invisible enemy. (. . .) Medical knowledge was of no help and medical treatment was of no value." As Cunningham sees it, Cipolla mistakenly projects our bacteriological conception of the plague back into on the seventeenth century (1992/2016: 240).

Cunningham defends his stance on retrospective diagnosis at greater length in a 2002-paper. The debate over retrospective diagnosis is concerned with four key questions: Is it legitimate to diagnose patients of the (more or less) distant past using today's medical categories and techniques? Does such procedure allow us to correct

---

[1] It would be an interesting exercise to relate Cunningham's views on this issue to positions taken in the "naturalism versus normativism" debate (cf. Boorse 1975, 1976; Engelhardt 1996: Ch. 5; Kingma 2014).

the diagnoses of doctors of the past? Does it permit us to say what diseases past patients 'really' were suffering from, and without being aware of it? And can we use retrospective diagnosis in order to write medical histories of the *longue durée*?

Cunningham answers 'no' to all four questions. One motivation we have already encountered: the insistence that diseases are not natural kinds. This is because diseases essentially involve patients' experiences and social expectations. And these experiences and expectations are inseparable from contingent and historically varying ways of sense-making. Such sense-making involves—but is not limited to—the causal explanation of the disease (2002: 13). Cunningham maintains that it "is this 'cause' dimension (. . .) which means that disease is always experienced socially, that it is not just a biological phenomenon" (2002: 14).

Most elements of diseases—their causes, their development through time, their characteristic ways of being experienced, their outcomes—are observable. They are observable either by the patients themselves or else by "bystanders": family members, friends, colleagues, or medical doctors. Medical doctors are recognized as experts on the *causes* of diseases. It is they who are responsible for diagnosis and therapy. As Cunningham sees it, it is the social settings of diagnosing and treating the patient that justify calling the causal dimension of disease a social practice (2002: 14).

Cunningham is fascinated by the social processes in and through which medical practitioners come to believe in timeless medical kinds. Trying to be faithful to the "modern scientific world-view", they interpret diseases as encounters between nature (conceived of as a-social and a-historical) and the human body (also thought of as a-social and a-historical). In and through this social-collective process of reconfiguring diseases, medical practitioners make the socially-culturally embedded character of diseases invisible (2002: 14). And once a disease is reconfigured as something a-social, a-historical, and thus purely biological, it seems plausible and tempting to search for this disease across cultures and historical periods. Diseases seem to be stable across time, and stably present in all cultures (2002: 15).

Cunningham does not simply deny the stability of diseases across time; he denies that such stability is "open to proof or disproof". This is due to the "incommensurability" of old and new disease concepts, as we already learnt in the last section. What is new in the 2002-paper is the thought that we should focus less on "disease concepts", and more on "how diagnosis happens" (2002: 16). "Disease identity" is always and everywhere an "operational identity", that is, an identity determined by the operations employed for establishing the presence of a disease. Cunningham is convinced that two ideas follow from viewing diseases in terms of their operational identity. First, "(. . .) the identity of any disease is made up of a compound of elements, of which the biological or medical is only one, and sometimes the least important one (. . .)", and second, "(. . .) you die of whatever your doctor says you die of" (2002: 17). Put differently, diseases are internally related to operations constituting social practices of diagnosis; operations of diagnosis change radically over time; one cannot belong to social practices at another historical time; hence: one cannot be diagnosed by the criteria of a distant historical period. Moreover, it is impossible to become enculturated into a social practice of which no members still exist. We are thus inescapably "outsiders" to past diagnostic practices and can at most try to offer partial and unverifiable translations and interpretations (2002: 17).

Cunningham offers two illustrations for his position. One is the case of his own father who died of cancer in 1987. Cunningham recounts the doctors' diagnostic reasoning and medical interventions. And he concludes that given today's operational identity of cancer, there was "no alternative diagnosis possible in this case" (2002: 19). We also learn of the father's unwillingness to learn of the cancer diagnosis. The latter is related to prevalent contemporary reactions to cancer, for instance as these have been studied by Susan Sontag (1978). This information is meant to convince us that the father's experience of suffering was inseparable from cultural tropes about cancer. This experience and these tropes were not ways of interpreting or framing the "real biological disease"; they were an essential part of the disease itself: "There is no 'real' disease, with an identity separate from its sufferers at any given time, which can be separated out as a timeless entity for us to give our modern labels to, years—centuries—after the events" (2002: 20).

The second, related, illustration is the juxtaposition of three tables listing causes of deaths, and their frequencies, at different times and places. The first one covers 1632-London, the second and third England and Wales in 1837 and 1999. There are striking differences between the lists and frequencies. For our context, it is worth highlighting in particular the position of "cancer": "The tables reveal that cancer is now the cause of death of 25% of the population of England and Wales. In 1632 in London just ten deaths out of over nine thousand were attributed to 'cancer or the wolf'. In [the 1837] statistics 'carcinoma' (...) killed (...) less than 1%" (2002: 33). This does not mean that cancer had become more prevalent or that the doctors of 1632 or 1837 failed to spot cancer patients. Such quick observations fail to do justice to the fact that "these three moments are separated not just in time but also in ways of thinking. (...) What counted as a disease and, more particularly, precisely how a disease was diagnosed, had changed beyond recognition" (2002: 34).

5.3 Jewson's Medical Cosmologies and the Modes of Production

The sociologist N. D. Jewson's paper "The disappearance of sick man from medical cosmology" is an early landmark in the "social-constructivist" history of medicine. It was first published in 1976, and republished in 2009. Jewson's historiography differs substantially from Cunningham's. Jewson aims to explain medical theories and practices in sociological terms; Marxist ideas are particularly central.

Jewson follows Erwin Ackerknecht (1967) in distinguishing, in Western Europe from the 1770s to the 1870s, between three "modes of the production of medical knowledge": "Bedside Medicine, Hospital Medicine and Laboratory Medicine" (1976/2009: 622). One core concept of Jewson's analysis is "medical cosmology". Medical cosmologies correspond roughly to Kuhnian paradigms: they contain metaphysical claims about the medical domain as a whole; and they direct research, methodologies, and testing. It is the role of sociology and history to analyse how medical cosmologies are "generated, sustained, and developed within a specific social group" (1976/2009: 622–3). Medical cosmologies structure all social relationships producing and maintaining medical knowledge (1976/2009: 623). Jewson combines the Kuhnian "medical cosmologies" with a Marxian rendering of the mode of production of medical knowledge. The central elements of this mode are "the patrons", "the medical investigators", "a system of patronage" (binding patrons and investigators together); assumptions about the domain of medical knowledge; the "occupational activity of medical investigators"; and the "product", that is, medical theoretical knowledge (1976/2009: 623).

In the period under investigations, Bedside Medicine came first. Its leading centre was Edinburgh. Bedside Medicine focused on the "sick-man" [sic!] as a "conscious human totality". Its main forms of inquiry were "phenomenological nosology and speculative pathology". Despite numerous disagreements and many conflicting theories, there existed a rough consensus on the thought that disease should be defined "in its external and subjective manifestations rather than its internal causes" (1976/2009: 623).

Bedside Medicine was challenged during the first few decades of the nineteenth century. A new medical cosmology emerged in Paris: "Hospital Medicine". Medical investigators now began to focus on the "morbid events" in the patients' bodies. Doing this in a systematic fashion became possible due to the fact that many patients were brought to hospital wards. The most important new product of Hospital Medicine were "correlations between external symptoms with internal lesions". Patients' symptoms were carefully recorded while the patients were alive, and autopsies performed upon their death. Patients' own reports of their sensations and feelings were secondary compared with the doctors' physical examinations. A further key innovation was the use of statistics to evaluate the effectiveness of different remedies (1976/2009: 624).

"Laboratory Medicine" was introduced in Prussian universities from around 1850. The main novelty was the introduction of natural-scientific ideas and instruments into medicine. Histology and physiology became central fields. Pathology was based on experimental physiology. Life was rendered as a process occurring in and between interacting cells; disease was a disturbance in these interactions. All of which means that "[m]edical practice became an appendage to the laboratory". Medical investigators advocated materialist ways of understanding biological phenomena and dismissed vitalism (1976/2009: 625).

Medical cosmologies differ in their orientation towards "person and object". In a person-orientation, the focus is on the specific attributes of individuals, and assessments of their health is regarded as open to discussion and negotiation. Individual needs are important and must be recognized. Moreover, life, disease and death are interpreted as processes that ultimately are mysterious and inscrutable. In consequence, there exist many different ways of making sense of one's own physical and mental experience. In object-oriented cosmologies, the category of the person is no longer in the central position. The role of doctor and patient are separated more sharply. Objective and quantifiable properties are at the centre of the medical investigators' attention. Phenomena of life and death are theorized as "physico-chemical processes (...) The study of life is replaced by the study of organic matter" (1976/2009: 626). It is obvious from the above, that Bedside Medicine was an instance of person-oriented medicine, and that Hospital and Laboratory Medicine leaned towards an object-oriented approach.

It remains for Jewson to link the different conceptions of medicine to the social-political orders. During the time of Bedside Medicine, the patrons came from the ruling feudal class, and they controlled the doctors' activities. The patrons were the patients that really mattered. And their social position allowed them to demand that doctors pay utmost attention to their individual needs. Doctors competed over the favour and recognition of their patron-patients. Medical theories reflected this constellation. The patron-patients expected medical theories to give them "a recognizable and authentic image of [their] complaint", and to recognize their "integrated psycho-somatic totality" (1976/2009: 627).

Hospital Medicine was an important step towards an object-oriented medical practice and theory. As far as social transformations were concerned, two developments were particularly important. The sick person was increasingly subordinated to the medical investigator; and medical doctors were controlled no longer by rich patron-patients, but by senior members of their profession. The French Revolution played a crucial role in these changes. Many medical institutions of the ancient regime were abolished and hospitals became central places for diagnosis, therapy, death, teaching and learning. Hospital doctors became the medical elite. The patients of the hospitals were poor and powerless vis-à-vis the doctors. The social changes in the relationships between doctors and patients left their marks on medical knowledge. Diseases were no longer couched primarily in terms of patients' experiences and surface changes; diseases were rendered as "a precise and objectively identifiable event occurring within the tissues, of which the patient might be unaware". Medical doctors came to regard themselves as a "homogeneous occupational group" and began to control recruitment, training and practice. The primary relations of a doctor were with their respective colleagues, not their patients. Medical practitioners also began to form associations and founded specialized medical journals (1976/2009: 627–8).

Laboratory Medicine went even further in the last-mentioned direction. Here the patient "was removed from the medical investigator's field of saliency altogether". The central medical investigator was not the elite clinician, but the "scientific research worker". The key steps towards Laboratory Medicine were first taken in Prussia. The central patron was the centralized state. There was competition between the medical practitioners and the laboratory research scientists, with the later controlling more resources and emerging as a "self-confident elite". For this elite, medicine became a mere chapter in a much wider interest in understanding organic matter. Medicine was envisaged to be something of an applied science. The laboratory medical scientist was at a greater spatial and social distance from the patient. At the same time, the training of medical doctors was further unified and controlled. And a new form of competition emerged: it was not a competition in which the patients declared the winners (as it had been in Bedside Medicine); it was a competition in which "cosmopolitan and disciplinary" criteria were decisive (1976/2009: 629–30).

5.4 Mol on Enactment and Ontological Politics

It might come as a surprise that I am treating Annemarie Mol's book *The Body Multiple* (2002) as paradigmatic of a third form of medical historiographical relativism. After all, Mol is an anthropologist and philosopher, and thus not a card-carrying historian. Still, the core of Mol's book is an *ethnography* in one Dutch hospital around the turn of the millennium, and as such willy-nilly historical. At the same time, there is no denying that Mol's writing are heavier with philosophical theorizing than texts by Cunningham and Jewson. I therefore have to begin with some philosophical preliminaries.

The key term in Mol's book is "ontology". But Mol is not doing ontology in the style of Heidegger or David Lewis. She is concerned with "ontological politics" (2002: 60, 1999). As she explains in an earlier paper with this title, "ontological" refers to "ontologies", that is, "ontology" in the plural. Different realities are "shaped", "enacted", "performed", or "done" within different "practices", that is, "realities have become multiple." "Politics" highlights the fact that the "process of shaping" (etc.) is "both open and contested" (1999: 75). Mol distinguishes between realities being "plural" and realities being "multiple". Talk of *plural* realities, or "pluralism", can take two forms. The first is "perspectivalism", or the view that of one and the same reality there can exist "mutually exclusive perspectives, discrete, . . . side by side." (For instance, the perspectives of doctors and patients are taken as "equals"; Mol 2002: 11, 20). The second form of pluralism is "constructivism"; it holds that whatever reality we find now, "alternative 'constructions of reality' might have been possible". Talk of realities being "multiple" is meant to say that realities do "not precede the mundane practices in which we interact with" them, and that realities are "enacted rather than observed" (1999: 75–77).

Not least because of Mol's writings, "enactment" has emerged as a crucial word in present-day "Science and Technologies Studies" (STS). Steve Woolgar and Javier Lezaum (2013) summarize the central ideas as follows. First, enactment theorists reject explanations in terms of "context" (2013: 323); that is, context is not an independent variable but itself "an emergent property of interaction available to its participants". Second, objects are "realized in the course of [. . .] certain practical activit[ies]", or, put differently, practices have "generative power . . . in the constitution of reality". Third, enactment theorists want their "ontological investigations" to be "empirical" and focus on "the practices of world-making". Fourth, "enactment" is meant to imply—more strongly than "social shaping" or "performing"—that there are no essences and that "entities" do not "pre-exist our apprehension of them" (2013: 324). Fifth, there is no *Ding an sich* beyond and below different enacted realities; rather rendering different enacted realities as "the same 'thing'" is a local achievement, "the upshot of active practical work" (2013: 325). (This seems what Woolgar and Lezaun refer to when they speak of enactment theory as a "situationism" (2015: 463)). Sixth, any type of object can be enacted. And seventh, the focus on ontology brings back political questions. Perspectivalism and constructivism typically end up with "irony, detachment or tolerance". But ontologies confront us with a "cosmopolitical choice: in which world would you like to live, and what can you do to bring such a world into being?" (2013: 325–6).

In *The Body Multiple* Mol takes herself to do "empirical philosophy" of medicine (2002: 1, 4). The empirical input comes from her own case study of how atherosclerosis in the lower limbs is diagnosed and treated in one Dutch university hospital (2002: 2). Mol starts with an ethnographic description of a consultation in the outpatient clinic. The patient talks to the doctor about pain in her lower leg, using a mixture of ordinary language and medical terms learnt from her GP and various other sources. Mol stresses that such reports are specific to time and place. For instance, the preserved doctor's notes on female patients' complaints in eighteenth-century Eisenach (a small town in Germany's Thuringia) show that lexica for capturing bodily experiences can change fundamentally over time (Duden 1991). Mol maintains that "the bodies of these women were different from those that we inhabit now" (2002: 26).

The interactions in the consultation room between the doctor, the patient, the patient's leg, a letter from the patient's GP, the doctor's desk, and the floor on which the patient walks, together "enact" an entity the doctor calls "intermittent claudication" (pain caused when the flow of blood to exercising muscles is diminished) (2002: 23). Mol emphasizes the role of the many different entities in the consultation room; again, this is a move away from philosophical epistemology and an insistence that knowledge is primarily located "in activities, events, buildings, instruments, procedures, and so on" (2002: 31). Intermittent claudication as a disease entity exists only in this setting. It is enacted only here. Ask a doctor why he moves from the patient's report to talking about "intermittent claudication" and they will refer back to all the procedures, instruments and events in the consultation room.

The textbook view of atherosclerosis focuses on pathology. It tells us that atherosclerosis is the build-up of substances (like fats or cholesterol) in the walls of arteries, a build-up that causes arteries to narrow. Mol asks us to put aside such textbook versions of medical knowledge and focus on medical practices instead. She goes so far as to say that, at least as far as atherosclerosis is concerned, pathology is not the "foundation" but something like "an afterthought": its typical place of enactment is the laboratory where the patient's amputated leg is studied with microscopes and surgical instruments (2002: 38). But note: the pathology is not "in" the amputated leg; it is rather enacted in the whole set-up and functioning of the laboratory (2002: 48).

The consultation room of the outpatient clinic and the laboratory are not the only places where different disease entities are enacted. Others are the X-ray departments where contrast dye is injected into specific arteries so that their narrowing becomes visible in X-rays, the hematology lab, the operating theatre, or the treatment room where the patient is given medical advice. All of these locations come with their specific enacted disease entities. The different techniques applied in the different settings lead to a "multiplication of reality". Mol rejects talk of incommensurability, but only if it is taken to refer to a "semantic notion". But she accepts incommensurability if it means that different instruments produce readings for which there is "no common measure" (2002: 72).

How do the different enacted disease entities hang together? How is "coherence" established between them? To ask these questions is already to accept Mol's central point, to wit, that the different disease entities are not simply phenomena related to an underlying noumenon, a disease *an sich*. Mol favourite way of putting her position is the slogan that a disease "more than one and less than many" (2002: 81). In the hospital, the various groups continuously negotiate how, in specific situations, the products of different techniques and devices are to be related to one another. Such attempts at relating objects to one another do not have to go as far as a single "shared, coherent ontology" (2002: 115). Such shared ontology is not necessary for therapeutic success. On the contrary: "That the ontology enacted in medical practice is an amalgam of variants-in-tension is more likely to contribute to the rich, adaptable, and yet tenacious character of medical practice" (2002: 115).

5.5   The Spectrum of Relativism

Why do I call Cunningham's, Jewson's and Mol's positions 'relativistic'? And how do their relativisms compare to each other, and relativistic positions in HPS-ST? To answer these questions, we need a few terminological distinctions.

Note first of all, that versions of relativism can be distinguished by the realm to which they apply: for instance, a relativism about *knowledge* is called 'epistemic', a relativism about *what there is* 'ontological', or a relativism about *concepts* 'semantic'. Relativisms can also differ by the contexts to which knowledge, ontology or concepts are relativized: this gives us, inter alia, *cultural* or *standpoint* relativisms.

A further important distinction is that between *descriptive*, *normative*, and *methodological relativisms*. Descriptive relativism makes the empirical claim that we find fundamentally different standards in different contexts. Forms of methodological relativism hold that we should investigate different phenomena *as if* they were of equal value. Finally, normative relativism demands that we regard the idea of absolute truths or absolute standards as flawed, absurd or incoherent.

We can now turn to the main elements of a relativist position. None of these elements, on its own, is sufficient as a definition of relativism; it's their various combinations or interpretations that define what one might call "spectrum of relativism" (Kusch 2021). I shall explain this spectrum using epistemic relativism as my example. Other forms of relativism can be generated by replacing 'epistemic' with 'conceptual', or 'ontological'.

> (DEPENDENCE) A belief has a status (e.g. 'knowledge') only relative to epistemic standards.

Such epistemic standards might be, for example, rules or *exemplary scientific achievements of the past*. (Think Kuhn on exemplars.)

> (PLURALITY) There is (has been, or could be) more than one set of epistemic standards; the standards of different sets can conflict. (I shall write 'S' for such sets.)

Relativism thus allows for the possibility that our own S is without an *existing* alternative.

> (CONFLICT) Epistemic verdicts, based on different S, sometimes exclude one another. This can happen either ...
>
> (a) because the two S license incompatible answers to the same question, or
> (b) because the advocates of one S find the answers suggested by the advocates of another S unintelligible.

(b) captures the cases of Kuhnian "incommensurability" (Kuhn 1962/2012, 2022)

> (CONVERSION) In some cases, switching from one S to another has the character of a 'conversion': that is, the switch is underdetermined by S, evidence or prior beliefs, and is experienced by the converting X as something of a leap of faith.

CONVERSION plays of course an important role in Kuhn, too (Kuhn 1962/2012: 338). "Conversion" is a fancy expression for something like a contingent and underdetermined assessment. In a scientific "crisis", scientists have to weigh up two score-sheets: the many successes and anomalies of the old paradigm against the promises and few successes of the new paradigm. There is no neutral set of criteria to compel the answer, and rational people can come to different decisions.

Note also that the switching from one S to another does not have the character of a 'clean sweep'. Not all standards are affected by such a transition. For example, Aristotelian geo-centrists and Copernican helio-centrists shared many epistemic norms or exemplars. What changed was thus a subset of the geocentric standards. But this subset was weighty enough to justify speaking of a 'fundamental change', or "change of paradigm."

> (SYMMETRY) Different S are symmetrical in that they all are:
>
> (a) based on nothing but local, contingent and varying causes of credibility (LOCALITY);
> (b) impossible to rank except on the basis of a specific S (NON-NEUTRALITY);
> (c) impossible to rank since the evaluative terms of one S are inapplicable to another S (NON-APPRAISABILITY);
> (d) equally true or valid (EQUAL VALIDITY).

SYMMETRY is, in many ways, the heart of relativism. It takes different forms, formulated by (a) to (d). LOCALITY runs directly counter to absolutist suggestions according to which there is a unique S that ought to be accepted by every rational being; enables us to capture truths that 'are there anyway'; or would be accepted by an ultimate, final, science.

LOCALITY allows that the proponents of the standards of one S may (legitimately) criticize the standards of another S. LOCALITY is naturally combined with NON-NEUTRALITY: when we rank different S, we must always rely on, or take our starting point from, some S other. I shall call the resulting relativism 'locality-relativism'.

A much stronger claim is advanced by NON-APPRAISABILITY. It insists that evaluative terms can only operate *within* an epistemic practice (as defined by a given S). This precludes the option of legitimately evaluating epistemic practices other than one's own. To give this form of relativism a name, I propose calling it a "relativism of distance"—a term introduced by Bernard Williams in a different context (1981). Williams emphasizes two central elements in this type of relativism. First, the confrontation with the other culture is merely "notional". That is to say, going over to the other side is not a real or "live" option for oneself. One cannot imagine adopting the view of the other side without making an endless number of changes to one's system of beliefs or values. And second, one's own "vocabulary of appraisal" seems out of place: "(. . .) for a reflective person the question of appraisal does not genuinely arise (. . .) in purely notional confrontation" (1981: 141–142). In other words, the ways of thinking and acting of the other side are so foreign that a reflective person finds them difficult to even evaluate.

EQUAL VALIDITY goes further still in declaring all S to be equally correct or valid. It is interesting to note that most criticisms of relativism assume that EQUAL VALIDITY is an essential component of relativism (Boghossian 2006; Seidel 2014; Baghramian and Coliva 2019). But it is hard to find any card-carrying epistemic or moral relativist who commits themselves to this element (Kusch 2019). Be this as it may, I shall speak of 'equal-validity relativism' to pick out positions that honor EQUAL VALIDITY.

5.6   Cunningham's Relativism of Distance

How do Cunningham's views on incommensurability and retrospective diagnosis relate to relativism? Cunningham himself acknowledges (in 2016) that his "approach looked relativist to some people, as if I was saying that all ideas and truths are only—merely—time-specific and contingent." In reply, Cunningham refuses the title of an "in-principle relativist"; he merely wishes to be an "open-minded historian" (2016a: ix). I think we can say more. And we can best do so, by showing that his views harmonize with some central contentions in Kuhn.

Kuhn is of course the first address when it comes to incommensurability. His views are the most sophisticated in his recently published "late writings" (Kuhn 2022). Five claims seem most pertinent here. First, the later Kuhn no longer speaks of incommensurability in the cases of "methods, problem-field, and standards of solution", but only in the case of "lexica" (2022: 7319). This is perhaps why the later Kuhn prefers the terms "untranslatable" (2022: 1501) or "simply unsayable" (2022: 1899) to the earlier "incommensurable". Second, the later Kuhn keeps the earlier provocative idea according to which a change in paradigm—or lexicon—amounts to a "world change". But he now emphasizes that such world change involves both the social structure of the relevant scientific community and the world around them

(2022: 1982). Third, untranslatability results from fundamental changes in taxonomic trees; what is important here are less the "features" that identify taxa, but the relations between taxa themselves. Kuhn's favourite examples are the different astronomical taxa in astronomic lexica before and after Copernicus (2022: 1906), or the different notions of motion in Aristotle and Newton (2022: 1553, 3190). It is impossible to translate Aristotle's claim about the status of the Sun in a truth-preserving way. Given our lexicon, the best we can say is "The Sun is a planet":

> But the translation is worse than imperfect. Using our lexicon, "The Sun is a planet" is false. The Greeks, we therefore suppose, were mistaken. But in the Greek lexicon, the Sun was a planet; it was, that is, more like Mars and Jupiter than like any of the stars. The corresponding Greek sentence was therefore true, not simply believed to be true. (2022: 1660–4)

Fourth, Kuhn therefore dismisses the thought that sentences of past lexica can be given a truth-value in our language. At least this is ruled out if we think of truth as correspondence (2022: 2655–2702). Kuhn therefore feels compelled to brush aside as "ill formed" questions concerning the existence of past taxa. He mentions "witches" and "phlogiston" as examples (2022: 1982). Fifth, and finally, Kuhn is happy to acknowledge that his work is based on a form of "methodological relativism" (2022: 2641–6). The expression refers to suspending his knowledge of today's science when aiming to grasp works of the past. Kuhn also believes that the choice between lexica is "interest-relative (...) instrumental and relativistic". When a community is faced with such choice, the "truth-value game" can no longer be played. The latter presupposes a shared lexicon and a high degree of "solidarity". Kuhn claims that his relativism is no "relativism with respect to truth" (2022: 1943–9). It is not truth but "effability" that is being relativized. It is not possible for the same proposition (involving taxa terms) to appear in two different communities with different lexica. It follows that such proposition cannot receive different truth values in the two communities (2022: 1897).

How does Kuhn relate to my "spectrum of relativism"? It seems that his position is close to a form of relativism I earlier called (following Williams): "relativism of distance". There are two elements in my spectrum of relativistic positions that are particularly important here. In the case of CONFLICT, relativism of distance opts for possibility (b), that is, the notion that advocates of different epistemic systems are not giving incompatible answers to the same questions, but rather find the answers of the other epistemic system unintelligible. And with respect to SYMMETRY, relativism of distance leans towards NON-APPRAISABILITY: it is impossible to rank epistemic systems since the evaluative terms of one S are inapplicable to another S.

We can now return to Cunningham. How do his views relate to those of Kuhn, and thus to relativism of distance? There is a considerable degree of commonality between Kuhn and Cunningham. Like Kuhn, Cunningham renders incommensurability primarily as an issue relating to categories. And in a way that parallels Kuhn's claims about "motion" or "phlogiston", Cunningham writes that many medical claims about plague or consumption can no longer be meaningfully used in today's bacteriological world. The pre-bacteriological position "is now a lunatic view"

(1992/2016: 239). Other than Kuhn, Cunningham says little about lexica; he instead focuses on one disease and its various identities. But it is not difficult to decide what Cunningham would say about lexica. Take Kuhn's claim that fundamental shifts in lexica lead to changed social and natural worlds. Cunningham would agree (as Cunningham 2007 shows clearly). After all, fundamental changes in lexica come with changes in medical diagnostics and forms of therapy, new authorities (e.g. the lab scientists), new architectural measures (think of hospital architecture, or means to promote hygiene), changing patterns and risks of infection, and changing populations of bacteria and viruses.

Admittedly, Cunningham says little about transformations in taxonomic trees, Kuhn's main focus in his late writings. But the similarities and differences between consumption/tuberculosis and other diseases changed of course radically as the central criteria shifted from symptoms to pathology to etiology. Cunningham's awareness of such issues is obvious from his 2002-paper where he reproduces the possible causes of death from the three different centuries. These lists differ radically. It is thus also natural to read Cunningham as agreeing with Kuhn on the limits of translation: there cannot be a straightforward translation of propositions formulated using past lexica into propositions formulated present lexica. Consumption understood in medieval or early-modern ways has no place in modern medicine; the further we go back in time, the more obviously this is true. Just think of the time when the ultimate cure was to be "touched and blessed" by a King or Queen (Bynum 2012: 39).

5.7 Is Cunningham Is Committed to Equal-Validity Relativism?

Nick Binney (unpublished, 2022) has taken issue with Cunningham's views on retrospective diagnosis. Binney speaks of Cunningham's position as an instance of "extreme" or "silly and unacceptable relativism". As Binney sees it, silly relativism "reduce[s] knowledge to belief, which destroys the possibility and purpose of inquiry (...) [It] makes it impossible for doctors to be wrong, or to make mistakes, or to improve their practice. (...) If they accepted this sort of epistemology, historians themselves would be unable to distinguish their own work from pure fiction" (2022: 18). Is Binney right?

To begin with, Cunningham's position does not rule out mistakes in diagnosis or therapy, or errors in historical work. For instance, when the doctors removed a large part of Cunningham's father's bowel, and when they joined up the healthy ends of the intestines, the doctors could have been careless and sloppy. They could have removed too much or too little of the bowel, and they could have joined up the intestines too loosely. Or, for another example, Cunningham reports that shortly before his father's death, the doctors had concluded that they were unable to rescue the 76-year-old. Again, an autopsy after the father's death might have revealed that a further operation could have saved his life. I do not find any indications that Cunningham would deny any of this.

And what goes for doctors also holds for historians. Cunningham does not reduce knowledge to mere belief, or truth to mere opinion. There are matters of fact about, say, the three tables of disease categories from different time periods. Maybe Cunningham overlooked that one of the three tables was an elaborate spoof, or that the third table was published in 1989 rather than 1999. He does not rule this out a priori. In other words, Cunningham's position is not the pseudo-Hamletian: 'Nothing is true or false, but thinking makes it so.'

Moreover, Cunningham does not claim that all talk of progress or regress in the medical history is unintelligible and best avoided. His position allows him statements like the following: infant mortality and the mortality of birthing mothers have decreased dramatically over the past 500 years; or bacteriology has enabled medicine to save numerous lives. Indeed, Cunningham says such things repeatedly in his BBC Radio series "The Making of Modern Medicine". And he is not contradicting himself in being enthusiastic about such advances. The position of the 2002-paper only rules out claims like: we now know what consumption really was and is; we have come to understand the plague ever better.

Needless to say, there are forms of relativism we should classify as "silly" and "unacceptable". To my mind, 'equal-validity relativism' in the realm of science— i.e., all scientific statements of different epistemic systems, paradigms or thought styles are equally valid—qualifies as silly and unacceptable. But Cunningham does not sign up to equal-validity relativism. Indeed, given his own account of incommensurability, he would not be allowed to claim that all systems of beliefs or norms, or all practices, are equally valid. It would involve him in taking a stance with respect to all possible systems or practices; it would be to deny or ignore the very incommensurability of systems and practices Cunningham insists on. 'Equally valid' implies 'comparable', 'measurable on the same scale'. And this is the very opposite of incommensurability.

5.8   Jewson's Locality-Relativism

In my analysis of Cunningham's relativism, it proved helpful to use Kuhn as a 'shoehorn' in order to 'squeeze' Cunningham into my spectrum of relativism. We need such 'shoehorn' also in Jewson's case. But here this role is performed by the "Strong Programme" in the "Sociology of Scientific Knowledge" (SSK). Barry Barnes and David Bloor, the architects of this position introduce relativism as follows:

> The simple starting-point of relativist doctrines is (i) the observation that beliefs on a certain topic vary, and (ii) the conviction that which of these beliefs is found in a given context depends on, or is relative to, the circumstances of the users. But there is always a third feature of relativism. It requires what may be called a 'symmetry' ... postulate [:] ... the incidence of all beliefs ... calls for empirical investigation and must be accounted for by finding the specific, local causes of this credibility. (1982: 22–23)

Note in particular the phrase "local causes of ... credibility": I used it above in order to specify one rendering of SYMMETRY: the thought that different sets of standards are based on nothing but local, contingent and varying causes of credibility.

Elsewhere, Bloor formulates the "Strong Programme" as follows:

(1) It would be causal, that is, concerned with the conditions which bring about belief or states of knowledge. Naturally there will be other types of causes apart from social ones which will cooperate in bringing about belief.

(2) It would be impartial with respect to truth and falsity, rationality or irrationality, success or failure. Both sides of these dichotomies will require explanation.

(3) It would be symmetrical in its style of explanation. The same types of cause would explain say, true and false beliefs.

(4) It would be reflexive. In principle its patterns of explanation would have to be applicable to sociology itself. (...) (1991: 7)

It is (3) that is generally seen as the central relativistic component of this programme. If we take rationality and irrationality, true and false beliefs, to have the same types of causes, are we not treating them as equals? And does this not amount to equal-validity relativism? (cf. Brown 2004). Barnes and Bloor disagree: their relativism does not treat all beliefs as equally true or equally valid. Beliefs are equal only insofar as they all can be studied by SSK (1982: 22).

Thus far, we have moved in the realm of methodological relativism. But Bloor is ready to also defend relativism as a philosophical position in its own right. Bloor writes: "(...) relativism is the negation of absolutism. To be a relativist is to deny that there is such a thing as absolute knowledge and absolute truth." Bloor explains what he means by "absolute knowledge" as follows: it would be "perfect, unchanging, and unqualified by limitations of time, space, and perspective. It would not be conjectural, hypothetical, or approximate, or depend on the circumstances of the knowing subject" (2011: 436–7).

Now we can return to Jewson. Jewson's study nicely exemplifies at least three of the four elements of the "Strong Programme": Jewson's study is *causal* in that it seeks to identify some of the causes that made it natural for different forms of medicine (practices and theories) to arise; Jewson's investigation is *impartial* in so far as it does not tell us which of the different medical cosmologies is closer to the truth, or is superior in rationality to the others; and Jewson's paper is *symmetrical* in its style of explanation: the same general types of social-political causes explain both true and false beliefs about health and disease. There are, in other words, no absolutes in Jewson's study, just historical contingencies of changing politics and medical cosmologies. The same form of methodological relativism seems characteristic also of other authors who follow Jewson's paradigm. This all fits nicely with *locality-relativism*.

It is inviting to compare and contrast Jewson's locality-relativism with Cunningham's and Kuhn's relativism of distance. Cunningham's periodization roughly fits with Jewson's; after all, the shifts in the disease identities of consumption/tuberculosis parallel the shifts from Bedside to Hospital to Laboratory Medicine. However, Jewson does not gesture towards semantic incommensurability as a by-product of medical Kuhnian revolutions. He does not claim that the practitioners of different forms of medical cosmology had difficulties understanding each other, or that historians today struggle to understand the medical practitioners of past cosmologies. To some limited degree this may be in line with SSK's critical view of Kuhnian revolutions. Barnes (1982) objects that Kuhn is wrong to restrict fundamental semantic change to revolutionary science. Barnes claims that the historical record proves otherwise: "any particular change which occurs in a revolutionary episode can occur equally in a period of normal science, whether it be meaning change, technical change, the inventions of new problem-solutions, or the emergence of new standards of judgements." Once this is recognized, Barnes claims, we have every reason to reject Kuhn's "functionalist" account of revolutions: "Hence there is nothing to compel a leap out of the system [that is, out of the old paradigm]: nothing makes it necessary to replace, rather than to develop, existing practice" (Barnes 1982: 86). Barnes' critique of the need for revolutions is not necessarily an attack on the very idea of incommensurability. Incommensurability may be the result of long-term gradual development.

5.9   Mol's Ontological Relativism: Between Equal Validity and Distance

What Kuhn is to Cunningham, or what Barnes and Bloor are to Jewson, Bruno Latour and Nelson Goodman are to Mol. Latour is highly critical of "social constructivism", that is, the views of authors like Barnes, Bloor or Jewson. To begin with, Latour denies that scientific contents and technologies can ever be explained by sociology (2005: 1672). One argument is that the social causes often invoked by social constructivists—"Society, Capitalism, Empire, Norms, Individualism, Fields, and so on" (2005: 2572)—are not specific enough to explain the peculiarities of specific contents (2005: 1958).

Moreover, Latour targets the "cultural relativism" of Claude Levi-Strauss, arguing that it was based on a "solid absolutism of natural sciences". For Levi-Strauss there were many different cultures, but one nature (2005: 2191). And he defended the equal validity of cultures with the thought: "give the primitives a microscope, and they will think exactly as we do" (1993: 1995). Different cultures are theorized as "so many more or less accurate viewpoints on that unique Nature. Rationalists will insist on the common aspects of all these viewpoints; relativists will insist on the irresistible distortion that social structures impose on all perception" (1993: 2101). Latour's master argument against cultural relativism is the claim: "There are no cultures. ... There are only natures-cultures" (1993: 2083, 2098). There is nothing that beliefs about nature can be *relative to*. Latour is scathing also about incommensurability and relativist standpoint epistemologies. Standpoints can always be modified and switched in and out of: "Don't believe all that crap about being 'limited' to one's perspective. All of the sciences have been inventing ways to move from one standpoint to the next, from one frame of reference to the next, for God's sake: that's called relativity" (2005: 2750). By the same token, the sciences work ceaselessly to make commensurable what at first was incommensurable (1993: 2276). Accordingly, Latour claims to be able to "travel from one frame of reference to the next, from one standpoint to the next" (2005: 1753).

Still, Latour happily answers "But of course, what else could I be?" to the question whether he is a relativist (2005: 2753). And there is a sense in which he too commits to an equality, if not of cultures, then at least of "collectives" of humans and nonhumans. The only dimension in which such collectives differ is "size"; it is this, and this alone, that allows some of them to "dominate" others (1993: 2173). In order to distinguish his relativism from that of Levi-Strauss or social constructivists, Latour qualifies it with the terms "empirical" (2005: 3479), "natural" (1993: 2135) and "relativist" (1993: 2264). Or he replaces "relativism" with the term "relationism" (1993: 2286). The thought is that relativism is more a strategy of research than a philosophical position; and it zooms in on the ways in which "objectivity" and the "relative universal" are created with local and contingent means. For instance, "[it] is possible to verify gravitation 'everywhere', but at the price of the relative extension of the networks for measuring and interpreting" (1993: 2399).

It is not easy to locate Latour relativism/relationism in my 'spectrum'. His focus on the local puts Latour in the proximity of locality-relativism. The remarks about 'size is all that matters', can be interpreted in different ways: if size is opposed to intellectual tools and evidence, then the result is equal-validity relativism. If 'greater size' means, or includes, 'better evidence' or 'better standards', then perhaps the result is a non-relativist position.

Finally, sometimes Latour expresses his differences with Bloor by suggesting that whereas Bloor is obsessed with epistemology, he (that is Latour) is focussed on "the question of ontology". That is to say, Latour claims not to be concerned with the representations but with "associations" or "networks" of "actors", human and nonhuman (2005: 2223).

This is not the place to debate Latour's position. Still, a couple of points are worth flagging since they help us to understand the relativistic spectrum in relation to historiography. Latour is right to object that some (early) social constructivists (like Jewson) were too quick in explaining scientific contents or technologies in terms of large and vague social causes like mode of production, feudalism, capitalism, and the like. And yet, I fail to see why this is an unavoidable vice rather than an accidental or contingent sin, and a sin *even by the lights of social constructivism itself.* Arguably, social constructivists over time have learnt to tell more sophisticated histories (e.g. Kusch 1999: Part I).

Latour's position vis-à-vis relativism is complex: he dismisses versions that involve incommensurability, cultures, perspectives, or standpoints. But Latour endorses relativism, or relationism, when it is rendered as the empirical programme of studying how local and relative means can bring about something quasi-universal (like gravity or the railroads). How good are his arguments? Concerning Latour's empirical programme of relativism, I can be brief. I find it important and I have learnt a lot from studying it.

I am less impressed with the ways in which Latour discusses incommensurability and the limitations of standpoints or perspectives. Latour insists that "all the sciences have been inventing ways to move from one standpoint to the next", that is, to make commensurable what at first was incommensurable. He also claims to be able to "travel from one … standpoint to the next". My first reply is that Latour has not made a publicly documented attempt to travel from the standpoint of a privileged white upper-class academic Frenchman to that of an exploited female migrant cleaner in Paris, or to that of a person with an intellectual disability. But even if Latour were able travel in this way, it would not show that such travel is possible for many others; people who are less well educated for example. Moreover, Latour has not shown for any of the Kuhnian cases of scientific revolutions that and how incommensurability was overcome.

We can now address Mol's position vis-à-vis relativism. She never comments on the topic explicitly. She does remark, however, that 'going ontological' means leaving behind "perspectivalism" with its "mutually exclusive perspectives, discrete, (…) side by side" on the one hand, and "constructivism" with its claim that "alternative 'constructions of reality' might have been possible", on the other hand (1999:76). These are obvious references to relativistic social constructivism and Kuhnian incommensurability. Has Mol overcome these positions with her ontological turn? I do not think so.

Focusing on "ways of worldmaking", as Mol does, has some parallels in earlier relativistic philosophy of science or metaphysics. One obvious parallel is Kuhn with his insistence that after a scientific revolution scientists live in a new world:

> (…) after Copernicus, astronomers lived in a different world (1962/2012: 117).
> (…) Lavoisier worked in a different world (1962/2012: 118).
> When [the chemical revolution] was done (…) the data themselves had changed. (…) we may want to say that after a revolution scientists work in a different world (1962/2012: 134).

An even more obvious parallel is Nelson Goodman's *Ways of Worldmaking* (1978). (Hacking once identified commonalities between Latour, Woolgar and Goodman. Hacking 1988) Here is an arresting passage:

> The physicist takes his world to be the real one … The phenomenalist regards the perceptual world as fundamental. … For the man-in-the-street, street, most versions from science, art and perception depart in some ways from the familiar serviceable world he has jerry-built from fragments of scientific and artistic tradition and from his own struggle for survival. The world, indeed, is the one most often taken as real; for reality in a world, like realism in a picture, is largely a matter of habit. Ironically, then… not only motion, derivation, weighting, order, but even reality is relative. (Goodman 1978: 20)

Goodman's and Kuhn's idioms of worldmaking are usually interpreted as instances of *ontological relativism*. We can approach such ontological relativism from the perspective of the spectrum of relativistic positions. The ontological relativist insists that claims about ontology are relative to ontological standards; that there can be more than one set of ontological standards; that ontological verdicts (on what there is) can conflict; that the switching from one set of standards to another can have the character of a conversion; and that different sets of ontological standards are symmetrical. Mol seems to accept versions of all five of these tenets. It is not totally clear which version of 'symmetry' best fits with her position. It is striking though that she makes no effort to distance herself from equal-validity relativism.

In explaining the parallels between Mol and Goodman, I do not mean to deny that Mol's position is highly original. The boldest new element in her ontology is "enactment". Different bodies are "enacted" in different practices or situations. Mol's strategy is reminiscent of the "bracketing of the world" in Husserlian phenomenology (see e.g. Kusch 1989). Phenomenologists aim to identify which acts of our "transcendental ego" make it possible for us to encounter different kinds of real and ideal objects. Phenomenologists claim that to do so we have to suspend our belief in the ready-made existence of these objects. Likewise, Mol wants to make visible how specific sets of interactions between humans and nonhumans make it so that a certain disease entity can be encountered as ready-made. And to make the role of the interactions visible, we must not assume that there is something fixed there from the start.

This is intriguing and suggestive and, to repeat, bold and original. Still, my main contention is this: the points at which Mol differs from, say Goodman, are not points that distance her from ontological historiographical medical relativism.

Mol's comments on incommensurability are also of interest here. On the one hand, the medical personnel in Mol's *The Body Multiple* seek to overcome incommensurability. Still, Mol goes out of her way to stress that such commensurability is a local and contingent achievement, no 'once and for all' translatability. Such "tinkering", such "bricolage" with terms seems to confirm rather than undermine the plurality of standpoints. Moreover, no defender of incommensurability has ever insisted that all instances of incommensurability are permanent. As Kuhn has it, they can be negotiated or mitigated if not by translation then at least by language learning.

The topic of incommensurability also connects Mol and Cunningham. There is a relevant connection between Cunningham's and Mol's discussions of the different identities of different diseases: that is, the symptomatological identity, the pathology, or the etiological identity. Whereas Cunningham talks about their 'successive' incommensurability, Mol analyses their co-existence in different parts of one and the same hospital. Does Mol's study refute Cunningham? After all, does she not show that the different conceptions of atherosclerosis can "live together"? I do not think that Mol's work refutes Cunningham's studies. The reason is the very emphasis in *The Body Multiple* on the local, temporary, contingent and isolated ways in which commensurability is achieved. Perhaps we should say that both Cunningham and Mol lean towards relativism of distance: temporal distance in Cunningham's case, spatial distance in Mol's.

5.10 Summary and Conclusions

In this paper, I have tried to show that one can find a variety of relativistic historiographical positions in the history of medicine. Cunningham, Jewson and Mol all lean towards epistemic or ontological historiographical relativism, but they do so in original and thought-provoking ways. Cunningham develops a medical analogue of Kuhn's incommensurability-relativism; Jewson's position is a counterpart of Barnes' and Bloor's sociology of scientific knowledge in the medical field; and Mol builds on, and pushes further, ontological-relativistic motifs familiar from the writings of Goodman and Latour. I used my spectrum of relativism as a grid for 'measuring' the variety of relativisms studied in this paper. Cunningham's historiography is a relativism of distance; Jewson's a locality-relativism. Mol's position has elements of an equality-relativism and a relativism of distance.

I have not attempted a systematic defence, criticism or refutation of any of the positions discussed here. I did so only insofar as I sought to put clear water between these position and straightforward versions of obviously unacceptable equal-validity (or "anything goes") relativism. This leaves open the questions whether relativism of distance is a defensible view, or whether the conceptual distinction between, say, relativism of distance and equal-validity relativism is ultimately justifiable. These are questions for another contexts.

References

Ackerknecht, Erwin. 1967. *Medicine at the Paris hospital 1794–1848*. Baltimore: The Johns Hopkins Press.

Baghramian, Maria, and Annalisa Coliva. 2019. *Relativism*. London: Routledge.

Barnes, Barry. 1982. *T.S. Kuhn and social science*. London: Macmillan.

Barnes, Barry, and David Bloor. 1982. Relativism, rationalism and the sociology of knowledge. In *Rationality and relativism*, ed. Martin Hollis and Steven Lukes, 21–47. Oxford: Blackwell.

Binney, Nick R. 2022. *Framing disease—Pictures or conversations*. Unpublished manuscript.

Bloor, David. 1991. *Knowledge and social imagery*. 2nd ed. Chicago: University of Chicago Press.

———. 2011. Relativism and the sociology of scientific knowledge. In *The Oxford companion to relativism*, ed. Steve Hales, 433–455. Oxford: Wiley-Blackwell.

Boghossian, Paul. 2006. *Fear of knowledge: Against relativism and constructivism*. Oxford: Clarendon Press.

Boorse, Christopher. 1975. On the distinction between disease and illness. *Philosophy and Public Affairs* 5: 49–68.

———. 1976. Health as a theoretical concept. *Philosophy of Science* 44: 542–557.

Brown, James Robert. 2004. *Who rules in science? An opiniated guide to the wars*. Cambridge, MA: Harvard University Press.

Bynum, Helen. 2012. *Spitting blood*. Kindle edition. Oxford: Oxford University Press.

Cunningham, Andrew. 1992/2016. Transforming plague: The laboratory and the identity of infectious disease. In Cunningham 2016b, 190–244. First published in 1992.

———. 2002. Identifying disease in the past: Cutting the Gordian knot. *Asclepio* 54: 13–34.

———. 2007. *The making of modern medicine*. Audible audiobook. London: BBC Worldwide.

———. 2016a. Introduction: 'We will not anticipate the past, our retrospection will now be all to the future'. In Cunningham 2016b, vii–xi.

———. 2016b. *The identity of the history of science and medicine. Variorum Collected Studies*. Kindle edition. London: Taylor and Francis.

Duden, Barbara. 1991. *Geschichte unter der Haut*. Stuttgart: Klett-Cotta.

Engelhardt, Hugo Tristram. 1996. *The foundations of bioethics*. Oxford University Press.

Goodman, Nelson. 1978. *Ways of worldmaking*. Indianapolis: Hackett.

Hacking, Ian. 1988. The participant irrealist at large in the laboratory. *British Journal for the Philosophy of Science* 39: 277–294.

Jewson, Nicholas D. 1976/2009. The disappearance of the sick man from medical cosmology, 1770–1870. First published in 1976. *International Journal of Epidemiology* 38: 622–633.

Kingma, Elselijn. 2014. Naturalism about health and disease: Adding nuance for progress. *Journal of Medicine and Philosophy* 39: 590–608.

Kripke, Saul. 1980. *Naming and necessity.* Oxford: Blackwell.

Kuhn, Thomas S. 1962/2012. *The structure of scientific revolutions.* 50th ed. Chicago: University of Chicago Press.

———. 2022. *The last writings of Thomas S. Kuhn: Incommensurability in science.* Kindle edition. Chicago: University of Chicago Press.

Kusch, Martin. 1989. *Language as calculus vs. language as universal medium: A study in Husserl, Heidegger and Gadamer.* Dordrecht: Kluwer.

———. 1999. *Psychological knowledge.* London: Routledge.

———. 2019. Relativist stances, virtues and vices. *Aristotelian Society Supplementary* 93: 271–291.

———. 2021. *Relativism in the philosophy of science.* Cambridge: Cambridge University Press.

Latour, Bruno. 1993. *We have never been modern.* Cambridge, MA: Harvard University Press.

———. 2005. *Reassembling the social: An introduction to actor-network theory.* New York: Oxford University Press.

Mol, Annemarie. 1999. Ontological politics: A world and some questions. *The Sociological Review:* 74–89.

———. 2002. *The body multiple. Science and cultural theory.* Kindle edition. Durham: Duke University Press.

Seidel, Markus. 2014. *Epistemic relativism: A constructive critique.* New York: Palgrave Macmillan.

Sontag, Susan. 1978. *Illness as metaphor and AIDS and its metaphors.* Kindle edition. New York: Penguin Books.

Williams, Bernard. 1981. The truth in relativism. In *Moral luck,* ed. Bernard Williams, 132–143. Cambridge: Cambridge University Press.

Woolgar, Steven, and Javier Lezaum. 2013. The wrong bin bag: A turn to ontology in science and technology studies. *Social Studies of Science* 43: 312–340.

Woolgar, Steven, and Javier Lezaun. 2015. Missing the (question) mark? What *is* a turn to ontology? *Social Studies of Science* 45: 462–467.

Chapter 6
Cultivate Your Own Garden—Some Reflections on Martin Kusch's Overview of Relativism in Medical History

Hans-Joerg Ehni

In the preface to the first edition of the *Critique of Pure Reason* Immanuel Kant compares sceptics to nomads, who despise the cultivation of soil and therefore dissolve the civic communities of their opponents from time to time (Kant 1929). This process of construction and destruction continues afterwards. Although scepticism and relativism are two different philosophical positions, relativists may show enough similarities with sceptics to count them among Kantian nomads. Historiographical relativists despise the claim that a settlement was always built on the same foundation. This raises a couple of questions: Who are these people? What are their reasons? What is more basically their general motivation? And how convincing are both—their reasons and their motivations?

As an answer to the first question Kusch distinguishes three basic forms of relativism: "locality relativism", "relativism of a distance", and "equal validity relativism". This distinction is based on the assumption that relativists attribute certain symmetries to epistemological standards, (or conceptual schemes, frameworks, paradigms, thought styles, etc.). These symmetries include locality, non-neutrality, non-appraisability, and equal validity. The first two forms of symmetry mean that sets of standards are firstly based on local and contingent causes of validity, and secondly that if we rank different sets of standards, we don't start from a neutral position, but from another specific set that we presuppose. Together these two forms of symmetry characterize what Kusch refers to as "locality-relativism". Non-appraisability means that a set of standards and its corresponding assumptions can only be evaluated from its own perspective, but not from the perspective of other sets of standards. This leads to a stronger form of relativism that Kusch names "relativism of a distance", using a concept of Bernard Williams. Finally, the strongest form of relativism assumes that sets of standards are symmetrical in the sense

H.-J. Ehni (✉)
Institut für Ethik und Geschichte der Medizin Gartenstr, Tübingen, Germany
e-mail: hans-joerg.ehni@uni-tuebingen.de

© The Author(s) 2024
M. Schermer, N. Binney (eds.), *A Pragmatic Approach to Conceptualization of Health and Disease*, Philosophy and Medicine 151,
https://doi.org/10.1007/978-3-031-62241-0_6

that they are equally valid and true. Kusch states that this is the only relativistic position which "qualifies as unacceptable [in the realm of science]", but that no relativists ascribe this position to themselves.

Kusch's main objective is to describe the positions at hand, not to defend or criticize them. Developing detailed critical arguments on the positions described by Kusch is also beyond this short commentary. But I want to briefly point out an aspect of medicine, which may deserve further consideration in the context of historiographical medical relativism. This aspect is medicine as an "art" (techne) with a practical orientation (Wieland 1993). The practical orientation is among other things expressed by the goals medicine pursues, for instance according to the formulation of an international project by the Hastings Centre ('The Goals of Medicine. Setting New Priorities' 1996). The last of the four goals the project report identifies is "The avoidance of premature death and the pursuit of a peaceful death" ('The Goals of Medicine. Setting New Priorities' 1996: Executive summary). How does the practical orientation relate to a possible criticism of Cunningham and in more general terms to the criticism of relativism in medical history?

Relativists may be confronted with the criticism that their position is incoherent, either because they defend self-contradictory claims, or because their theoretical approach is contradictory if applied in a reflexive way. Here, I only want to address the first problem and take Cunningham as an example. His key term according to Kusch is "incommensurability". This is explained by the circumstance that medical claims in one system of medical knowledge are no longer meaningful in another and that according to 'non-appraisability' one system of medical beliefs cannot be evaluated in terms of another. Diseases are not natural kinds, but they are composed of the patients' experience, and social expectations. Another expression of incommensurability is that the stability of diseases over time is 'not open to prove or disprove'. Based on these assumptions Cunningham criticizes the idea of a retrospective diagnosis. Despite of the 'non-appraisability' of different sets of standards in different systems of medical beliefs, according to Kusch, Cunningham assumes that there has been medical progress in some respects reflected in Cunningham's statements such as "bacteriology has enabled medicine to save numerous lives". This partial belief in medical progress—again following Kusch—is no contradiction to statements which imply that we cannot say that we have understood the plague ever better.

Does this also hold considering understanding and/or reaching the goals of medicine such as avoiding a premature death? At least this would add some complexity to Cunningham's relativism and make it more difficult to defend. According to 'non-appraisability' he would have to claim that medicine of the seventeenth century cannot be understood and evaluated from the perspective of today's medicine. It would be plausible to extend this assumption to the goals of medicine since these goals include key concepts of the respective medicine itself such as 'health' and 'disease', and also 'death' or 'premature'. Beyond this 'non-appraisability' in terms of understanding could also extend to the concept of a 'medical goal' itself. This also seems to follow from Cunningham's criticism of Cipolla's opinion that the treatment seventeenth century medicine had to offer against plague was of no value. For Cunningham, according to this quote we cannot judge the earlier medical model in terms of "value of a treatment" from the perspective of today's bacteriological knowledge.

But this implies that we neither can properly understand the goals of seventeenth century medicine nor measure these goals against the goals of today's medicine. That is a coherent extension of 'non-appraisability' to the goals of medicine. Consequently, it is also not possible to say whether seventeenth century medicine has been better or worse in terms of reaching the goals mentioned above (or if it had the same goals or what they meant). Would it be plausible to argue that today's medicine couldn't have avoided premature deaths in a plague outbreak in the seventeenth century? It seems to be hardly defensible to argue that a 'premature death' and the corresponding goal would have meant something so different for seventeenth century medicine (and indeed for the people living at the time), that we cannot understand this from today's (medical) perspective. People and medical doctors in the seventeenth century certainly had the ability to grasp, when a death was 'premature' considering the age that people could expect to live to. They also likely understood that medicine was not of much use to avoid such a 'premature death' when somebody has contracted the disease that was called 'plague' at the time.

Maybe to defend this assumption that seventeenth century medicine was at least of some value compared to today's medicine and that we cannot really compare the two different values, Cunningham could argue that seventeenth century medicine might have been better in respect of this goal, but not in respect of another one which may have been particular to the medicine of the time, and this goal which would be different from any of today's medical goals couldn't simply be summarized with the degree in which it and other goals have been reached in an overall appreciation. But what should such a potential goal be, which is missing or has changed? What value might define the X contained in this goal, which cannot be translated in this context? This can be illustrated by an example and a heated discussion Martha Nussbaum reports from a conference (Nussbaum 1999). A French anthropologist according to Nussbaum regretted the disappearance of the cult of an Indian goddess, Sittala Devi, who was supposed to protect her believers from smallpox. This happened when smallpox vaccination was introduced to India by the British. When confronted with the criticism that after all the vaccine was better for health and survival, the anthropologist replied that only if one would give up this perspective, one could begin to understand the otherness of Indian traditions. This example across cultures shows what 'non-appraisability' across time may mean: That something is lost to today's perspective evaluating how well medical goals are met in terms of survival or premature death. Such a view can hardly be combined with the notion of medical progress as Cunningham does.

Of course a medical historian could finally give up the idea of medical progress to avoid this incoherence. But as Nussbaum shows in her report, this leads to another version of relativism: ethical relativism.[1] And this may add more apparent or real incoherences to the positions at hand since their motivation at least partially seems to be an ethical one: Doing justice to different perspective including those of the past. For historiographical medical relativists, at least for those who defend a version of a 'relativism of a distance', it may consequently be more difficult to cultivate their own gardens, even only temporarily.

References

Kant, Immanuel. 1929. *Immanuel Kant's critique of pure reason.* Trans. Norman Kemp Smith. London: Macmillan. A different edition (trans. J. M. D. Meiklejohn) is available online at: https://www.gutenberg.org/files/4280/4280-h/4280-h.htm#chap01.
Nussbaum, Martha C. 1999. *Sex and social justice.* Oxford/New York: Oxford University Press.
The Goals of Medicine. Setting New Priorities. 1996. *The Hastings Center Report* 26 (6): P1–27.
Wieland, Wolfgang. 1993. The concept of the art of medicine. In *Science, technology, and the art of medicine: European-American dialogues*, Philosophy and Medicine, ed. Corinna Delkeskamp-Hayes and Mary Ann Gardell Cutter, 165–181. Dordrecht: Springer. https://doi.org/10.1007/978-94-017-2960-4_10.

---

[1] A discussion of the extent to which Jewson or Mol have to face the same problem is beyond the scope of this article. The criticism of Cunningham should not imply that epistemological or ontological relativists also have to be ethical relativists.

Chapter 7
Is There an Epistemic Role for History in Medicine? Thinking About Thyroid Cancer

Nicholas Binney

7.1 Introduction

Doctors and other medical researchers frequently express a strange intuition—that the history of medicine has an important epistemic role to play in their practice. On this view, the history of medicine can and should be used to inform the evaluation of medical knowledge in the present. For example, thyroid cancer researchers have reflected on the overdiagnosis of this disease in the last few decades, in which a large number of patients were diagnosed with thyroid cancers that would not have harmed them. Such patients were subjected to unnecessary surgeries and radiotherapy, interventions which themselves carry significant risks. Thyroid cancer researchers lay the blame for this overdiagnosis on the development of diagnostic categories such as the 'follicular variant of the papillary thyroid carcinoma' (FVPTC), because instances of this category and others were understood to be dangerous when they were not. However, they do not simply call for more empirical research into the prognosis of patients with such conditions to address these false beliefs. Rather, they call for *historical* research into the development of such categories, because they see the powerful role such histories might play in contemporary medical practice. "By understanding the history of FVPTC, future classification of tumors will be greatly improved" (Tallini et al. 2017: 15). They take it upon themselves to write articles entitled "The History of the Follicular Variant of Papillary Thyroid Carcinoma" (Tallini et al. 2017), and the "Evolution of the Histologic Classification of Thyroid Neoplasms and its Impact on Clinical Management" (Xu and Ghossein 2018). Such historical work is seen by these and other researchers as highly relevant to

N. Binney (✉)
Section Medical Ethics, Philosophy and History of Medicine, Erasmus MC University Medical Center, Rotterdam, The Netherlands
e-mail: n.binney@erasmusmc.nl

© The Author(s) 2024
M. Schermer, N. Binney (eds.), *A Pragmatic Approach to Conceptualization of Health and Disease*, Philosophy and Medicine 151,
https://doi.org/10.1007/978-3-031-62241-0_7

understanding and improving contemporary medical practice (Jones et al. 2015; Steere-Williams et al. 2023; Mackowiak et al. 2017; Gale 2001).

This intuition about the epistemic role for medical history in medical practice is opposed by another widely held intuition, that in philosophy goes by the name of the *distinction between the context of discovery and the context of justification*. According to this distinction, "It is one thing to understand how a scientific claim was generated and accepted and another to ask whether it is justified, in light of the available evidence" (Arabatzis 2006: 227). On this view, knowledge of the history of medicine has no important epistemic role to play in contemporary medical science. I challenge this distinction here.

There are many versions the context distinction, most of which have been criticized and may have collapsed (Hoyningen-Huene 1987, 2006; Schickore and Steinle 2006). One version of it, however, remains—what Paul Hoyningen-Huene (2006: 120) has called the "lean" distinction. On this version, two *questions* should be distinguished. For any given claim p, on the one hand we might ask "How did someone come to accept that p?", and on the other hand we might ask "Is p justified?" (Sturm and Gigerenzer 2006: 134). The first question belongs to the context of discovery, whilst the second belongs to the context of justification. If we accept that these contexts are distinct, then answering one of these questions does not inform the answer to the other. They are just different questions, which need answering separately. This is not to say that the answers to these questions have to be *different*. The reasons that convinced historical actors to believe that p might well justify belief in p. Even so, accepting this lean distinction, it is not because they are the answer to the question 'How did someone come to accept that p?' that those reasons justify belief in p. Hoyningen-Huene (2006) argues that this lean distinction should be acceptable to everyone. This, then, is a version of the distinction between the contexts of discovery and justification that is still widely accepted. This lean distinction will be challenged here by showing how history does indeed have an epistemic role to play in medical practice.

This is a difficult task. The historian of science Lorraine Daston, in a recent interview, laments the current state of disintegration between history, philosophy and science, especially the lack of interest shown by many scientists in the histories of their disciplines (Loncar 2022). Daston distributes the blame evenly across history, philosophy and science: "There are three parties who have to pull up their socks" (Loncar 2022). Philosophers do not take the historical contingency of knowledge seriously enough. Historians have focused too much on local histories, making it difficult to see how such histories are relevant to contemporary science. By emphasizing the historical contingency of knowledge, historians also tend to under-cut scientists' aspirations of discovering genuine knowledge. Scientists tend to dismiss both historical and philosophical work as "blather". "So we need a philo-sophical remake of the concept of truth that does justice to the historical dynamism of science" (Loncar 2022).

Working in the traditions of integrated history and philosophy of science pioneered by N.R. Hanson (1958) and Ludwik Fleck (1979), I can meet this challenge. One of Hanson's most profound insights is that when scientists observe they do not simply *see*, they *see as* and *see that*. Scientists need to see objects *as* members of a certain kind of object in order to see *that* objects of that kind have, and will continue to have, certain properties. This involves seeing some properties of objects as sufficiently important to see the objects that share them as of the same kind because of these similarities. It also involves seeing other properties as sufficiently unimportant to see objects that do not share them as of the same kind despite these differences. These similarities need to be understood as allowing scientists to see that objects of that kind will behave in a certain way. Unless such connections between isolated experiences made, experiences can have no significance for scientific work.

Fleck had a similar insight and referred to the creative activity of seeing as the *active element of knowledge* (Fleck 1979). This creative act allowed further observations to be formulated, which Fleck called the *passive element of knowledge*. What Fleck referred to as active elements Hanson called *patterning statements;* and what Fleck called passive elements Hanson called *detail statements* (Hanson 1958: 87–88). It is these passive elements or detail statements that prevent these epistemologies from collapsing into pernicious forms of relativism (Binney 2023). By tracing the development of a field of knowledge in terms of a network of shifting and changing active and passive elements, I can show how facts are a human creation, the contents of which are only explicable in the light of their history, without thereby becoming a work of fantasy. This history is presentist and yet not whiggish, as its purpose is to play a productive role in scientific practice (Chang 2021). According to Fleck, who studied the history of syphilis, "It is not possible to legitimize the "existence" of syphilis in any other than a historical way" (Fleck 1979: 23). The same goes for thyroid cancer.

As the historical story told below can be technical, it is useful to spoil the story, so that readers can see where it is going. Applying insights about seeing as and seeing that to the case of thyroid cancer, it is immediately clear that a vast range of properties could be used to classify patients with thyroid growths. To name a few, thyroid growths can be large or small, encapsulated or unencapsulated, be organized into follicles or papillae, and have normal looking nuclei or abnormal looking nuclei. In the mid-twentieth century, thyroid tumours tended to be classified by how their cells organized histologically, into follicles or into papillae. In the last decades of the twentieth century, oncologists had come to classify thyroid growths with abnormal looking nuclei as papillary carcinomas of the thyroid, even if these growths had a follicular organization. Thus, the appearance of the nucleus was taken as much more important to tumour classification than histological organization. Patients with growths with a follicular organization but abnormal nuclei, were seen as the same as patients with growths with papillary organization with abnormal nuclei. This way of seeing patients as members of a kind led to the development of the 'follicular variant of the papillary thyroid carcinoma' (FVPTC) as a way of classifying patients. Oncologists did look to see whether the FVPTC was as dangerous as the classical papillary carcinoma and found that it was. Patients with FVPTC had the same poor prognosis and responded to the same treatment as patients with the classical papillary carcinoma. For thirty years, doctors recommended that papillary carcinomas, including the FVPTC, be treated aggressively with surgery and radiotherapy. However, eventually doctors saw that thyroid cancer was substantially overdiagnosed. Part of the reason for this was that some patients with the FVPTC had invasive tumours, many of which were dangerous, but other patients had encapsulated and non-invasive tumours, which are not dangerous. Doctors learned to see patients as the same according to whether their growths were invasive. Looking at patients like this, the FVPTC is a composite object, made up of a kind patient with a comparatively poor prognosis and another kind of patient with a good prognosis, as opposed to a single kind of patient with the same prognosis.

The story of the overdiagnosis of thyroid tumours is obviously much more complicated than the simple story below. However, there is an important lesson that can be learned even from this simple story. *Different ways of seeing have a profound influence over what doctors think they are observing, and thus over what observations are taken to justify.* If the FVPTC is seen as a single kind, and as the same as other classic papillary carcinomas, then observing that such patients have a poor prognosis can be used to justify the general conclusion that such patients in the future will have a poor prognosis. However, if the FVPTC is seen as a composite object, then observing that such patients in one study had a poor overall prognosis does not say very much about how patients with FVPTC will behave in the future. The overall prognosis of these patients will depend on the proportions of the different kinds of patients that will make up the composite object of FVPTC in the future. This composition may vary in the future, and the behaviour of the group as a whole may change as a consequence.

Nobody would deny that, in science, observations are relevant to the justification of scientific claims. However, what scientists observe is not simply a function of what is in front of them in the present. What scientists observe is a function of how they *see as* and how they *see that*. How scientists *see as* and *see that* is a function of the history of their field. It follows from this that what scientists observe is a function of the history of their field, which makes this history relevant to the justification of scientific claims. History, on this view, does not just provide one way of marshalling the evidence that can function autonomously from its history in the justification of scientific claims. On this view, history is an integral part of the observations themselves, and thus part of the justification of scientific claims. The answer to the question 'how did someone come to accept that p?' is part of the answer to the question 'is p justified?'. Even the lean distinction between discovery and justification collapses.

Ultimately, this is a story about what cancer, or carcinoma, is. Broadly, it is a story about a dialogue between two views: (1) That cancer is a malignant tumour, and (2) that cancer is a tumour with a certain microscopic structure. These two views were adopted iteratively over the nineteenth and twentieth centuries, gradually being fleshed out and specified in different ways, each one forming the other. Often, both views were adopted at once. The following sections describe how concepts of cancer, malignant, benign, carcinoma and adenoma were formed in the nineteenth century (Sect. 7.2), before following the development of thyroid carcinoma in particular over the twentieth century. I describe the how the concepts of papillary carcinoma and malignant adenoma emerged (Sect. 7.3), to be replaced by those of papillary and follicular carcinoma and undifferentiated carcinoma (Sect. 7.4). I describe how tumours came to be seen as needing only small amounts of papillary tissue to count as papillary carcinoma (Sect. 7.5), before being seen as needing no papillary tissue at all (Sect. 7.6). This is the origin of the FVPTC. I describe how seeing papillary carcinoma in this way lead to the overdiagnosis of thyroid cancer, following the widespread introduction of ultrasound guided fine needle aspiration in the 1990s (Sect. 7.7). I close with a discussion of how knowing this history could have informed what physicians saw when they looked at their patients and how it could inform their practice (Sect. 7.8).

This work is incomplete. It provides little more than a scaffold on which to build future work. There is much it leaves out. For example, the crucial story of microcarcinoma is entirely absent, as is an account of how wider societal concerns influenced medical practice. These are important. For example, the problem of overdiagnosis was much less pronounced in Japan than in America, and this has been explained as a consequence of America being the much more litigious society, causing physicians to practice more defensively, producing more diagnoses (Kakudo et al. 2012). This chapter is limited to describing how the network of active and passive elements was "tuned" over time (Fleck 1979: 86) to allow the malignant behaviour of tumours to be predicted from their microscopic appearance—a highly pragmatic project. Incomplete as it is, this history will show how researchers built the expectation of homogeneity, of kindhood, into their classifications. This is what gives history its epistemic function. Even though this is a creative process, the interaction of active and passive elements preserves the integrity of science.

7.2   Cancer and Malignancy in the Nineteenth Century

Cancer is an ancient term, as is carcinoma. Over the centuries, they have been used to refer to a wide variety of ulcerating, spreading swellings (Skuse 2015; Walshe 1846: 6–7). Since the early modern period, the term malignant was used to mean "evil", and "likely to rebel against God or authority" (Skuse 2015: 76). By the middle of the nineteenth century, malignant disease had come to refer to tumours with certain clinical behaviour: they spread, came back after surgery, could not be cured, and killed. Cancer was understood by many to be a malignant growth of cells (Arnold-Forster 2021; Timmermann 2013: 1–33).

> Designating, then, by the terms *Malignant Growths*, or *Cancer*, those growths which constantly possess, in a greater or less degree, the clinical attributes enumerated above, we may again subdivide them into so many species, according to their anatomical peculiarities (Laurence 1858: 3).

Others, however, disagreed. These physicians argued that malignancy was a hopelessly confused concept, referring to far too many different things to be useful for defining a disease (Walshe 1846, 1853). In any case, malignant was a clinical designation, and not fit for use when defining diseases anatomically (Bennett 1849: 170; Lebert 1851: 4–5; Vogel 1847: 190).

> The term "cancer" has been so commonly applied indefinitely to any growth possessing malignant properties, that "cancerous" and "malignant" have come to be regarded by many as synonymous terms. It is important, however, clearly to distinguish between them. A cancer is a growth possessing the above-named definite structure; a malignant growth, on the other hand, is one which, independently of its structure, is infectious. (See "Malignancy.") "Cancerous," is an anatomical term; "Malignant," is a clinical one (Green 1871: 153).

The opposition of these different ways of classifying growths was well expressed by the British surgeon Robert Druitt (1854), who made a valuable analysis of the concept of malignancy in an article entitled "On the modern philosophy of cancer". "The vital characteristics of cancer may be described in about fourteen terms, the sum of which equals the word malignant" (Druitt 1854: 33). According to Druitt, malignancy indicated a *constitutional* condition, as opposed to a local one. The whole body is involved in the condition, not just a part. Malignant tumours grow *quickly* and *constantly*. They are *painful*, can *ulcerate*, and produce *cachexia*. They can *degenerate and soften,* rather than remaining firm. Malignant tumours are *invasive*, in that they grow beyond their tissue of origin into surround tissue. They can *invade the lymphatics*. They can also undergo *diffusion*, producing *secondary deposits*, causing new tumours to grow at distant parts of the body. Malignant tumours are *resistant to treatment* and *return following extirpation* (surgical removal). They show *heterology of structure*, in that their gross and microscopic structure is different to the surrounding tissue. Druitt (1854) closes his list with a fifteenth characteristic: malignant tumours cause *death*.

Druitt made his analysis of malignancy to show that it is a multifaceted and often vague concept. As so many different clinical phenomena were grouped under the heading 'malignant', he did not see that any real entity would be picked out by such a classification. Classifying tumours according to clinical phenomena was no more likely to capture the 'real elements' of a disease than classifying tumours according to whether they looked like a cabbage or a cauliflower. "But this mode of classification cannot be trusted to; because masses of structure, widely differing in their real elements, may be nearly alike to the naked eye" (Druitt 1854: 31). Druitt held that classifying according to the microscopic physical structures of a tumour was the only way to capture the real elements of the disease.

> It [classifying according to physical structure] divides according to the forms, structures, and chemical composition exhibited at various periods of development, and studies the vital properties which exist together with them. It does not take mode of life as its basis, and consider physical properties as accidents (Druitt 1854: 31).

As classifying according to "vital qualities" such as malignancy was, for Druitt, to classify according to accidental properties, attempts to produce stable facts about the relationships between such accidentally defined objects and other things was bound to fail. This difficulty was augmented by the vagueness of the concept of malignancy, making statistics collected about tumours "worthless" (Druitt 1854: 31). The struggle between champions of "vital qualities" and champions of "physical structure" would continue to obstruct the art of healing until the correct method of classification was adopted. Druitt noted that other doctors held the opposite view, "that we ought to "choose modes of life rather than structure for determining the affinities of morbid products, and for arranging them under generic names. As of all tumors, so especially of cancers, the true nature is to be apprehended only by studying them as living things."" (Druitt 1854: 32). Even so, he held that this was a mistake, claiming that "before any fact can be held as certain, or can be estimated statistically, a reform of nomenclature and classification, and the disuse of that most noxious word "malignant", are essential" (Druitt 1854: 36). "But for all this it is absurd to erect the malignant into a separate class, from the innocent tumors of the same structure. It is like putting all the fatal cases of a disease into one class and the unfatal into another" (Druitt 1867: 111).

Rather than arguing about which of these two approaches, 'physical structure' or 'mode of life', is the real element of cancer, here I will make a Fleckian analysis. Rejecting the notion that there is a *real element*, Fleck instead considered what was the *active element* of knowledge for a group of researchers. The active element is an association that is taken for granted by a group of researchers, which then allows them to generate facts (Binney 2023). Here, one group of researchers, including Druitt, took for granted the association between a certain physical structure and cancer: if this structure was present then there was cancer, and if there was cancer this structure was present (see Fig. 7.1). Others took for granted the association between malignancy and cancer: if malignancy was present then there was cancer, and if there was cancer then malignancy would be present. These active elements were constitutive of cancer for these differing schools of thought. Rather than needing either of these elements to capture the "real element" of cancer, which Fleck thought was unintelligible, Fleck's epistemology allows that these active elements formed part of two different scientific cultures. Sometimes, *both* micro-scopic structure *and* malignant behaviour were seen as constitutive for cancer, constituting yet another way of understanding cancer. As active elements are adopted by groups of researchers, they are in a sense chosen by these researchers: active elements are responsive to the will of these researchers.

Knowledge is not limited to its active elements. The adoption of the active element allows researchers to generate further elements, which Fleck called *passive elements*. He called these passive because they resisted the will of researchers. So, adopting active elements such that cancer is some physical structure of a tumour, and such that to be malignant is to invade surrounding tissue, then researchers will be able to make observations about how often cancers are malignant. What cancer is may be for researchers to decide, and what malignancy is may be for researchers to decide, but once these decisions are made, whether cancers are malignant is not for researchers to decide. It is a mater to be decided by observation. These empirical relationships between culturally created objects do not exist unless certain culturally malleable active elements are adopted. Nevertheless, these relationships resist the will of researchers, which is why Fleck called them passive.

The analysis of elements into active and passive is useful, but it is also rather too abrupt. The resistance of the passive element is not necessarily solid, or even firm.

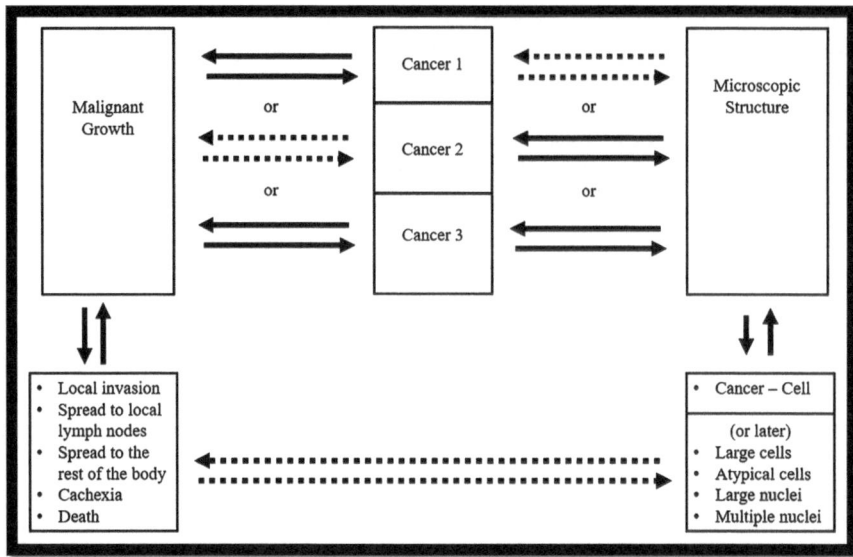

Fig. 7.1 Different Ways of Understanding Cancer. Cancers might be seen as malignant tumours, and malignant tumours might be seen as cancer. By accepting this, it becomes possible to observe the microscopic structures (e.g. a cancer-cell or (later on) enlarged and atypical cells) that are empirically associated with cancer (malignant tumours). This is cancer 1. Alternatively, cancers might be seen as certain microscopic structures (e.g. a cancer-cell or (later on) enlarged and atypical cells), and these microscopic structures might be seen as cancer. By accepting this, it becomes possible to observe that some cancers are not malignant, and some malignant tumours are not cancers, as Druitt did. This is cancer 2. Finally, cancers might be seen as *both* malignant tumours *and* certain microscopic structures, and malignant tumours and those microscopic structures might be seens as cancers. If this is accepted, then it must be the case that malignant tumours and the relevant microscopic structures will covary. This is cancer 3. Finding that they do not covary precipitated changes to how microscopic strucures were understood, in order to better predict malignant behaviour. This involved using cancer 1, 2 and 3 in an interative process, developing how cancer was understood over time
Key: A solid arrow represents an active element of knowledge. Read these, for example, "malignant growth → cancer 1", roughly as "if malignant growth then cancer 1". A dotted arrow represents a passive element of knowledge. Read these, for example, "malignant growth ˙ > cancer 2", roughly as "an observed proportion of malignant growths are cancer 2".

For example, tumours may recur following surgical removal, but this may not mean that they are malignant. It may be that the surgeon did not manage to remove all of the tumour. Researchers have to use many active elements in concert to specify what malignancy, physical structure, recurrence, etc. are before firm resistance can be generated. Such specification also involves tacit elements of knowledge, such as how to properly carry out a surgical operation. Solid resistance is only produced in dense networks of active and passive elements, and passive resistance can be much less constraining than this.

Physicians and pathologists set about adjusting and readjusting the network of active and passive elements in an effort to find those microscopic structures that would tell them about the malignant behaviour of a tumour. Following Herman Lebert (1845, 1851), Druitt (1867: 123) and others thought that there was a specific "cancer cell" that was present in all cancers and unique to them (Loison 2016; Turner 1863). As Druitt thought that malignancy was not constitutive of cancer, he did not require that malignant tumours have a cancer-cell, or that putative instances of the cancer-cell be malignant. Others, however, disagreed. Malignancy was just too important a phenomenon to be downplayed in his way. Repeated findings that malignant tumours had no cancer cell, and that benign tumours (and even healthy tissue) did, led to the rejection of the cancer cell theory. "Firstly, in some tumours, of an undoubtedly cancerous nature, cells are absent: secondly, that they have been found in tumours of a non-malignant character" (Henry 1855: 415). The passive relationship between the culturally constructed objects of cancer, malignancy, and cancer-cells was not as expected or desired. "Cells precisely similar to these are met with in other morbid growths, and even in the normal tissues. There is thus no specific 'cancer cell.'" (Green 1871: 154).

Careful microscopic work traced to origin of cancer of the breast to the epithelial cells lining mammary glands. Eventually, the cells of these carcinomas were actively associated with growths of glandular epithelial tissue (Green 1871: 153; Thiersch 1865; Waldeyer 1872). Carcinomas were epithelial growths. Instead of there being a universal cancer-cell, there was one type of tissue from which carcinomas originated. However, not all growths of glandular epithelium were cancer, on this view. In addition to the *carcinomas*, some tumours were *adenomas*. Adenomas were formed of cells in "a condition similar to that already described as occurring in the development of an ordinary glandular tumor" (Green 1871: 155). Adenomas could be thought of as growths of normal glandular epithelium in an organ (Green 1871: 150). The carcinomas, however, were growths of abnormal looking epithelial cells: misshapen cells of various sizes with large or even multiple nuclei. "The cells are characterized by their large size, by the diversity of their forms, and by the magnitude and prominence of their nuclei and nucleoli" (Green 1871: 153). For some, that carcinomas and adenomas were tumours with certain microscopic structures, rather than tumours with certain clinical behaviour, was an active element of knowledge. Hence, the observation of "malignant adenoma" was common in the late nineteenth century (Russell 1890; Sutton 1894: 497). For others, however, that adenomas were benign tumours, and that carcinomas were malignant, was an active element of knowledge (Kelynack 1899). This was also true for growths originating from non-epithelial tissue. Sarcomas, for example, were seen as malignant growths of connective tissue, whilst fibroma were seen as benign growths of connective tissue (Kelynack 1899). "The classes which we now recognize as malignant are the sarcomata and the carcinomata" (Mann 1883: 302).

It can easily be forgotten that this knowledge of the microscopic structures and behaviour of cancer was a contingent part of an ongoing process of adjustment and readjustment. Too much faith might be placed in the power of the microscope to reveal the 'real' microscopic structures of cancer. Both English and American physicians warned against this overconfidence (Mann 1883; Marsden 1874). "It has been thought by the uninitiated that we would see, with our lenses, the word "cancer" printed on every cell and every fibre" (Mann 1883: 306).

The same could be said for the contrast between the malignant and the benign. Malignancy, being associated historically with rebellion from God's authority, was understood as a sort of rebellion against the organizing authority of the body. Malignancy was seen by some as was the reversion of cells back to being single-celled organisms, which then competed with the body of which they were previously part. "[I]t further seems permissible to regard this "cancer—process" as essentially consisting in a local "cell—rebellion," certain cells casting off their allegiance to the central authority" (Snow 1893: 4). On this view, *all* malignant growths were in *full* rebellion (Arnold-Forster 2020). But why should this be so? Druitt's analysis of malignancy revealed fifteen *separate* components, which might be found in different combinations to produce very different malignant behaviours. Is a growth that spreads to local lymph nodes, but does not cause cachexia or kill the patient really malignant in the same sense as one that does? Aren't there different, and perhaps lesser, forms of malignancy? Indeed, that a growth could be *semi-malignant* was suggested many times in the nineteenth century. However, this view was largely rejected, as all malignant growths were seen as in full rebellion. "Some of the benign tumors – myxoma, chondroma, and some forms of fibroma – have received the reputation of being *semi-malignant* on account of their occasional recurrence after extirpation. *A tumor is either benign or malignant: there is no connecting – link between them*" (Senn 1895: 74).

This limited exploration of knowledge of cancer in the nineteenth century shows that facts about things like cancer, malignancy, carcinoma and adenoma are the product of historically contingent decision making. And yet, these facts are not the pure inventions of the little scientific cultures that produce them. It is worth keeping these contingencies in mind as we consider the development of knowledge of thyroid cancer in the twentieth century.

7.3   Bringing Order to Chaos: 1900–1950

Turning now to the diagnosis of thyroid cancer, we see that in this field the commonly held views about how to identify carcinomas quickly got into trouble. The thyroid is a butterfly shaped gland that makes and secretes iodine-based hormones affecting metabolism. It sits wrapped around the front of the trachea, just under the larynx. Iodine deficiency can cause swelling of the thyroid, as can thyroid tumours. Patients might present to doctors with a swelling of their thyroid, and the doctor might take a biopsy of this swelling to examine it under the microscope to see if it was a malignant tumour. One troubling phenomenon was finding thyroid tumours that looked like benign adenomas under the microscope, which turned out to be malignant in behaviour. Such tumours were sometimes referred to as malignant adenoma, which some physicians thought was a contradiction in terms.

> Another unusual form of malignant disease is that which is commonly known by the objectionable term of "malignant adenoma." By this term is meant a tumour which at first sight, and even after careful microscopical examination, appears to be innocent but which nevertheless is found to reproduce itself in distant parts of the body (Berry 1902: 229).

The term 'malignant adenoma' was objectionable because, to some physicians, that adenomas were benign tumours was an active element of knowledge. That adenoma had a certain microscopic structure, resembling that of normal tissue, was also an active element of knowledge to these doctors. They therefore expected that tumours with that structure would be benign. Whether they liked it or not, doctors found tumours with the microscopic structures that defined adenomas were sometimes malignant. The passive resistance generated by the network of elements undercut the active elements that brought it into being. This threw research into thyroid cancer into a state of chaos.

> Those who have had to deal with malignant tumours of the thyroid pathologically know from personal experience some of the difficulties encountered. Those who have sought to solve these difficulties by reference to the literature no doubt have been impressed by the great state of confusion, the endless conflict of authoritative opinion, the hopelessly involved terminology, the lack of satisfactory classification, and the inadequacy of the criteria by which to recognize malignancy in a fairly high percentage of cases (Graham 1925: 30).

The American doctor Allen Graham (1925), in what proved to be an influential study, sought to try to create some order in the chaos surrounding thyroid tumours. Graham and colleagues (1925) studied 108 cases of thyroid tumours gathered from their hospital between 1905 and 1922. He recognized two main types of thyroid tumours: *papillary carcinoma* and *malignant adenoma*. Papillae are finger-like projections of connective tissue that are covered in epithelial cells. They are quite a different structure to follicles, sacs of connective tissue lined with epithelial cells, found in normal thyroid tissue. Whilst the papillary carcinoma was largely composed of the papillary structures, the malignant adenoma could be composed of a huge variety of cell types and structures, including regions of papillary structures. "For this reason there is an endless variety of histological pictures to be encountered in the group and in individual tumors" (Graham 1925: 39). Graham did not recommend trying to separate this endless variety into different kinds based on their microscopic structure, as others had done. He claimed these two groups, papillary carcinoma and malignant adenoma, had different malignant behaviour and that this should be the basis for their classification. Papillary carcinoma metastasized to local lymph nodes, whilst malignant adenoma metastasized via the blood to the rest of the body. The papillary carcinoma showed low grade malignancy, in that they were not especially likely to metastasize, whereas the malignant adenoma were highly malignant. "For the foregoing reasons, it is important to preserve the distinction between these two types of lesions" (Graham 1925: 37). Thus, the sub-classification of thyroid carcinoma was not based on similarity in microscopic structure, but rather on those microscopic structures that were informative about malignant behaviour.

For both papillary carcinoma and malignant adenoma, he argued that the more invasive they were, the greater their malignancy. The structure of their cells did not matter a great deal, what mattered was the invasiveness of the tumour. For the papillary carcinoma: "The pathological diagnosis of carcinoma in this group of cases rests on the demonstration of local invasion, and not on the character of the cells primarily" (Graham 1925: 37). Note that for Graham malignant adenoma was a form of carcinoma: it was not a benign tumour. "[T]he term malignant adenoma is used to designate a type of thyroid carcinoma" (Graham 1925: 38). What distinguished a benign from a malignant adenoma was the degree to which it invaded blood vessels. For the malignant adenoma: "the histological appearance of cells and tissue is not a reliable basis for the differentiation of benign and malignant adenomata. At the same time, it was proposed that invasion of blood-vessels be utilized as the most reliable means of making the distinction" (Graham 1925: 41–42).

Graham thus introduced new active elements, classifying tumours in new ways. Papillary tumours were actively associated with papillary histological structure, and adenoma were actively associated with any other structure. Each was actively associated with a distinctive malignant behaviour: papillary carcinoma metastasized to lymph nodes, whereas malignant adenoma metastasized through the blood. Graham actively associated carcinoma with invasiveness, and benign tumours with the absence of invasiveness. However, he only did this after he had observed, *passively*, that in his case series invasiveness went along with malignant behaviour. He observed a passive relationship between invasiveness and malignancy, and then elevated this passive relationship to an active one to define carcinoma. In effect, Graham had adjusted the active elements that distinguished carcinoma (including malignant adenoma) from benign adenoma to allow the prediction of malignant behaviour.

Accepting these active elements, researchers should expect that tumours composed of papillae and are invasive to metastasize to local lymph nodes; and that tumours composed of anything else to metastasize through the blood to the rest of the body. They should also find that tumours that are not invasive should not metastasize. However, this was not up for these researchers to decide. That was a matter of passive resistance, and only further empirical work would tell how well these expectations faired.

7.4   Differentiated and Undifferentiated Tumours

Important modifications to the classification of thyroid tumours were made in the early 1950s. An "Atlas of Tumor Pathology" for the thyroid, published by the U.S. Armed Forces Institute of Pathology, was produced by pathologists Shields Warren and William Meissner (1953). This atlas contained many photographs of histological preparations, and importantly, a revised classification of thyroid tumours. These revisions were deemed necessary because previous efforts at classification had not managed to capture the relevant aspects of tumour behaviour.

The first great distinction made in this new classification was between *differentiated* and *undifferentiated* tumours. Many malignant adenomas were reclassified as undifferentiated tumours, and the term 'malignant adenoma' was no longer used.

The differentiated tumours had recognizable, mature types of cells, organized into recognizable structures, such as papillae and follicles. *Undifferentiated* tumours had no such structures, and their cells took on a jumble of different immature forms. Undifferentiated tumours were observed to be highly malignant and dangerous, and thus worthy of recognition as a distinct type of cancer (Warren and Meissner 1953: 78–87). Again, subdivisions of cancer were based on the clinical behaviour of the tumour, not on the identification of specific microscopic structures.

The differentiated tumours were divided into *papillary* and *follicular* types. Papillae are connective tissue fingers covered in the epithelial cells that would normally line the follicles of the thyroid gland. To be a papillary tumour, the pattern had to be predominantly papillary, but some follicular structures were allowed. The follicular pattern formed follicles: little sacs of connective tissue lined with epithelial cells. These sacs contained secreted substances and could resemble normal thyroid tissue quite closely. To be follicular, the predominant pattern had to be a follicular one, although follicles of different levels of maturity were permitted. Tissue resembling embryonic thyroid, with quite solid tissue, or foetal thyroid tissue, with very small follicles, were acceptable, as was tissue with normal or distended looking follicles. Even so, the follicular tumours in this atlas classification had rather more follicular organization than the malignant adenoma of Graham's classification. The follicular tumours were almost always encapsulated in a connective tissue sheath, whilst the papillary tumours tended not to be. Warren and Meissner (1953) thought that this distinction was readily apparent histologically and also happened to be useful clinically. Here, that papillary tumours had a papillary structure and that follicular tumours had a follicular structure were active elements of knowledge. Whether benign or malignant, the different sorts of differentiated thyroid tumours were defined according to these supposedly apparent histological patterns.

By contrast, Warren and Meissner (1953) thought that benign and malignant differentiated thyroid tumours were difficult to tell apart. The cells and structures could look very similar. Like Graham (1925) had advised, this atlas classification also distinguished between adenoma and carcinoma using the invasiveness of the tumour. Invasion of the blood vessels, the fibrous capsule or the lymphatics distinguished carcinoma from adenoma.

Warren and Meissner (1953) were keen to emphasize that these tumours were malignant because they have this microscopic property of invasion, and not because they would produce metastases.

> All tumors showing the basic pathological criteria of cancer must be called malignant, no matter whether they are growing slowly or rapidly, or whether they are early or late. That early thyroid cancers usually are curable by local excision and that some thyroid cancers grow slowly does not disprove the fact that they are malignant (Warren and Meissner 1953: 40).

Intriguingly, here even malignancy is defined according to microscopic structures. This association between microscopic signs of invasion and malignancy, carcinoma and cancer are all active for Warren and Meissner (1953). If these structures are present, then the tumour is cancerous and malignant, no matter how it behaves clinically. It is not evident from their descriptions that all of these microscopic structures were chosen as cancerous *precisely because* they were informative about the clinical behaviour of tumours.

If a tumour was comprised largely of papillary tissue and was invasive then it was a papillary carcinoma. If a tumour was comprised largely of follicular tissue and was invasive then it was a follicular carcinoma. With these active elements in place, Warren and Meissner (1953) hoped their classification could be used to predict whether and how tumours would spread. As Graham (1925) had done, the atlas classification noted that papillary carcinoma tended to spread through the lymphatics to local lymph nodes, whereas follicular carcinoma tended to spread through the blood vessels to distant sites in the body. The relationship between differentiated thyroid carcinoma and these aspects of the biological behaviour of a tumour was passive.

7.5 Papillary Carcinomas Comprised Mostly of Follicles

Modifying Warren and Meissner's (1953) classification, Woolner et al. (1961) described four main classes of thyroid carcinoma: *papillary, follicular, anaplastic* and *solid with amyloid stroma*. This latter class of tumour is an addition following the description of an apparently new type of tumour first described a few years earlier (Hazard et al. 1959). The anaplastic tumour corresponded to the undifferentiated tumours in the Warren and Meissner's (1953) atlas. Papillary and follicular tumours also appeared in this new classification, but they were modified (Doniach 1963).

In Warren and Meissner's (1953: 69) classification, papillary carcinomas were allowed to have a few "foci of follicular growth", but follicular tissue was not allowed to be the dominant pattern. Papillary carcinomas needed to be predominantly composed of papillary tissue. This was not so for Woolner, who argued that papillary carcinoma could be mostly composed of non-papillary tissue, or even mostly of follicular tissue (Woolner et al. 1961; Woolner 1971).

Why classify tumours like this? On this view, papillary carcinomas could have a greater proportion of follicular tissue than follicular carcinomas. Why abandon the active association between different types of thyroid carcinoma and their microscopic structure? Well, Warren and Meissner (1953) expected that the tumours comprised mostly of follicular tissue would metastasize via the blood and would be more dangerous than the tumours comprised mostly of papillary tissue. Woolner et al. (1961) and Woolner (1971) found that this was not so. They found that many tumours comprised mostly of follicular tissue metastasized to lymph nodes and were not especially dangerous—they behaved like papillary tumours. Most tumours were a mixture of different patterns, but this did not seem to matter—the presence of any papillary tissue was what was informative about tumour behaviour (Chen and Rosai 1977; Vickery 1983). Now, Woolner and colleagues could have just preserved Warren and Meissner's (1953) classification, by accepting that some follicular tumours behave like papillary tumours; but they did not. Instead, they argued that it was the clinical behaviour of the tumour that should determine how it is classified, not its microscopic structure. "In effect, tumors of similar biologic behavior are placed in the same category, regardless of the details of microscopic architecture" (Woolner 1971: 500).

Woolner and colleagues (1961) and Woolner (1971) also provided graphs showing the survival of patients over time and found differences in the papillary and follicular groups. In both groups, the invasiveness of the tumour is important for the patient's prognosis (as Graham (1925) had suggested). Follicular tumours were almost always encapsulated, and invasion could be assessed by the degree of invasion of their capsule. Papillary tumours, by contrast, were most often not encapsulated, and their invasion could be assessed by whether they were contained within the thyroid, or if they had escaped the thyroid gland to invade nearby tissues. The prognosis following surgery was worse for the invasive follicular tumours than for the invasive papillary tumours. Consequently, Woolner (1971) did not think that tumours with such different prognoses should be classified together, even though they resembled each other histologically.

> In this papillary category may show 1 to 2% of papillary formation and 95 to 98% follicular structure. Such predominantly follicular variants have the same mode of spread and excellent prognosis after operation as other, more papillary examples. By contrast, an encapsulated angioinvasive cancer called "follicular" in our classification has no papillary component microscopically, does not spread to nodes typically, and kills by metastasis to the lungs or bone. Such a tumor may be composed largely of small follicles or may show considerable follicular differentiation in a cellular or solid background. Since highly invasive angioinvasive cancer kills approximately 50% of its victims and, by contrast, papillary carcinoma with a strongly follicular component is regularly cured, it is a mistake, I believe, to include both types of tumor under one heading in any useful classification (Woolner 1971: 501).

Thus, for Woolner (1971), the connection between follicular carcinoma and its biological behaviour was active. Follicular carcinoma metastasized via the blood to distant organs, and invasive ones killed half of the patients suffering with it. Similarly, the connection between papillary carcinoma and its biological behaviour was active. Papillary carcinoma metastasized via the lymphatic system and had a much better prognosis following surgical removal. The microscopic structure of these tumours was treated in this research as a passive element and changed to match the biological behaviour. Once suitable microscopic structures had been found, they were elevated into active elements and used to define papillary and follicular tumours.

7.6   Papillary Carcinoma with no Papillary Structures at All

In the 1970s, the microscopic structures used to identify papillary carcinomas were modified again. Even before Woolner's work (1971) and Woolner et al. 1961), other researchers had noticed that some tumours comprised mostly of follicular tissue behaved like the papillary tumours. These tumours were called "follicular variant of papillary carcinoma" (Lindsay 1960: 43). Researchers also noted that many papillary carcinoma cells had odd looking nuclei. These nuclei had a "ground-glass" appearance, which distinguished them from normal thyroid cells, and from many other thyroid carcinomas (Lindsay 1960; Franssila 1973). They noticed that many tumours composed entirely of follicular tissue had nuclei with this same ground-glass appearance. Studying these tumours, researchers found that they behaved like papillary tumours, even though they had no papillary tissue in them (Franssila 1973).

The American doctors Karl Chen and Juan Rosai (1977) reviewed cases of thyroid carcinoma at their hospital and had found six cases of thyroid carcinoma composed of follicular tissue that had these odd-looking nuclei. None of these tumours was encapsulated. These patients had presented for goitre or lymph node swelling and had been treated surgically and, in some cases, with radiotherapy. Most had significant local invasion of the tumour and metastases to local lymph nodes. None developed distant metastases to bone or lung, even after over a decade of follow up in three cases. The biological behaviour of these tumours was like that of papillary carcinoma, and not like that of the follicular carcinoma. "The findings of this study suggest that "papillary" and "follicular" types of thyroid cancer, which are regarded by most people as indicators of two sharply contrasting biologic behaviors, can no longer be taken literally as synonymous with papillae and follicles" (Chen and Rosai 1977). Chen and Rosai (1977) argued that these nuclear features should be used to distinguish papillary and follicular carcinomas, and not the structure of the tissue.

> The single most important criterion to differentiate this variant of papillary carcinoma from a bona fide follicular carcinoma is the nuclear pattern which is characterized, in the former, by uniformity and ground-glass appearance. Other features, such as presence or absence of encapsulation or psammoma bodies, are helpful in the differential diagnosis, but should not be taken as absolute since a certain degree of overlapping does occur (Chen and Rosai 1977: 129).

Again, researchers tuned their classification to help them predict biological behaviour. Papillary carcinomas were actively associated with the tendency to spread to lymph nodes and have a good prognosis. Follicular carcinomas were actively associated with the tendency to spread to distant sites and to have a less favourable prognosis. Papillary carcinomas were also actively associated with having at least some papillary tissue. Follicular carcinomas were actively associated with having some follicular tissue and no papillary tissue. These active elements, together, produce the expectation that carcinomas with some follicular tissue and no papillary tissue would not tend to metastasize to regional lymph nodes and have a good prognosis. However, this passive resistance did not manifest. A group of patients with follicular tissue and no papillary tissue were found to have a tendency to metastasize to local lymph nodes. This passive resistance undercut the network of active elements that brought it into being. Something had to change. These rogue carcinomas were observed to have ground-glass nuclei, as many papillary carcinomas did. So, the microscopic morphology of papillary carcinoma was adjusted to preserve its biological behaviour. The biological behaviour of papillary carcinoma was not adjusted to preserve its microscopic morphology. Papillary carcinoma was still defined according to microscopic morphology, but biological behaviour took conceptual priority.

As discussed, papillary carcinomas were most often not encapsulated. However, sometimes, perhaps as much as 14% of the time, they were. Furthermore, it was soon observed that encapsulated papillary carcinoma could metastasize even though invasion of their capsule was not apparent microscopically. "As lymph node metastases were observed in two primaries with neoplastic papillae and without evidence of capsule invasion, the latter is not a necessary criterion for the diagnosis of EPC [encapsulated papillary carcinoma] which show presence of unequivocal cytologic hallmarks of papillary thyroid carcinoma" (Schröder et al. 1984: 92). For papillary carcinoma, invasion of the capsule was not the guide to malignancy that it was in follicular tumours. Studies investigating the malignant potential of encapsulated papillary carcinoma presented small case series of around ten to twenty-five patients, sometimes with tumours composed entirely of papillary tissue (Oyama et al. 1993), and sometimes including follicular variants as well (Carcangiu et al. 1985). Overall, about 25% of encapsulated papillary thyroid carcinoma were found to metastasize to local lymph nodes, with distant metastasis being very uncommon (Chan 2002). The encapsulated papillary thyroid carcinoma was considered to be a less aggressive form of the papillary carcinoma, or perhaps a precursor for it. "[T]he encapsulated papillary carcinoma can in fact be seen as precursor of the widely invasive papillary carcinoma" (Schröder et al. 1984: 93).

Notice, that in order to investigate the malignant potential of the papillary thyroid carcinoma, pathologists needed to be able to recognize these tumours *and then* look to see how they behaved clinically. Indeed, studies reported that as many as 10–20% of papillary thyroid carcinoma produced blood borne metastases to distant sites such as the lung or bone (Carcangiu et al. 1985). When working in this mode, that papillary thyroid carcinoma metastasized exclusively to local lymph nodes was no longer an active element of knowledge. Here, the association between carcinoma and microscopic structure was active, whilst the association between carcinoma and clinical behaviour was passive. Again, it might be easy to forget that the microscopic structures that defined thyroid carcinoma where chosen, through a long historical process, precisely because they predicted the clinical behaviour of these tumours.

Even though the follicular variant was defined using these new nuclear morphological criteria, there was still considerable leeway for different pathologists to interpret criteria differently and come to different judgements about whether a case was a carcinoma. Did the ground glass nuclei have to be found in all parts of the tumour, or only in regions? How much coverage was required? Other nuclear changes other than ground-glass appearance were also used to identify follicular variants, such as cytoplasmic invagination into the nucleus, overlapping nuclei and nuclear grooves. These were all seen as important when distinguishing cancer from non-cancer. One study found that the judgement of even expert pathologists was quite variable (Lloyd et al. 2004). The network of active and passive elements did not produce sufficiently firm resistance to strictly determine whether cancer was present or not. Even so, pathologists commonly made the diagnosis of the follicular variant of the papillary thyroid carcinoma in their work.

This way of seeing papillary carcinoma quickly caught on (Thompson et al. 1978). In the 1980s, 1990s, and in the new millennium, numerous studies found that the follicular variant of the papillary thyroid carcinoma had very similar biological behaviour to the papillary thyroid carcinoma with some papillary tissue in them (Carcangiu et al. 1985; Lin and Bhattacharyya 2010; Passler et al. 2003; Tielens et al. 1994; Zidan et al. 2003). These include case series of up to 100 thyroid carcinoma patients, who were diagnosed because of a thyroid swelling, a swelling of the lymph nodes in their neck, because of distant metastases, or because of previous operation which needed further surgery due to recurrence, in Europe and the U.S.A. (Carcangiu et al. 1985; Passler et al. 2003; Tielens et al. 1994; Zidan et al. 2003). They compared patient characteristics and clinical course of the follicular variant of the papillary thyroid carcinoma with that of the follicular carcinoma and the classic papillary carcinoma. They looked to see how wide array tumour characteristics influenced clinical behaviour, including: the ratio of papillary to follicular tissue; the sex of the patients; the degree of encapsulation; the amount of lymphocytic infiltration; and whether the tumour had pushing or infiltrating margins (Carcangiu et al. 1985). In many cases the follow up was not especially long, often as low as three years. Others followed patients for more than 20 years (Zidan et al. 2003). These studies would often re-examine slides of pathological specimens collected years ago. One made use of the U.S. 'Surveillance, Epidemiology and End Results' (SEER) database to investigate tens of thousands of papillary thyroid cancers registered between 1988 and 2006 (Lin and Bhattacharyya 2010; see also Yu et al. 2013). About one third of these were follicular variants and two thirds were classic papillary carcinomas. All reached a fairly similar conclusion: follicular variants should be seen as papillary thyroid carcinoma and treated as such.

> Our findings with the follicular variant of PTC amply confirm or previous results and the belief expressed by several authors that this tumor belongs to the papillary group. Its clinical behavior and the cohort of morphologic features that accompanied it were clearly that of PTC [papillary thyroid carcinoma] (Carcangiu et al. 1985: 825).

> Although debate has surrounded the differences in the clinical courses and outcomes of patients with FV-PTC [follicular variant of the papillary thyroid carcinoma], our data suggest that the prognoses of patients with FV-PTC is basically similar to that of the C-PTC [classic-papillary thyroid carcinoma], and patients should be treated and counseled accordingly (Lin and Bhattacharyya 2010: 715).

7.7 Lumps that Needed Splitting

When ultrasound guided fine needle aspiration became widely available in the 1990s, doctors began finding large numbers of the follicular variant of the papillary thyroid carcinoma. Despite this, death rates from the thyroid cancer remained flat. These tumours were almost all small tumours of less than two centimeters in diameter. In the early 2000s, researchers began suggesting that thyroid cancer was being substantially overdiagnosed, especially in South Korea and the U.S.A. (Davies and Welch 2006, 2010; Ahn and Welch 2015). Researchers feared that patients were being harmed with unnecessary surgery and radioactive iodine therapy. Even though these tumours appeared to be harmless, they were still *seen as* carcinoma, and that is important.

As they had ground-glass nuclei, many tumours composed entirely of follicular tissue had come to be seen as papillary carcinoma. For other follicular tumours, which were almost always encapsulated, whether or not the tumour invaded this capsule was often the basis for deciding whether a tumour was a carcinoma, a cancer, or an adenoma, a benign tumour. All papillary thyroid tumours were considered carcinoma. It did not matter whether they were encapsulated or not, and it did not matter whether the capsule was invaded or not. Thus, the only difference between a papillary carcinoma and a follicular adenoma could well be the presence of these ground-glass nuclei. As the encapsulated, non-invasive follicular variants were seen as the same as any other papillary thyroid carcinoma, pathologists did not distinguish between them and other follicular variants when collecting statistics about their malignant potential. Benign behaving follicular variants, which might have been classed as adenoma if it were not for their ground-glass nuclei, were grouped with tumours with more malignant behaviour, making these benign variants appear more dangerous than they were.

> Analyzing E-CPTC [encapsulated-classic papillary thyroid carcinoma] and E-FVPTC [encapsulated follicular variant of the papillary thyroid carcinoma] under the same umbrella can therefore mask the behavior of either entity and lead clinicians to believe that these are similar entities with regard to their metastatic spread, prognosis, and classification. For example, one often quoted study reports a 24% metastatic nodal rate in EPTC without correlating the growth pattern of the tumor with the presence of metastases. It is therefore impossible to know which EPTC is responsible for nodal disease. Is it the E-CPTC or the E-FVPTC? Do noninvasive E-CPTC behave in a fashion similar to noninvasive EFVPTC? Unfortunately, it is very difficult to answer these questions on the basis of the currently published literature (Rivera et al. 2009: 120).

One of the drivers of overdiagnosis was finding many more encapsulated, non-invasive follicular variants of the papillary thyroid carcinoma, which as it happens have an almost entirely indolent behaviour. This indolent behaviour was masked by lumping these tumours together with more troublesome tumours, as they were seen as the same. If these tumours were seen as minimally invasive follicular carcinoma or adenoma, many doctors would not have recommended complete thyroidectomy and radioactive iodine therapy, as they would if they were seen as papillary thyroid carcinoma (Rivera et al. 2009).

Several alterations to practice have been recommended to help prevent overdiagnosis of thyroid cancer. Some have recommended referring to lesions that currently qualify as cancerous as "indolent lesions of epithelial origin" (or IDLE), instead of as cancer or as carcinoma (Esserman et al. 2014: e234). One could develop stricter criteria for malignancy, for example by tightening up guidance on what counts as ground-glass nuclei. Another suggestion is to invent a category of intermediate or uncertain risk for malignancy, to give pathologists space to admit to their uncertainty (Renshaw and Gould 2002). Indeed, just openly discussing that even expert pathologist disagree might help make practice less defensive (Renshaw and Gould 2002).

The classification of thyroid cancers has shifted since the overdiagnosis this disease was identified. Pathologists had started to refer to encapsulated, non-invasive, follicular variant of the papillary thyroid carcinoma, but because this still recognizes the follicular variant as a type of thyroid carcinoma, the American Thyroid Association has recommended a new classification: the "noninvasive follicular thyroid neoplasm with papillary-like nuclear features" (NIFTP) (Haugen et al. 2017). Another publication recommends collecting the terms used to describe these problematic tumours with follicular tissue, such as "follicular adenoma with equivocal papillary thyroid carcinoma nuclei" and "minimally invasive follicular thyroid carcinoma" into one class: the "well differentiated tumor of uncertain behavior" (WDT-UB) (Kakudo et al. 2012: 155). Risk stratification for thyroid tumours today integrates information about tumour size, encapsulation, invasion, patient demographics in addition to histological type (Tallini et al. 2017). Management strategies are also changing. For example, researchers in Japan have explored what happens if patients with low-risk tumours, especially microcarcinomas, are left without any therapeutic intervention at all, an approach called "active surveillance" (Ito et al. 2018; Lohia et al. 2020; Sutherland et al. 2021). In such management strategies, patients only go to surgery if they develop local lymph node metastases, which are then removed along with the primary tumour. Outcomes for this strategy are reportedly quite successful. Reengineering of medical concepts and practices proceeds apace. The story of thyroid cancer continues.

Many of the innovations designed to prevent the overdiagnosis of thyroid cancer rely on being able to identify the pathology that places patients at high risk of some negative outcome. One might think, then, that the problem of overdiagnosis can be solved by attending carefully to the risks that certain pathologies pose. One of the main lessons of this history is that this is not the case. Researchers investigating the follicular variant of the papillary thyroid carcinoma studied the risk this pathology passed to patients carefully over several decades. They used data from case series and huge databases such as SEER to show, over and over again, that the follicular variant of the papillary thyroid carcinoma was not benign. Around 20% of cases would produce metastases to local lymph nodes, and it would also, on occasion, metastasize to distant sites and perhaps kill the patient. It was a slow growing cancer, but a cancer nonetheless, and should not be ignored. All this attention to the risks it posed was not enough to prevent its overdiagnosis.

7.8 Conclusion

Try as I might, I cannot see any of the ways of classifying thyroid tumours encountered in this history as capturing a mind and culture independent reality about human tissue. Back in the nineteenth century, Druitt held that microscopic structure captures the real elements of tumours, whilst their biological behaviour, their 'mode of life', was artificial and born of clinical interest. I do not see why one should be more real and less artificial than the other. The elements that constitute the relevant microscopic structures and the relevant—malignant—modes of life, need to be actively chosen by people. Once these active elements are in place, how they relate to each other is much more constrained. This passive resistance, which only exists as a result of human activity, and yet is not pure fantasy because it resists human volition, is then used to shape the active elements that brought it into being.

Why did physicians see certain patients as having the follicular variant of the papillary thyroid carcinoma? Why did they see these patients as a homogeneous kind, with clinical behaviour similar to other papillary tumours? Why did they not see the encapsulated follicular variants as follicular adenoma—as entirely benign? This way of seeing was not simply a function of what was in from of them, it was a function of the history of thyroid carcinoma. Papillary and follicular structures were found to align with distinctive forms of malignant behaviour: spreading to local lymph nodes and spreading to the rest of the body. As this was of clinical interest, these microscopic structures were chosen to define kinds of thyroid tumour. Invasiveness was found to align with malignant behaviour in follicular tumours and was used to distinguish follicular carcinoma from follicular adenoma. Papillary tumours were found to metastasize whether they were encapsulated or invasive, and thus were all deemed carcinoma. Tumours with no papillary structures, but with ground-glass nuclei, were found to behave like papillary tumours by tending to metastasize to local lymph nodes and were classified as such. Someone ignorant of this history, if presented with these as kinds of tumour and shown evidence of their behaviour, would have no reason to believe that they would not continue to behave in that way. Armed with some history, that person could see why the kinds are seen as they are. They might say that these follicular variants were once considered to be follicular tumours, and we know that invasiveness of the capsule is important for predicting the behaviour of follicular variants. Might this not also be the case for these tumours? They could see from the history that the cases used to seed the idea that ground-glass nuclei should define papillary carcinoma were all invasive tumours, with advanced disease. Might this putative kind be better seen as an aggregate of at least two different sorts of tumour, one malignant and the other benign? If so, why believe that this composite object would behave in the same way following the introduction of ultrasound guided fine needle biopsy, which might find more of the benign type of tumour than before? Without this history, we lose the ability to justify why we see as we do.

Whether a belief is justified depends upon observations, and observations depend upon their history. Thus, historical premises, such as 'ground-glass nuclei were adopted as an active element of papillary carcinoma for these reasons', can inform present day conclusions, such as 'that result is the aggregated effect of two different sorts of tumour'. This collapses the lean distinction between discovery and justification and gives history an epistemic role in medical practice.

Physicians today debate whether unencapsulated follicular tumours with papillary nuclear features should be seen as cancer. Some say yes, because the relevant microscopic structures are present. Without this history, physicians might feel that this settles the matter, as microscopic structures are the real elements of cancer after all. This history reveals, however, that seeing these structures as the real elements of cancer is a product of a contingent history. We can choose to change how we see, if it serves our purposes. If these microscopic structures do not predict malignancy, then perhaps we should change what we see as cancer, as those before us did. Some might reply that these tumours are malignant, as they spread to local lymph nodes in around 20% of cases. But, why should we see this as malignancy? Even if physicians wait until after these metastases have occurred, surgery to remove these tumours is highly successful. Given this, why not see this as a different, less severe, kind of malignancy? The reasons we do not are in the nineteenth century, with the rejection of semi-malignant as a useful concept. It was there that we learned to see cells as loyal or rebellious, but nothing else. Why not recover from the past the idea that malignancy is many different phenomena, perhaps with many different mechanisms? Why not say that this is a different sort of malignancy, a different sort of cancer, for which it is safe and prudent to wait and see if metastases occur? This is the rationale used by those advocating active surveillance in these types of us. However, it may be especially hard to convince patients of this. Patients know that malignancy is an evil. Waiting to see whether that evil harms you might seem like waiting until a gunman actually fires the gun before springing into action. Physicians and pathologists are by no means free to change what cancer and malignancy are, because the patients know what they are, and know that they are terrible. "This indicates the dominance of the mass over the elite in a democratic thought collective" (Fleck 1979: 124). But cancer and malignancy need not be seen like this. What cancer and malignancy are needs to be negotiated between physicians, pathologists and patients, and history should be part of that negotiation.

Acknowledgement The research for this chapter was supported by the Dutch Research Council (NWO) as part of the project 'Health and disease as practical concepts', project number 406.18. FT.002.

References

Ahn, Hyeong Sik, and H. Gilbert Welch. 2015. South Korea's thyroid-cancer "Epidemic"—Turning the Tide. *New England Journal of Medicine* 373 (24). Massachusetts Medical Society: 2389–2390. https://doi.org/10.1056/NEJMc1507622.

Arabatzis, Theodore. 2006. On the inextricability of the context of discovery and the context of justification. In *Revisiting discovery and justification*, ed. Jutta Schickore and Friedrich Steinle, 215–230. Springer.

Arnold-Forster, Agnes. 2020. "A Rebellion of the Cells": Cancer, modernity, and decline in Fin-de-Siècle Britain. In *Progress and Pathology*, 173–193. Manchester University Press. https://www.manchesterhive.com/display/9781526147547/9781526147547.00015.xml.

———. 2021. *The cancer problem: Malignancy in nineteenth-century britain*. Oxford/New York: Oxford University Press.

Bennett, John Hughes. 1849. *On cancerous and cancroid growths*. Sutherland and Knox.

Berry, James. 1902. An address on the diagnosis and treatment of the various forms of Goitre. *The Lancet*, Originally published as Volume 1, Issue 4105 159 (4105): 1227–1231. https://doi.org/10.1016/S0140-6736(01)83681-3.

Binney, Nicholas. 2023. Ludwik Fleck's reasonable relativism about science. *Synthese* 201 (2): 40. https://doi.org/10.1007/s11229-022-04018-w.

Carcangiu, Maria Luisa, Giancarlo Zampi, Alberto Pupi, Antonio Castagnoli, and Juan Rosai. 1985. Papillary carcinoma of the thyroid. A clinicopathologic study of 241 cases treated at the university of florence, Italy. *Cancer* 55 (4): 805–828. https://doi.org/10.1002/1097-0142 (19850215)55:4<805::AID-CNCR2820550419>3.0.CO;2-Z.

Chan, John K.C. 2002. Strict criteria should be applied in the diagnosis of encapsulated follicular variant of papillary thyroid carcinoma. *American Journal of Clinical Pathology* 117 (1): 16–18. https://doi.org/10.1309/P7QL-16KQ-QLF4-XW0M.

Chang, Hasok. 2021. Presentist History for Pluralist Science. *Journal for General Philosophy of Science* 52 (1): 97–114. https://doi.org/10.1007/s10838-020-09512-8.

Chen, Karl T.K., and Juan Rosai. 1977. Follicular variant of thyroid papillary carcinoma: A clinicopathologic study of six cases. *The American Journal of Surgical Pathology* 1 (2): 123.

Davies, Louise, and H. Gilbert Welch. 2006. Increasing incidence of thyroid cancer in the United States, 1973–2002. *JAMA* 295 (18): 2164–2167. https://doi.org/10.1001/jama.295.18.2164.

———. 2010. Thyroid Cancer Survival in the United States: Observational Data From 1973 to 2005. *Archives of Otolaryngology–Head & Neck Surgery* 136 (5): 440–444. https://doi.org/10.1001/archoto.2010.55.

Doniach, I. 1963. Carcinoma of the Thyroid: The Pathology of Thyroid Carcinoma. *Proceedings of the Royal Society of Medicine* 56 (5). SAGE Publications: 354–356. https://doi.org/10.1177/003591576305600513.

Druitt, Robert. 1854. On the modern philosophy of cancer. *Association Medical Journal* 2 (54): 31–36.

———. 1867. *The principles and practice of modern surgery*. Blanchard and Lea.

Esserman, Laura J., Ian M. Thompson, Brian Reid, Peter Nelson, David F. Ransohoff, H. Gilbert Welch, Shelley Hwang, et al. 2014. Addressing overdiagnosis and overtreatment in cancer: A prescription for change. *The Lancet Oncology* 15 (6): e234–e242. https://doi.org/10.1016/S1470-2045(13)70598-9.

Fleck, Ludwik. 1979. *Genesis and development of a scientific fact*. University of Chicago Press.

Franssila, Kaarle O. 1973. Is the differentiation between papillary and follicular thyroid carcinoma valid? *Cancer* 32 (4): 853–864. https://doi.org/10.1002/1097-0142(197310)32:4<853::AID-CNCR2820320417>3.0.CO;2-2.

Gale, Edwin A.M. 2001. The discovery of Type 1 diabetes. *Diabetes* 50 (2): 217–226. https://doi.org/10.2337/diabetes.50.2.217.

Graham, Allen. 1925. Malignant tumors of the thyroid. *Annals of Surgery* 82 (1): 30–44.

Green, Thomas Henry. 1871. *An introduction to pathology and morbid anatomy*. Henry C. Lea.

Hanson, Norwood Russell. 1958. *Patterns of discovery: An inquiry into the conceptual foundations of science*. Cambridge University Press.

Haugen, Bryan R., Anna M. Sawka, Erik K. Alexander, Keith C. Bible, Patrizio Caturegli, Gerard M. Doherty, Susan J. Mandel, et al. 2017. American Thyroid Association Guidelines on the Management of thyroid nodules and differentiated thyroid cancer task force review and recommendation on the proposed renaming of encapsulated follicular variant papillary thyroid carcinoma without invasion to noninvasive follicular thyroid neoplasm with papillary-like nuclear features. *Thyroid* 27 (4). Mary Ann Liebert, Inc., Publishers: 481–483. https://doi.org/10.1089/thy.2016.0628.

Hazard, John B., William A. Hawk, and George J.R. Crile. 1959. Medullary (Solid) carcinoma of the thyroid – A clinicopathologic entity. *The Journal of Clinical Endocrinology & Metabolism* 19 (1): 152–161. https://doi.org/10.1210/jcem-19-1-152.

Henry, Alexander. 1855. on the ancient and modern doctrines of cancer. *Association Medical Journal* 3 (122): 413–416.

Hoyningen-Huene, Paul. 1987. Context of discovery and context of justification. *Studies in History and Philosophy of Science Part A* 18 (4): 501–515. https://doi.org/10.1016/0039-3681(87)90005-7.

———. 2006. Context of discovery versus context of justification and Thomas Kuhn. In *Revisiting Discovery and Justification*, ed. Jutta Schickore and Friedrich Steinle, 119–131. Springer.

Ito, Y., A. Miyauchi, and H. Oda. 2018. Low-risk papillary microcarcinoma of the thyroid: A review of active surveillance trials. *European Journal of Surgical Oncology (SI: Thyroid Cancer Management)* 44 (3): 307–315. https://doi.org/10.1016/j.ejso.2017.03.004.

Jones, David S., Jeremy A. Greene, Jacalyn Duffin, and John Harley Warner. 2015. Making the case for history in medical education. *Journal of the History of Medicine and Allied Sciences* 70 (4). Oxford University Press: 623–652.

Kakudo, Kennichi, Yanhua Bai, Zhiyan Liu, and Takashi Ozaki. 2012. Encapsulated papillary thyroid carcinoma, follicular variant: A Misnomer. *Pathology International* 62 (3): 155–160. https://doi.org/10.1111/j.1440-1827.2011.02773.x.

Kelynack, T.N. 1899. The Pathology of Renal Tumours: Substance of an address introductory to a discussion at the annual meeting of the British Medical Association, Portsmouth, Aug. 4, 1899. *Edinburgh Medical Journal* 6 (3): 233–242.

Laurence, John Zachariah. 1858. *The diagnosis of surgical cancer*. Churchill.

Lebert, Hermann. 1845. *Physiologie pathologique, ou, Recherches cliniques, expérimentales et microscopiques: sur l'inflammation, la tuberculisation, les tumeurs, la formation du cal, etc.* Baillière.

———. 1851. *Traité pratique des maladies cancéreuses et des affections curables confondues avec le cancer*. Baillière.

Lin, Harrison W., and Neil Bhattacharyya. 2010. Clinical behavior of follicular variant of papillary thyroid carcinoma: Presentation and survival. *The Laryngoscope* 120 (4): 712–716. https://doi.org/10.1002/lary.20828.

Lindsay, Stuart. 1960. *Carcinoma of the thyroid gland*. Charles C Thomas Publisher. https://search.worldcat.org/title/644050824.

Lloyd, Ricardo V., Lori A. Erickson, Mary B. Casey, King Y. Lam, Christine M. Lohse, Sylvia L. Asa, John K.C. Chan, et al. 2004. Observer variation in the diagnosis of follicular variant of papillary thyroid carcinoma. *The American Journal of Surgical Pathology* 28 (10): 1336. https://doi.org/10.1097/01.pas.0000135519.34847.f6.

Lohia, Shivangi, R. Martin Hanson, Michael Tuttle, and Luc G.T. Morris. 2020. Active surveillance for patients with very low-risk thyroid cancer. *Laryngoscope Investigative Otolaryngology* 5 (1): 175–182. https://doi.org/10.1002/lio2.356.

Loison, Laurent. 2016. Forms of presentism in the history of science. Rethinking the project of historical epistemology. *Studies in History and Philosophy of Science Part A* 60 (December): 29–37. https://doi.org/10.1016/j.shpsa.2016.09.002.

Loncar, Samuel. 2022. Does science need history? A conversation with Lorraine Daston. *Marginalia*. https://themarginaliareview.com/daston-interview-p1/.

Mackowiak, Philip A., Donna Parker, and Lindsay D. Croft. 2017. The case for medical history in physicians' education: A survey of what physicians and physicians-in-training think. *The American Journal of Medicine* 130 (4). Elsevier: 494–497. https://doi.org/10.1016/j.amjmed.2016.11.025.

Mann, M.D. 1883. Some points of the histology of malignant Tumors*Read before the Buffalo Microscopical Club, January 9, 1883. *Buffalo Medical and Surgical Journal* 22 (7): 300–306.

Marsden, Alexander. 1874. *A new and successful mode of treating certain forms of cancer*. Churchill.

Oyama, Tetsunari, Tsunehiro Ishida, Kaoru Ishii, Shinji Sakurai, Takashi Joshita, Atsuhiko Sakamoto, and Takashi Nakajima. 1993. Encapsulated papillary carcinoma of the thyroid gland: Clincopathological and cytoflurometric study in comparison with non-encapsulated papillary carcinomy. *Pathology International* 43 (9): 516–521. https://doi.org/10.1111/j.1440-1827.1993.tb01165.x.

Passler, Christian, Gerhard Prager, Christian Scheuba, Barbara E. Niederle, Klaus Kaserer, Georg Zettinig, and Bruno Niederle. 2003. Follicular variant of papillary thyroid carcinoma: A long-term follow-up. *Archives of Surgery* 138 (12): 1362–1366. https://doi.org/10.1001/archsurg.138.12.1362.

Renshaw, Andrew A., and Edwin W. Gould. 2002. Why there is the tendency to "overdiagnose" the follicular variant of papillary thyroid carcinoma. *American Journal of Clinical Pathology* 117 (1): 19–21. https://doi.org/10.1309/CJEU-XLQ7-UPVE-NWFV.

Rivera, Michael, R. Michael Tuttle, Snehal Patel, Ashok Shaha, Jatin P. Shah, and Ronald A. Ghossein. 2009. Encapsulated papillary thyroid carcinoma: A clinico-pathologic study of 106 cases with emphasis on its morphologic subtypes (Histologic Growth Pattern). *Thyroid* 19 (2). Mary Ann Liebert, Inc., publishers: 119–127. https://doi.org/10.1089/thy.2008.0303.

Russell, William. 1890. An address on a characteristic organism of cancer. *British Medical Journal* 2 (1563): 1356–1360.

Schickore, Jutta, and Friedrich Steinle. 2006. *Revisiting discovery and justification: Historical and philosophical perspectives on the context distinction.* Springer.

Schröder, Sören, Werner Böcker, Henning Dralle, Karl-Bernd Kortmann, and Claudia Stern. 1984. The encapsulated papillary carcinoma of the thyroid a morphologic subtype of the papillary thyroid carcinoma. *Cancer* 54 (1): 90–93. https://doi.org/10.1002/1097-0142(19840701)54:1<90::AID-CNCR2820540119>3.0.CO;2-0.

Senn, Nicholas. 1895. *The pathology and surgical treatment of tumors.* Saunders.

Skuse, Alanna. 2015. *Constructions of cancer in Early Modern England: Ravenous natures,* Wellcome Trust–Funded Monographs and Book Chapters. London: Palgrave Macmillan. http://www.ncbi.nlm.nih.gov/books/NBK547257/.

Snow, Herbert. 1893. *A treatise, practical and theoretic on cancers and the cancer-process.* London: J. & A. Churchill.

Steere-Williams, Jacob, Justin Barr, Claire D. Clark, and Raúl Necochea López. 2023. Remaking the case for history in medical education. *Journal of the History of Medicine and Allied Sciences* 78 (1): 1–8. https://doi.org/10.1093/jhmas/jrac049.

Sturm, Thomas, and Gerd Gigerenzer. 2006. How can we use the distinction between discovery and justification? On the weaknesses of the strong programme in the sociology of science. In *Revisiting discovery and justification*, ed. Jutta Schickore and Friedrich Steinle, 133–158. Dordrecht: Springer.

Sutherland, Rosie, Venessa Tsang, Roderick J. Clifton-Bligh, and Matti L. Gild. 2021. Papillary thyroid microcarcinoma: Is active surveillance always enough? *Clinical Endocrinology* 95 (6): 811–817. https://doi.org/10.1111/cen.14529.

Sutton, J. Bland. 1894. *Tumors, innocent and malignant: Their clinical features and appropriate treatment.* Cassell & Company.

Tallini, Giovanni, R. Michael Tuttle, and Ronald A. Ghossein. 2017. The history of the follicular variant of papillary thyroid carcinoma. *The Journal of Clinical Endocrinology & Metabolism* 102 (1): 15–22. https://doi.org/10.1210/jc.2016-2976.

Thiersch, Carl. 1865. *Der Epithelialkrebs namentlich der Haut: eine anatomisch-klinische Untersuchung.* Engelmann.

Thompson, Norman W., Ronald H. Nishiyama, and Jay K. Harness. 1978. Thyroid carcinoma: Current controversies. *Current Problems in Surgery* 15 (11): 1–67. https://doi.org/10.1016/S0011-3840(78)80004-5.

Tielens, Emile T., Steven I. Sherman, Ralph H. Hruban, and Paul W. Ladenson. 1994. Follicular variant of papillary thyroid carcinoma. A clinicopathologic study. *Cancer* 73 (2): 424–431. https://doi.org/10.1002/1097-0142(19940115)73:2<424::AID-CNCR2820730230>3.0.CO;2-I.

Timmermann, C. 2013. *A history of lung cancer: The recalcitrant disease.* Springer.

Turner, William. 1863. On the present aspect of the doctrine of cellular pathology: A lecture delivered at an evening meeting of the Royal College of Surgeons of Edinburgh, 27th February 1863. *Edinburgh Medical Journal* 8 (10): 873–897.

Vickery, Austin L., Jr. 1983. Thyroid papillary carcinoma: Pathological and philosophical controversies. *The American Journal of Surgical Pathology* 7 (8): 797.

Vogel, Julius. 1847. *The pathological anatomy of the human body.* Lea & Blanchard.

Waldeyer. 1872. Die Entwickelung der Carcinome. *Archiv für pathologische Anatomie und Physiologie und für klinische Medicin* 55 (1): 67–159. https://doi.org/10.1007/BF01937199.

Walshe, Walter Hayle. 1846. *The nature and treatment of cancer.* Taylor and Walton.

———. 1853. *Practical treatise on cancerous diseases and on curable affections confounded with cancer.* Vol. 11. https://www.ncbi.nlm.nih.gov/pmc/articles/PMC5192898/.

Warren, Shields, and William A. Meissner. 1953. *Tumors of the Thyroid Gland, Atlas of Tumor Pathology.* Washington, DC: Armed Forces Institute of Pathology.

Woolner, Lewis B. 1971. Thyroid carcinoma: Pathologic classification with data on prognosis. *Seminars in Nuclear Medicine (Current Concepts of Thyroid Disease Therapy)* 1 (4): 481–502. https://doi.org/10.1016/S0001-2998(71)81042-5.

Woolner, Lewis B., Oliver H. Beahrs, B. Marden Black, William M. McConahey, and F. Raymond Keating. 1961. Classification and prognosis of thyroid carcinoma: A study of 885 cases observed in a thirty year period. *The American Journal of Surgery* 102 (3): 354–387. https://doi.org/10.1016/0002-9610(61)90527-X.

Xu, Bin, and Ronald Ghossein. 2018. Evolution of the histologic classification of thyroid neoplasms and its impact on clinical management. *European Journal of Surgical Oncology: The Journal of the European Society of Surgical Oncology and the British Association of Surgical Oncology* 44 (3): 338–347. https://doi.org/10.1016/j.ejso.2017.05.002.

Yu, Xiao-Min, David F. Schneider, Glen Leverson, Herbert Chen, and Rebecca S. Sippel. 2013. Follicular variant of papillary thyroid carcinoma is a unique clinical entity: A population-based study of 10,740 cases. *Thyroid* 23 (10): 1263–1268. https://doi.org/10.1089/thy.2012.0453.

Zidan, Jamal, Drumea Karen, Moshe Stein, Edward Rosenblatt, Walid Basher, and Abraham Kuten. 2003. Pure versus follicular variant of papillary thyroid carcinoma. *Cancer* 97 (5): 1181–1185. https://doi.org/10.1002/cncr.11175.

Chapter 8
A Plea for More History

Timo Bolt

In his *Is There An Epistemic Role For History In Medicine? Thinking About Thyroid Cancer* Nicholas Binney (Chap. 7, this volume) aims to show how historical work can have its epistemic function, that is: a role in understanding and therefore justifying medical knowledge and practice. From my perspective as a medical historian, Binney's programmatic text raises three questions: (1) Is this epistemic role for history important?; (2) Is Binney's argument convincing?; (3) How can a plea for history *appear* so ahistorical, or more positively put: how can we make Binney's promising approach even more historical? In what follows, I will try and answer these questions.

8.1 Is it Important?

Why would it be important for medical history to have an epistemic role? Great medical histories are and can be written without having any epistemic function. Most medical historians do not even strive for their work to have such a function. In this sense, the 'lean distinction' between the contexts of discovery and justification, although convincingly challenged by Binney, does correspond with everyday scholarly practice. Medical historians tend to ask different questions than philosophers of medicine or biomedical scientists and usually answering them does not require history to have an epistemic function. A similar situation is at hand as the one Monica Greco describes, in her Chap. 17, in this volume, with respect to social research. She points at an 'agnosticism' regarding the nature of symptoms, which

T. Bolt (✉)
Section Medical Ethics, Philosophy and History of Medicine, Erasmus MC University Medical Center, Rotterdam, The Netherlands
e-mail: t.bolt@erasmusmc.nl

© The Author(s) 2024
M. Schermer, N. Binney (eds.), *A Pragmatic Approach to Conceptualization of Health and Disease*, Philosophy and Medicine 151,
https://doi.org/10.1007/978-3-031-62241-0_8

"reflects an implicit methodological commitment to a bifurcated division of labour that allocates social scientist to the study of (discourse-mediated) 'culture', while forbidding them from entertaining hypotheses about 'nature', such that any truth about bodily symptoms as such is left as a matter for medicine or natural science to research and establish."

Binney's chapter encourages medical history to move beyond this agnosticism and its corresponding self-limiting role within the division of labour between disciplines. Embracing an epistemic role for history would do just that. It would also be important, for example in those instances where it is necessary to be able to distinguish between good and bad science, or between scientific knowledge and mere belief. Moreover, it could enrich the historical 'framing disease' literature, which is somewhat unbalanced: on the one hand highly sophisticated in analysing the role of cultural and social factors, and political and economic interest in the social construction of disease, but on the other hand relatively unsophisticated, sometimes even naively realistic, when it comes to the medical 'content' or the 'biological event' that is being framed (Aronowitz 2008; Cooter 2010; Rosenberg and Golden 1992). Like Binney, I feel addressed by Lorraine Daston's lamentations about the disintegration between history, philosophy, and science. I also agree with Daston that historians have made it difficult to see how their work is relevant to contemporary science and medicine. Fortunately, there is hope, as Binney notes that doctors and medical researchers frequently express the intuition that history of medicine can and should be used to inform the evaluation of medical knowledge in the present. This suggests that adopting an epistemic role would increase the relevance of medical history and further its collaboration with other fields.

8.2   Is it Convincing?

In my view, Binney is highly original and analytically sophisticated in his use of relatively 'old' work by Hanson and Fleck. Hanson's most profound insight is that when scientists observe, they do not simple *see,* they see *as* and *see that.* Scientists need to see objects *as* members of a certain kind of class, in order to see *that* objects of that kind have certain properties. This implies that the starting point of analysis should not be some ultimate reality, but the way people shape order by categorizing things and attributing 'kindhood'. In Hanson's terminology this involves the "patterning statements" which make people see things *as*. This in turn allows further observations which enable us to make "detail statements". Quite ingeniously, Binney relates Hanson's "patterning statements" to Flecks "active elements" of knowledge, and Hanson's "detail statements" to Flecks "passive elements" of knowledge. According to Binney, it is the passive elements, and the detail statements, that prevent these epistemologies from collapsing into "silly relativism". Although they are dependent on and produced by the active elements, and the patterning statements, they are not fully determined by them: they "resist the will" of the researchers involved.

In a practical and methodological sense Hanson and Fleck 'work'. They enable Binney to trace the development of a field of medical knowledge in terms of its shifting and changing active and passive elements. He reconstructs how criteria for the diagnosis and classification of thyroid cancer were "actively" defined and regularly modified in the face of "passive resistance" since the mid-nineteenth century. The result is an insightful "genealogy" of a "problematic situation": the substantial overdiagnosis of the so-called follicular variant of the papillary thyroid carcinoma (FVPTC).

The objective of his paper, however, is not to give a full account of the specific case of thyroid cancer, but to make a more general point about how history can have an epistemic function by adopting a Fleckian analytic approach. Amongst others, he shows that historical knowledge helps to make sense of medical ideas and practices that, at first glance, seem to be illogical. The most striking example is how some tumours with a follicular histological organisation and without any papillary tissue, nevertheless, came to be classified as 'papillary' tumours. Moreover, he convincingly argues that what doctors and medical scientists observe is a function of the history of their field. This means that history "is an integral part of the observations themselves and thus part of the justification of scientific claims", or in other words: history has an epistemic function.

8.3   How to Make it Even More Historical?

Binney does not intend nor claim to provide a full historical case-study. His chapter has a different purpose: "Incomplete as it is, this history will show how researchers built the expectation of homogeneity, of kindhood, into their classifications. This is what gives history its epistemic function here. Even though this is a creative process, the interaction of active and passive elements preserve the integrity of science." This is what Binney's chapter is all about: preserving the integrity of science and being able to conclude: "What are created are *genuinely* scientific facts". Thus, it addresses one of the key issues of this volume (Kusch Chap. 5, this volume; Binney et al. Prologue Chap. 2, to this volume): that it is possible to reject simplistic forms of realism and to frame concepts of health and disease as pragmatic and historically contingent, *without* slipping into "pernicious forms of relativism". In his view, this is also a prerequisite for overcoming the disintegration between history, philosophy, and science.

Given these objectives, it is understandable that Binney leaves out a great deal of historical context. He explicitly mentions that the development and implementation of cancer screening programs, fears of litigation, changes to surgical and radiological practice, and influence of patient attitudes, and other 'contextual' issues are *not* discussed in his chapter. He also explains why: "*Before* these elements are explored, it is important to show how they can be relevant to the production of pathological facts without compromising pathological science." Binney's use of the word 'before' suggests a hierarchy: epistemological analysis comes first, is primary, historical context is only secondary. This is at odds with most work in medical history, which usually prioritises contextual over epistemic issues in their analysis of the framing of diseases. Moreover, after almost half a century of 'social history of medicine' and 'history from below', Binney's selection of historical actors and sources will seem problematically one-sided to medical historians. He only discusses medics and biomedical scientists, and only their ideas seem to matter. In contrast with this, social scientist Wadman (2023: 5) recently argued: "The delineation of disease is not merely an epistemic activity. It relies on practical work, social negotiations, and material infrastructures."

Fortunately, Binney's own argument points the way for bridging this gap with medical historiography and making his approach even more historical. The contextual issues he chooses not to discuss, as well as the "practical work, social negotiations, and material infrastructures" mentioned by Wadman, are part of the network of active elements of knowledge, and therefore, by definition, *epistemic*! This is highly important because, as Binney repeatedly stresses, the passive elements of knowledge are dependent on *and* generated by the network of active elements (but not fully determined by them). In personal communications he has sometimes used the metaphor of a trampoline to explain this: a trampoline does not resist us unless we are jumping on it, and how we jump on it will change how it resists us. This implies that more is at stake than just the issue of passive resistance being generated and facts being 'genuine'. It is also relevant to ask: Who are doing the jumping, how are they jumping and what kind of 'genuine' facts are thus produced?

Here, medical historical studies about the 'framing' of diseases, their 'framing *effects*', and the cycles of feedback between them can be of great help in answering questions such as: Who are setting the agenda and formulating the objectives of medicine and biomedical research? And how and why are networks of active elements 'chosen', and in many cases maintained or reinforced, even despite passive resistance? (see a.o.: Aronowitz 2008). In addition, in the specific case of the history of the classification of thyroid cancer, some of the who?'s, how?'s, and why?'s only make sense in the light of what historians call 'the Western biomedical tradition'. Binney rightly starts his story in the mid-nineteenth century, when Western biomedicine started to look for, and make visible, the pathologies, abnormal structures and disturbed processes within the body, which were regarded as *underlying* the symptoms and signs of disease (Bynum et al. 2006; Kleinman 1993; see also Fangerau's Chap. 3, in this volume). It is against this backdrop that the historical actors in Binney's story tried to predict the prognosis, clinical behaviour over the course of time, and mode of metastasis of tumours on the basis of what kind of abnormal microscopic structures were present. It was also line with the Western biomedical tradition that they, at a certain point, decided to take the appearance of cell nuclei as more important to tumour classification than histological organization.

This example suggests that we could do well with *more* Fleck. Within the necessarily narrow scope and objectives of Binney's chapter it is understandable that he concentrates on the interplay between active and passive elements of knowledge. But integrating other Fleckian concepts, such as 'thought style', 'thought collective', 'apriori's', 'esoteric circles', and 'exoteric circles', would make his promising approach more historical and better aligned to the existing medical historiography. Admittedly, however, that would require a book rather than a chapter.

References

Aronowitz, Robert. 2008. Framing disease: An underappreciated mechanism for the social patterning of health. *Social Science & Medicine* 67: 1–9. https://doi.org/10.1016/j.socscimed.2008.02.017.

Bynum, W.F., Anne Hardy, Stephen Jacyna, Christopher Lawrence, and E.M. Tansey. 2006. *The Western medical tradition: 1800–2000.* Cambridge: Cambridge University Press.

Cooter, Roger. 2010. The life of a disease? *Lancet* 375: 111–112. https://doi.org/10.1016/S0140-6736(10)60034-7.

Kleinman, Arthur. 1993. What is specific to Western medicine? In *Companion encyclopedia for the history of medicine*, ed. W.F. Bynum and R. Porter, 15–23. London: Routledge.

Rosenberg, Charles E., and Janet Golden. 1992. *Framing disease: Studies in cultural history.* New Brunswick: Rutgers University Press.

Wadman, Sarah. 2023. Disease classification: A framework for analysis of contemporary developments in precision medicine. *SSM—Qualitative Research in Health* 3: 100217. https://doi.org/10.1016/j.ssmqr.2023.100217.

Chapter 9
Scope Validity in Medicine

Lara Keuck

9.1   Introduction

If a test measures what it means to measure, it is deemed "valid." First defined in this way in psychological research (Kelley 1927), the concept of validity has pursued a steep career. Since at least the mid-twentieth century, the ideal of validity has been theorized, debated, translated into methods, and used to regulate and (de-)legitimate knowledge concerning health and disease. For instance, a specific rodent model of chronic mild stress was considered one of the best validated animal models of depressive disorder in humans according to existing concepts of validity (e.g. Willner and Mitchell 2002). However, clinical trials on therapeutics that had been successfully tested in the animal model failed. The reason for this failure in the human context has been attributed to the fact that only a small portion (and therefore a financially uninteresting market) of patients who are diagnosed with depression suffer from a subtype of the disorder for which this model is a good predictor (Belzung 2013). The clinical trial population was not stratified in a way that allowed to test whether or not the drug works. Put differently, the experimental design of the preclinical model restricted the successive domain of application of the research results. This case of translational failure can be analyzed in several ways: we can question the meaning of 'best validated model' if the animal model cannot be adequately extrapolated to clinical trials on depression. Or we can blame the pharmaceutical company's marketing-oriented selection of too broad inclusion

L. Keuck (✉)
History and Philosophy of Medicine; Faculty of History, Philosophy, and Theology, and Medical School OWL, Bielefeld University, Bielefeld, Germany

Max Planck Research Group "Practices of Validation in the Biomedical Sciences", Max Planck Institute for the History of Science, Berlin, Germany
e-mail: lara.keuck@uni-bielefeld.de

© The Author(s) 2024
M. Schermer, N. Binney (eds.), *A Pragmatic Approach to Conceptualization of Health and Disease*, Philosophy and Medicine 151,
https://doi.org/10.1007/978-3-031-62241-0_9

criteria for undermining the model's validity. Both approaches are fair enough, yet the blame game that often results can easily overshadow that validity is never unmediated, never absolute. Mismatching of scopes are not just (though also) a problem of polemics and the rhetoric of big pharma or overpromising biomedical research. We lack an understanding of the scientific activities involved in capturing and evaluating how well the scope of an experiment—the *actual domain of application* of the results of preclinical research—fits to its *intended target domain of application* in the clinical context.

This chapter analyzes mismatching disease operationalizations as challenges to validity in biomedicine, and introduces the new concept of scope validity to capture this problem. It combines an adequacy-for-purpose view towards validity (e.g. Alexandrova and Hybron 2016; Feest 2019; Parker 2020) with a pragmatist and particularistic perspective on disease concepts (e.g. Demazeux and Keuck 2023; see also Binney et al., Chap. 2, in this volume for a pragmatist perspective on disease concepts; Kusch, Chap. 5, in this volume for differentiating pragmatism from relativism; Binney, Chap. 7, in this volume for conceptualizing change in disease operationalizations). The chapter proceeds as follows: the second section focuses on mismatching disease operationalizations as a missing link in the evaluation of animal models of human mental disorders. Against this background, I clarify how my notion of scope validity differs from existing concepts of validity, in particular construct validity, external validity, and predictive validity. In the third section, I advocate much in the spirit of practical concepts of disease for a relational epistemology to biomedical objects of inquiry. I argue for relational epistemology as a philosophical framework for capturing the extent to which (and the conditions under which) the relata of a specific animal model, a clinical trial design, and the diagnosis in clinical guidelines match. This line of argument builds on my particularistic perspective, which side-steps all-encompassing validity theories and general philosophical theories of disease, while being attentive to the diversity of validity and disease theories that are at work in every single study design. Against this background, I argue for the potential of a philosophy of science in practice approach to identify existing medical scientific methods that could be analyzed as responding to problems of scope validity. For instance, some forms of retrospective epidemiological studies and reverse translation trials in animal models (testing effective clinical interventions in animals) might be understood as instances of 'scoping methods,' which provide us with information on the (mis-)matching of disease operationalizations in different research and application contexts. In the concluding section, I address the functions we might ascribe to scope validity: as a tool for evaluating study designs in translational medicine, as a description of how knowledge generation within one biomedical context conditions the way in which a medical problem needs to be identified in another context, and as an analytic category for studying scientific methods of matching scopes across research contexts. I conclude with a common thread between the philosophical questions that scope validity raises: the adequacy of approaches to medical research.

9.2 Validity, Scope, and Scope Validity

This section introduces scope validity. I first analyze the role of abstract targets (or constructs) in validity concepts (Sect. 9.2.1). I will then examine the limitations of this approach for evaluating animal models of human diseases (Sect. 9.2.2). Against this background, I discuss the representational scope of models in biomedical research and present scope validity as a complementary conceptual tool to identify the target population to which a research result might be best generalizable (Sect. 9.2.3).

9.2.1 Validity Concepts and the Guiding Ideal of a Construct

Validity has been debated for almost a century, especially in the psychological sciences. Most validity theorists take the educational psychologist Truman Lee Kelley's 1927 dictum as point of departure: "The question of validity would not be raised so long as one man uses a test or examination of his own devising for his private purposes, but the purposes for which schoolmasters have used tests have been too intimately connected with the weal of their pupils to permit the validity of a test to go unchallenged (. . .) *The problem of validity is that of whether a test really measures what it purports to measure*" (Kelley 1927: 14, my italics). Validity seems to involve "the acceptance of a set of operations as an adequate definition of whatever is to be measured" (Bechtoldt 1951, 1265, quoted in Cronbach and Meehl 1955, 282). Or at least this is the case for a specific understanding of validity. Indeed, this was the worry of Lee Cronbach and Paul Meehl, the heads of the Committee of the American Psychological Association that was tasked to formulate *Technical Recommendations for Psychological Tests*. They suggested an elaborate terminology of different kinds of validity, naming their "chief innovation" the introduction of a new term that they called "construct validity": "Construct validity is not to be identified solely by particular investigative procedures, but by the orientation of the investigator. (. . .) When an investigator believes that no criterion available to him is fully valid, he perforce becomes interested in construct validity because this is the only way to avoid the 'infinite frustration' of relating every criterion to some more ultimate standard" (Cronbach and Meehl 1955, 282). They suggested a new concept, namely that of construct validity, to give "investigators" a possibility to address a specific kind of doubt: not a doubt about the performance of a test, but about its informativeness about an abstract target.

The concept of construct validity becomes more intelligible when taking into account the nature of 'constructs.' Ken Schaffner (Forthcoming: 1) defines constructs as concepts that "refer to entities that are general, abstract, and putatively explanatory. Examples include notions such as intelligence, working memory, gamma frequency oscillation circuits, normal and abnormal personality types, disorders such as schizophrenia, and even the 'self.'" If a test has a high construct validity, it is highly informative about the abstract entity in question. A valid test can be understood as providing evidence for the reality of the construct (if we can measure intelligence, it exists), and/or as being a good way to test the manifestation (e.g. of intelligence) in an individual that allows for drawing conclusions that are also of relevance outside of the test context.

Psychometricians, who were the first to introduce and broadly apply notions of *construct validity*, for instance with regards to psychological testing of personality traits or intelligence, have developed a nuanced terminology. Keith Markus and Denny Borsboom (2013: 3) define a *construct* as a "property tested or intended for testing," which "assumes a substantive interpretation of this property." The semantic representation of this property is then the *construct label*. Since the "researchers do not directly observe" the property, the psychometricians treat it as a *latent variable*, which allows them "to represent statistical relationships with some latent variable, whatever it may be, without specifying the substance of that variable."

According to Schaffner (2012, Forthcoming), the introduction and use of validity concepts in psychometric, psychiatric, and animal model research contexts have given rise to quite different discussions with varying underlying philosophical commitments to laws, pragmatism, and reductionism. However, Schaffner also stresses that the notion of a construct as an abstract entity has been central to all three of these contexts, even if, for instance, Robins and Guze's (1970) criteria on how to assess whether a diagnosis of schizophrenia was valid did not at all refer to Cronbach and Meehl's term of construct validity. Moreover, he seems to agree with Cook and Campbell (1976) who "asserted that C[onstruct] V[alidity] was involved *whenever* one dealt with causes and outcomes." (Schaffner Forthcoming: 2). It is a fair assumption that construct validity served a regulatory function for the many other validity concepts — internal, external, predictive, descriptive, aetiological, face, etc. (Sect. 9.2.2) — that had been introduced and discussed in the last 65 years. At stake was the question of how well a certain model or test hit the abstract target of inquiry, be it with respect to representing its pathophysiology, determining its relationship to a latent variable, or to developing a screening device for drug testing. While the plurality of validity concepts indicate an awareness of the various aims and interests in assessing the hitting of the target disease entity, the practical definition of the target mark, and the fitting of different definitions within contexts of experimentation and contexts of application remained undertheorized. Discussions on the validity of animal models for human diseases illustrate exemplarily why this concern matters.

9.2.2 The Logic of Validation in Animal Models of Human Diseases

Within the field of animal-based modelling of human diseases, most researchers have used variants of three suggested validity concepts: predictive validity, face validity, and construct validity. Yet, there is no homogenous use of these concepts and their derivatives—not even in a comparably confined field, such as that of animal models of human depressive disorders (see Belzung and Lemoine 2011). In general, a high *predictive validity* denotes a high "human-animal correlation of therapeutic outcomes," that is to say, pharmacological (or other interventionist) therapeutic effects in humans can be reproduced in the animal and vice versa (ibid.: 3). A high *face validity* of an animal model means that it exhibits a "phenomenological identity" to the human disorder, which is mostly understood in terms of "an attempt to mimic diagnostic criteria of the psychiatric conditions, such as those listed in the American Psychiatric Association's Diagnostic and Statistical Manual of Mental Disorders." (ibid.: 4). A high *construct validity* means that the animal model is informative about the human disease in the sense that the model can be used to gain knowledge about the disease entity in question.

But how do we know what qualifies as a legitimate instance of this construct? Models often fulfil a seemingly paradoxical role, the trained psychologist turned animal researcher Paul Willner noted while revising the face validity criteria of animal models of depression that were first proposed by McKinney and Bunney (1969). Along with his revisions, Willner introduced new measures of predictive and construct validity (Willner 1984), which he argued were necessary updates because, for one, "in relation to animal models of depression, similarity of aetiology and biochemistry are unsuitable as validating criteria since they are themselves the subject of intense research and speculation." (ibid.: 1). For another, Willner's update was motivated by his perception that scientific progress in depression research had led to new hypotheses regarding the interrelation between environmental factors and endogenous depression as well as more elaborate experimental set-ups to induce and test behaviors, for instance, the animals' reactions to 'chronic stress.'

This example, and, more generally, the plethora of validity concepts that scientists, psychiatrists, and philosophers have elaborated in the last decades reflect the manifold interdependences between determining the explanatory role, the predictive power, and the representational scope of a given test or model (see, e.g., Kendler and Parnas 2012). Perhaps best known and most discussed in philosophy of science is the differentiation between internal and external validity (e.g., Cook and Campbell 1979): research results are *internally valid*, when they are reproducible and significant within the confined parameters of a controlled test, but need to prove their *external validity* outside of the controlled, experimental setting in real-world contexts (see also Guala 2003; Cartwright 2009). Extrapolation and external validity have been the subject of many philosophical inquiries into application-oriented sciences, some of which have motivated normative conclusions on how science should work in order to be useful for society (e.g., Kitcher 2003; Cartwright 2009). Yet, the chronological and epistemic order that presumes that internal validity always precedes external validity is challenged in biomedical research, which operates in a more iterative mode (Huber and Keuck 2013). Biomedical research does not start at the bench and end in the clinics; the material and conceptual transfers are multidirectional.

This iterative go-between of clinical and laboratory demands and insights is particularly evident within animal models of diseases that are thought to occur only in humans, such as Alzheimer's disease, a neurodegenerative disease leading to dementia and death. The establishment of a mouse model begins with a reverse translation from bedside to bench, often including transfer of genetic material from human patients to laboratory animals. It necessitates the selection of clinical symptoms (e.g., memory deficits, but not personality changes) and their translation into test procedures for animals (e.g., behavioral testing of mice's memory deficits in the Morris Water Maze). After establishing and characterizing the animal model, pharmaceuticals are tested in these in-bred animals. The conclusions in the lab legitimate whether the drug should be tested in clinical trials on humans.

The zigzagged logic of animal modelling has implications for thinking about what it means to 'hit' the target of inquiry. The resort to an abstract disease construct has clouded rather than facilitated the assessment of the representational scope of an animal model with regards to human patients. Alzheimer's disease is perhaps a particularly strong case in point with its unknown aetiology and its ambiguous definition (Huber and Keuck 2013; Keuck 2020; Daly and Keuck 2024). The first mouse model that exhibited both a (nowadays debated) histopathologic hallmark of the disease (amyloid beta plaques) and memory deficits (Hsiao et al. 1996) had been established through the transfer of genetic material of the so-called Swedish mutation. This genetic mutation had been characterized within a human genetic field study that had traced families in which severe, early onset forms of dementia had occurred throughout generations. The geneticists that had isolated (and later patented) the Swedish mutation acknowledged that Alzheimer's disease was "genetically heterogeneous" (Mullan et al. 1992: 345). However, the mouse model was not presented and evaluated as a model that might provide more insights into the devastating illness of this Swedish family, but as a model for Alzheimer's disease in general. In the past 25 years, several hundred further mouse models for Alzheimer's disease have been established and elaborately validated, but in terms of translational research this approach did not provide for successful extrapolations. Just as in depression research, Alzheimer's researchers working with mouse models have blamed the clinical trial designers for redefining the medical target: "The nosology of A[lzheimer's] D[isease] keeps shifting, the consequence of not knowing its etiology. This situation makes it difficult to place mouse models precisely into human context and demands an adaptive framework for utilizing mice as models of the human disease." (Ashe and Zahs 2010). In other words, the clinical redefinitions have made the animal researchers' validation work of Alzheimer mouse models invalid.

The zigzagged logic of validating animal models for human diseases may remind us of the philosophical characterization of so-called looping effects. Originally, Ian Hacking (e.g., 2007) has described looping effects that stem from classified people's reactions to the way they have been classified, which can result in a change of this classification (e.g., of autism or of homosexuality in psychiatric manuals). Such 'moving targets' could be seen as one cause for a subset of classificatory shifts. The problem that I address in this paper, however, encompasses many more kinds of mismatches between an implicit or explicit definition of a target of inquiry in one setting (e.g., a particular lab) and in another (e.g., in a clinical trial, or in a general physician's practice). Jackie Sullivan (2009) has argued in a similar vein that it might turn out to be difficult to assess what neuroscientific studies can tell us about memory in general, when the protocols that are used in different laboratories to operationalize memory differ so strongly from each other that it is no longer clear whether they actually relate to the same phenomenon.

With regards to Alzheimer's disease, one strategy — that is currently propagated by the National Institute of Aging — to solve this problem of the shifting target is to bind the construct label Alzheimer's disease to a measurable variable like the occurrence of amyloid plaques in the brain (Jack et al. 2018). However, this strategy has several problems: as it reduces the mental illness to a biomarker, it is likely to be overinclusive with respect to false positives, because heightened values of amyloid beta also occur in people who never develop clinical symptoms. It also deprivileges alternative aetiological hypotheses, which might, according to some epidemiologists, account for a significant proportion of cases that are contemporarily diagnosed as Alzheimer's disease (e.g., Glymour et al. 2018). The overarching problem is that neither contemporary epidemiological nor biomarker approaches provide sufficient grounds for defining Alzheimer's disease unequivovally: similar to what Paul Willner described with respect to the challenges of modelling depression in mice when we do not really know what qualifies depression in humans, we are faced also in the case of Alzheimer's disease with an epistemological underdetermination of the target of inquiry (Daly and Keuck 2024). What does it mean in such cases to deem a model valid?

9.2.3 Scope Validity

My suggestion is to take a step back from the definitory muddle (or warfare, in some cases) that surround many abstract constructs, and think about a measure that better qualifies the actual scope of a given test or model. With respect to biology and biomedicine, most scholars have identified the representational scope of a model with the degree to which a model and its target share essential properties or functional processes and therefore are instances of the same 'general biology' (see e.g., Burian 1993; Schaffner 1998; Keller 2000; Ankeny and Leonelli 2011; see also Steel 2008 for a defense of 'comparative process tracing' to grant successful extrapolation even if properties between the model and target differ). However, extrapolation and representation might, at least in some cases, work significantly differently within models for general biology as compared to biomedicine: in biomedicine, the relationship between experiment and application is one of substitution (i.e. animals replace human patients) rather than, necessarily, an exemplification of general biology (i.e. animals represent general patterns of interest; see Rheinberger 2006a; Huber and Keuck 2013; Germain 2014; Green 2024). For example, when xenografts or human genetic material are used to generate humanized animal models (as is the case in many Alzheimer mouse models), the question is not only one of how conclusions drawn from an animal disease can be extrapolated to sick humans, but also which aspects of the human disease can be instantiated in the animal not least since Alzheimer's disease is thought to not naturally occur in mice.

In biomedicine, the target of the representational scope must be qualified not just regarding the comparability (be it the similarity or the possibility for comparative process tracing) of animal and human physiology, but also with respect to two further dimensions. First, we need to consider the degree to which this model can account for relevant aspects of human illness, recovery, or even the side effects of pharmaceuticals. For instance, weight gain as a side effect of a person taking antipsychotic drugs might be observable in animals but not the development of depressive symptoms due to the experience of the social stigma of obesity. To assess the psychological harmfulness, aspects of social (human) life need to be taken into consideration that are abstracted away in most experimental settings. This dimension relates to a model's face validity, i.e., the phenomenological similarity between model and target, whereby here face validity includes so-called patient-relevant outcomes. The second additional dimension that needs to be considered when determining the representational scope of a model in biomedicine is connected to the "reference class problem" (see, e.g. Hájek 2007): the Alzheimer's mouse mentioned above was potentially a much better model for the disease running in the Swedish family from which the mutated genetic material was transferred than it was for all people diagnosed with Alzheimer's disease in the 1990s. Similarly, the reference class of depression that the chronic mild stress rodent model is best compared to does not comprise all incidences of major depressive disorder, but a subclass that consists of people who developed depressive symptoms associated with stress. This dimension has some commonalities with variants of aetiological validity. However, it does not necessarily need to be based on a causal hypothesis of disease manifestation, which sometimes, but not always motivates stratification practices of delineating diagnostic groups. In psychiatry, so-called transnosographic and theranostic approaches have been adopted to suggest some re-classifications of mental illnesses (for an example, see Guessoum et al. 2020; for a critical assessment of a "precision psychiatry" approach, see Tabb and Lemoine 2021). Besides psychiatry, such re-grouping practices are much discussed and used in oncology, where, for example, umbrella or basket trials of cancer treatments cut across the traditional organ-specific classifications of neoplasms, and group traditional diagnoses together in new ways — with new chances and challenges for study trial designs and their implementation (e.g., Strzebonska and Waligora 2019). One of the challenges is akin to the missing link between assessments of the validity of animal models and successful translations into general health care practice that I elucidate in this paper: we need a measure of adequacy to ascertain how good the fit is between a given model or study population and the diagnoses in the "real world" patient population. This means to move from the abstract idea of a disease entity to concrete practices of identifying diseases in a given local context — and moving from a model's or trial design's construct validity to their scope validity in relation to a concrete context of application.

The idea of validity as a relational concept is not new, but to my knowledge it has not yet been elaborated in the medical sciences without assuming that there is an abstract disease entity that can be better or worse hit. For instance, Paul Willner once defined construct validity as "a theoretical account of the disordered behavior in the model, a theoretical account of the disorder itself, and a means to bring the two theories into alignment" (Willner 1994, quoted in Belzung and Lemoine 2011: 5). Other researchers even "mentioned the similarity of etiology, but also an interesting criterion that was unfortunately abandoned: the precision of the sub-nosographic entity ('Does the laboratory model describe (...) a naturally occurring psychopathology or only a subgroup?')" (Belzung and Lemoine 2011: 3, quoting Abramson and Seligman 1977). In the past two decades with the advance of -omics, big data analysis, and dimensional approaches to psychiatric classification, most prominently the Research Domain Criteria (RDoC) of the National Institute of Mental Health, the focus of many researchers has shifted away from abstract disease entities and towards modelling more fine-grained in-group differentiations, for instance between 'good' and 'poor responders' to antidepressants in mouse models (e.g. Herzog et al. 2018; for RDoC as a challenge to the predominance of diagnostic kinds in psychiatric theory, see Tabb 2017; Solomon 2022; Demazeux and Keuck 2023). Thus, there has been an increasing interest in refining a model's representational scope in practice, but this has not been theorized vis-à-vis the scientific validity concepts that still guide the choice of and warrant the extrapolation from animal models.

Concepts of validity that resort to an abstract entity cloud the fact that we lack a conceptual tool to capture mismatches between the scope of an animal model and the scope of a clinical trial. Catherine Belzung's (2013) frustrating conclusion of a failed clinical trial that had not been able to reproduce the effects of a pharmaceutical in human patients, which had been preclinically tested in her best-validated animal model of a specific form of depression, could be re-read as a call for taking the implications more seriously that the representational scope of the model has for defining appropriate inclusion and exclusion criteria in the clinical trial design.

Let me close this section with a working definition of scope validity:

(SV) *Scope validity denotes the matching between the target as operationalized in the setting of experimentation and the target as operationalized in the setting of application.*

In this understanding, the scope validity of an animal model in relation to a clinical study would include an assessment of the conditions of the particular clinical trial, which are best fitted to allow for testing the preclinically tested drug's mode of action in the human context. Importantly, scope validity is not identical to external validity, but rather an additional tool to *refine the frame under which external validity is assessed*.

In other words, we could describe a failure due to problems of scope validity as occurring when the domain of successful applicability is different from the domain of application that the intervention was tested on. Consider this definition from biomedical researchers Bert 't Hart et al. (2018) who apply validity terms to animal models of Multiple Sclerosis (MS): "External validity: represents the extent to which the observed effect of a treatment in an animal model can be generalized to the *total* MS patient population." (ibid.: 263, my italics). In contrast to this total generalizability of external validity, scope validity would then represent the matching between *MS as it is diagnosed* in the *studied* patient population and *MS as it is modelled* in the animal. Importantly, I do not assume one specific form of relevant similarity between a target as it is modelled and a target as it is diagnosed, when I refer to the matching of targets. The question of what makes a good match rather is a central question that the focus on scope validity helps to address (see the next sections). In general, if the scope validity is high, this should imply that the translational set-up is, given the current scientific knowledge, well suited to test whether the preclinical results could be extrapolated to the human patient study group. Scope validity does in and of itself not capture the total generalizability of the intervention's effect, but rather the matching of a model and a specific clinical trial design. This has important consequences, because if we take 't Hart and colleagues' definition of external validity as stand-alone measure, we would need to disqualify animal models that only allow for generalizability for a small set of the total patient population as having a weak external validity. Yet, as has been argued in the case of clinical trials on antidepressants, there might be good candidates that could be effective for a subset of the patient population. If we took this group as reference class for evaluating external validity, the external validity would be presumably much higher than if we took the group of all people who receive a diagnosis of depression. At the same time, this need not mean that we should change the diagnostic criteria of a given disease altogether. In contexts beyond drug testing, for instance the assessment of socioeconomic factors that impact (mental) health, it might be more adequate to sample the patient population in a different way. As I have argued elsewhere in more detail, there are good reasons for taxonomic and explanatory pluralism in medicine, but it demands additional measures that check what transferred data or translated results exactly refer to (Kutschenko 2011a, b). Scope validity responds to this task.

9.3   Towards a Relational Epistemology

In this section, I contextualize my approach to scope validity within a relational epistemology that is based on a particularistic perspective on disease (Sect. 9.3.1). I argue that this approach allows to ask new philosophical questions about specific scientific methods that respond to the challenge of scope validity (Sect. 9.3.2).

9.3.1   A Particularistic Perspective on Disease

My definition of scope validity is grounded on a relational approach to the objects of medical research: this means to not compare how well a given practice hits an ideal disease entity, but how well the target of one practice fits the target of another practice for which it attempts to provide a solution. This implies, for example, to not take for granted that practices of diagnosing disease or pursuing clinical trials carve nature at similar joints. The goal of a relational approach is to critically assess the extent and ways in which, for instance, the disease target that is operationalized when testing a drug in the highly-controlled setting of a clinical trial fits to the disease target that is reflected in the diagnostic practices of a primary health care setting (i.e. the context of application for which the context of testing attempts to provide a solution). In the trial, patient groups are often selected with expensive diagnostic technologies, such as PET-neuroimaging, but these technologies of identification are not available, affordable, and possibly desirable in all health care settings in which the experimental knowledge shall be put to practice. This is in those cases problematic, in which the ways of carving out the target in the experimental setting leads to a meaningfully different patient group composition than the one to which the knowledge is translated. In these cases, the practices identify different types though they are said to refer to the same disease. A relational theory acknowledges that in every single context, in every single laboratory, on every single occasion, a concrete manifestation (in an individual patient, in a clinical population, or in a model system of biomedical experiment) must be newly attributed to an abstract phenomenon-of-interest (like 'disease x' or 'memory', see Sullivan 2009; Feest 2011; Meunier 2012; Hauswald and Keuck 2017; Huber and Keuck 2017 for examples that are, however, not analyzed with explicit reference to a relational theory). Much in line with Gaston Bachelard's concept of a *phénomènotechnique*, the technologies and experimental procedures that are applied to make the phenomenon-of-interest examinable within the confines of a given research context impact the delineation of the target object (see Rheinberger 2006b). The very particular target objects of biomedical experiments therefore do not precede research although they aim to answer a question that is raised by a reality that exists outside of the laboratory.

The degree to which this particularistic perspective matters for successful translation of medical research results across local settings is a case-by-case empirical question. Some positively tested interventions into medical issues may require thorough knowledge and strict adherence to the precise rules of operationalization applied to the study. In other cases, there might be more tolerance.

The general approach of turning philosophical attention to practices of research is in line with a methodological development to characterize the generation, translation, and assessment of scientific knowledge in the real (read: social, complex, messy) world (see e.g., Wagenknecht et al. 2015). A main strand of research within this field of study has been the examination of how value judgements and divergent interests define the aim of a given research enterprise and thereby affect the design and evaluation of scientific studies (e.g., Longino 2002; Carrier 2004; Douglas 2009; Solomon 2015). "Identifying these features of a local epistemology, particularly the assumptions and values that link methods to kinds of knowledge sought, is a matter not just of picking out the methods and standards that link data to hypotheses in research articles but of reconstructing them from an analysis of the context of

inquiry: correspondence; accounts of controversy and of interventions in controversy; study of institutional settings, priorities, and constraints." (Longino
2002: 187).

The turn to local epistemologies of medical research has given rise to the
acknowledgement that the multitude of sub-disciplines (e.g., anatomy, epidemiology, pharmacy) within medicine as well as the scientific approaches to medicine
make use of various epistemological frameworks and metaphysical assumptions
regarding theories of disease(s) (see e.g., Lemoine (2011) for an elaboration on the
general claim with respect to explanations of disease). Anya Plutynski (2018)
recently inquired with respect to the conclusions of Marta Bertolaso's (2016)
study on the multitude of understandings of cancer, whether "we should consider
giving up the very idea of general theories of cancer." It is not clear if the different
theories relate to the same object. Preclinical studies might give rise to disease
ontologies in the plural — just as Annemarie Mol has argued with respect to
arthrosis within medical practice (Mol 2002). If this is the case, the much-discussed
epistemological question of whether explanations that are yielded from different
experiments will result in an integrated pluralism (as advocated prominently by
Mitchell 2009) becomes second to questioning the very conditions for identifying
disease and translating knowledge based on site-specific identification practices
across different domains of medical research and practice.

9.3.2   Scoping Methods

There have been some suggestions to apply a relational account to capture the
interdependencies between world, data, data models, and theory in the life sciences
(e.g., Leonelli 2019). Scope validity takes this route even further: it side-steps
general metaphysical assumptions about diseases though acknowledging that different practices of diagnosing and defining disease come with ontological implications.
This perspective urges us to ask in every case study and in every context how exactly
the target object is framed and what strategies are applied to evaluate in how far a
given research setting conditions the scope of application. The relational approach is
well suited to make differences in scientific practices, theoretical assumptions, value
judgements, and interests of various actor groups explicit. A relational approach is
well-compatible with approaching disease as historical and practical concepts (see
e.g., van der Linden and Schermer, Chap. 19, this volume, Binney, Chap. 7, this
volume, Fangerau, Chap. 3, this volume). However, it puts a lot of normative weight
on the assessment of the *adequate* identification of the target in local settings as well
as the evaluation of their matching across contexts of experimentation and application. The relational perspective thus allows us to ask new philosophical questions
regarding scientific methods. From a philosophical point of view, we can ponder the
dimensions of adequacy in medical research. From a scientific point of view, we can
probe methods to assess and increase the scope validity of a model or test regarding
its intended use.

From a philosophy of science in practice perspective, it can be a useful first step to turn the attention to methods that are already used by scientists and that can be related to scope validity in some way or another — though they have so far not been analyzed as responding to the same meta-methodological issue. Examples could include backward or reverse translation of animal models and retrospective epidemiological studies. Reverse translation denotes the testing of an intervention that is known to work (or not to work) in humans in animal models. They aim to check for and characterize failures of animal studies to reproduce the human effects, to refine measures of outcome parameters in animals, and/or to compare the validity of different animal models with respect to their capacity of mimicking the proven (in-)effectiveness in human trials. Indeed, good candidates for *scoping methods* seem to be connected to discussions of dissatisfaction with the current structure and practice of biomedical research, and attempts to remedy the experienced shortcomings. For instance, Bert 't Hart and colleagues (2018) define reverse translation as, "when a promising new treatment fails to show efficacy in clinical trials, the reason(s) for failure are investigated by retesting in a relevant animal model (clinic to lab)." They describe reverse translation as an important step to better understand species- (or strain-)specific pathophysiological mechanisms of a disease and problems that result thereof for extrapolation. The scientists echo the complaint from Belzung and Lemoine (2011: 1) that too little research has been funded that applies "the back-translational approach. . . . going from the bedside to the bench." Experimental designs that employ reverse translation could be used as "a learning principle" ('t Hart et al. 2018: 267) for drawing conclusions about the pathophysiological mechanisms that led to a failure of extrapolation from animals to humans. I propose that reverse trials could be used as a scoping method to investigate which target a model might best fit and how to improve and assess the matching.

Conceptualized in this way, we can compare methodological strategies like backward-trials with methods from other subdisciplines such as epidemiology. I mentioned above that some epidemiologists have been very critical of the new biomarker-based research framework to investigate Alzheimer's disease. This framework builds on evidence from longitudinal studies that identified a population according to their performance in neurocognitive tests, and then followed this cohort of people with 'mild cognitive impairment' over years to ascertain their heightened risk to develop symptoms of dementia due to Alzheimer's disease. In contrast, the skeptical epidemiologists argue that evidence from retrospective studies have not been taken seriously enough (Amieva, Glymour, personal communication). In these studies, the starting point is not a putative risk population, but people who have already developed severe symptoms of dementia and received a clinical diagnosis. The epidemiologists then backtrack the patients' medical (and biographical) records for commonalities in their midlife, years before they received the diagnosis. Such studies have shown that a low body mass index was significantly correlated with a dementia diagnosis two decades later (e.g., Qizilbash et al. 2015). This method does not mean to discard neurocognitive testing as tool to identify a population at risk, but it would help to quantify how many patients who receive a dementia diagnosis at the end of the study were overlooked by neurocognitive testing, because cognitive problems did not occur as early signs in their cases. Again, such retrospective study designs have mostly been discussed as testing a hypothesis (here: the falsification of a potential correlation between obesity and dementia) or to generate alternative aetiological hypotheses (that high metabolic rates might be involved in dementia development) for additional testing in longitudinal epidemiological or laboratory studies. However, we could also analyze and apply such retrospective studies as a *scoping method* for testing what proportion of the clinical diagnoses of Alzheimer's disease did not previously fall into the category of mild cognitive impairment (this has been suggested to me by epidemiologist Hélène Amieva).

9.4   Conclusion

In this paper, I have shown that traditional validity concepts assess the informativeness of a model or test by deploying some abstract concept of the target of modelling. These validity concepts have proven to be ill-suited to assess and refine how well the target of experimentation matches the target of application, not in general, but within the particular local context. To fill this gap, I have introduced a relational approach to medical objects of inquiry that side-steps metaphysical questions about what disease is in general and that is apt for investigating how knowledge generation within one biomedical context conditions the way in which a medical problem is identified in another context. Scope validity does not contradict other validity measures but elucidates a dark spot, and thus could be used complementary to other validity concepts. In contrast to variants of construct validity that assess how close the test hits the abstract entity, scope validity captures how well the target in the experimental test fits to the target in the application setting. The process of forming an ideal of a given disease entity in modern medical sciences (Rosenberg 2002) puts medical scientists and philosophers in the position of having to judge the right way to delineate disease(s). My alternative, relational approach focuses instead on the differences between the relata of animal models, research populations, and the group of people who receive a diagnosis in general health care. Instead of prioritizing one way of delineating disease according to an assessment of how close the given operationalization comes to the idealized disease entity, scope validity addresses the matching (and mismatching) of identifying disease types across concrete contexts.

This tasks researchers in philosophy and science to identify, first, which approaches should be legitimately included in such an analysis; second, how to assess practices of identifying disease within a given context; third, how to examine their matching; and fourth, how to guide the assessment of the matching towards the values we want to see instantiated in a good health care system. Each of these steps, and perhaps the last one most of all, will undoubtedly raise many discussions and concerns, because it might lead us to question the freedom and disinterestedness of scientific inquiry. It is important to raise these (and other) concerns and to examine in detail under what conditions they are warranted. However, it is equally important to keep in mind that when animal scientists ask for funding from medical research organizations and when pharmaceutical companies run experiments on human beings, there needs to be some accountability for how this research can benefit humans. The resort to abstract concepts of disease has at least in some cases deprivileged attempts to improve scope validity, as exemplified in the case of a pharmaceutical company's strategic choice to not test a potential antidepressant in a better matching, but much smaller subpopulation. If there was a regulatory require-ment to assess (and publish this assessment of) the scope validity of a clinical trial design in relation to the animal models that were used to provide mechanistic evidence for a drug's mode of action, such strategic choices would at least be more difficult to advocate. Scientists from various subdisciplines have already developed methods, such as reverse translation, that could be used to examine the matching of scopes across research contexts. Scope validity can serve as a meta-methodological category for identifying, collecting, comparing, and analyzing such *scoping methods*, thereby bringing attention to the epistemological work done in these subdisciplines.

My philosophical account of a relational approach to medical issues and the focus on scientific methods of assessing adequacy bears certain assumptions and limitations with regard to using scope validity as a conceptual tool. It does not provide a fixed set of criteria of adequacy that can serve as a universal standard for evaluating medical research. Rather, the next step would be to provide a more nuanced vocabulary for weighing the premises of local operationalizations of disease within a given experi-mental design against its intended scope of application. An assessment of a research trial's scope validity neither privileges a certain definition of disease, nor does it necessarily entail that only research should be funded that fits best to received diag-nostic criteria. This means that researchers who detail their experiment's or model's scope validity need to question what their disease operationalization implies for the use of their research results in other contexts. Given the social organization of biomedical research as a highly segregated, multi-professional enterprise, the answers to the question of how the premises of research designs in different labs, clinical studies, and application contexts fit to each other will remain underdetermined in many cases. There are too many variables in the process of translating research. I want to argue that this should not be seen as a shortcoming, but rather as an indicator that the concept of scope validity might indeed be of use as a tool for science. The assessment of scope validity will generate questions, which can be made productive when directing philo-sophical and scientific research into applying and possibly inventing or improving methods to better qualify the adequacy of translational medical research.

Acknowledgement I would like to thank the organizers, my commentator Frank Wolters, and the participants at the workshop on "health and disease as practical concepts" at Erasmus University in Rotterdam, as well as the participants of the Philosophical Club at Bielefeld University, and the commentators and organizer of the Scope Validity in Medicine Panel of the American Philosophical Association's Meeting in Philadelphia 2020 for reading earlier drafts of this paper and giving me very helpful input for my revisions.

References

Abramson, Lyn Y., and Martin E.P. Seligman. 1977. Modeling psychopathology in the laboratory: History and rationale. In *Psychopathology: Experimental models*, ed. J.D. Maser and M.E.P. Seligman, 1–26. San Francisco: WH Freeman.

Alexandrova, Anna, and Daniel M. Haybron. 2016. Is construct validation valid? *Philosophy of Science* 83 (5): 1098–1109.

Ankeny, Rachel A., and Sabina Leonelli. 2011. What's so special about model organisms? *Studies in History and Philosophy of Science Part A* 42 (2): 313–323. https://doi.org/10.1016/j.shpsa.2010.11.039.

Ashe, Karen H., and Kathleen R. Zahs. 2010. Probing the biology of Alzheimer's disease in mice. *Neuron* 66 (5): 631–645. https://doi.org/10.1016/j.neuron.2010.04.031.

Bechtoldt, H.P. 1951. Selection. In *Handbook of experimental psychology*, ed. S. S. Stevens, 1237–1267. New York: Wiley.

Belzung, Catherine. 2013. Innovative drugs to treat depression: Did animal models fail to be predictive or did clinical trials fail to detect effects? *Neuropsychopharmacology* 39 (5): 1041–1051. https://doi.org/10.1038/npp.2013.342.

Belzung, Catherine, and Maël Lemoine. 2011. Criteria of validity for animal models of psychiatric disorders: Focus on anxiety disorders and depression. *Biology of Mood & Anxiety Disorders* 1 (1): 1–16. https://doi.org/10.1186/2045-5380-1-9.

Bertolaso, Marta. 2016. *Philosophy of cancer: A dynamic and relational view*. Dordrecht: Springer.

Burian, Richard M. 1993. How the choice of experimental organism matters: Epistemological reflections on an aspect of biological practice. *Journal of the History of Biology* 26 (2): 351–367. https://doi.org/10.1007/bf01061974.

Carrier, Martin. 2004. Knowledge and control: On the bearing of epistemic values in applied science. In *Science, values, and objectivity*, ed. P. Machamer and G. Wolter, 275–293. Pittsburgh: University of Pittsburgh Press.

Cartwright, Nancy. 2009. Evidence-based policy: What's to be done about relevance? *Philosophical Studies* 143 (1): 127–136. https://doi.org/10.1007/s11098-008-9311-4.

Cook, Thomas D., and Donald Thomas Campbell. 1976. The design and conduct of true experiments and quasiexperiments in field settings. In *Handbook of industrial and organizational psychology*, ed. M.D. Dunnette, 223–326. Chicago: Rand McNally.

———. 1979. *Quasi-experimentation: Design and analysis issues for field settings*. Chicago: Rand McNally.

Cronbach, Lee J., and Paul E. Meehl. 1955. Construct validity in psychological tests. *Psychological Bulletin* 52 (4): 281–302. https://doi.org/10.1037/h0040957.

Daly, Tim and Lara Keuck. 2024. Alzheimer's Disease. Engaging with an unstable category. In *Handbook of the Philosophy of Medicine*, 2nd Edition. Dordrecht: Springer.

Demazeux, Steeves, and Lara Keuck. 2023. Comment peut-on être précis les yeux fermés? In *Promesses et limites de la psychiatrie de precision*, ed. C. Gauld, E. Giroux, and S. Demazeux, 201–230. Paris: Hermann.

Douglas, Heather E. 2009. *Science, policy, and the value-free ideal*. Pittsburgh: University of Pittsburgh Press.

Feest, Uljana. 2011. What exactly is stabilized when phenomena are stabilized? *Synthese* 182 (1): 57–71. https://doi.org/10.1007/s11229-009-9616-7.

———. 2019. Why replication is overrated. *Philosophy of Science* 86 (5): 895–905.

Germain, Pierre-Luc. 2014. From replica to instruments: Animal models in biomedical research. *History and Philosophy of the Life Sciences* 36 (1): 114–128. https://doi.org/10.1007/s40656-014-0007-0.

Glymour, Medellena Maria, Adam Mark Brickman, Mika Kivimaki, Elizabeth Rose Mayeda, Geneviève Chêne, Carole Dufouil, and Jennifer Jaie Manly. 2018. Will biomarker-based diagnosis of Alzheimer's disease maximize scientific progress? Evaluating proposed diagnostic criteria. *European Journal of Epidemiology* 33 (7): 607–612. https://doi.org/10.1007/s10654-018-0418-4.

Green, Sara. 2024. *Animal Models of Human Disease*. Cambridge: Cambridge University Press.

Guala, Francesco. 2003. Experimental localism and external validity. *Philosophy of Science* 70 (5): 1195–1205. https://doi.org/10.1086/377400.

Guessoum, Sélim Benjamin, Yann Le Strat, Caroline Dubertret, and Jasmina Mallet. 2020. A transnosographic approach of negative symptoms pathophysiology in schizophrenia and depressive disorders. *Progress in Neuro-Psychopharmacology and Biological Psychiatry* 99: 109862.

Hacking, Ian. 2007. Kinds of people: Moving targets. *Proceedings of the British Academy* 151: 285–318.

Hájek, Alan. 2007. The reference class problem is your problem too. *Synthese* 156 (3): 563–585. https://doi.org/10.1007/s11229-006-9138-5.

't Hart, Bert A., Jon D. Laman, and Yolanda S. Kap. 2018. Reverse translation for assessment of confidence in animal models of multiple sclerosis for drug discovery. *Clinical Pharmacology & Therapeutics* 103 (2): 262–270. https://doi.org/10.1002/cpt.801.

Hauswald, Rico, and Lara Keuck. 2017. Indeterminacy in medical classification: On continuity, uncertainty, and vagueness. In *vagueness in psychiatry*, ed. Geert Keil, Lara Keuck, and Rico Hauswald, 93–116. Oxford: Oxford University Press.

Herzog, David P., Holger Beckmann, Klaus Lieb, Soojin Ryu, and Marianne B. Müller. 2018. Understanding and predicting antidepressant response: Using animal models to move toward precision psychiatry. *Frontiers in Psychiatry* 9. https://doi.org/10.3389/fpsyt.2018.00512.

Hsiao, Karen, Paul Chapman, Steven Nilsen, Chris Eckman, Yasuo Harigaya, Steven Younkin, Fusheng Yang, and Greg Cole. 1996. Correlative memory deficits, a elevation, and Amyloid Plaques in transgenic mice. *Science* 274 (5284): 99–103. https://doi.org/10.1126/science.274.5284.99.

Huber, Lara, and Lara Keuck. 2013. Mutant mice: Experimental organisms as materialised models in biomedicine. *Studies in History and Philosophy of Science Part C: Studies in History and Philosophy of Biological and Biomedical Sciences* 44 (3): 385–391. https://doi.org/10.1016/j.shpsc.2013.03.001.

———. 2017. Philosophie der biomedizinischen Wissenschaften. In *Grundriss Wissenschaftsphilosophie: Die Philosophien der Einzelwissenschaften*, ed. Simon Lohse and Thomas Reydon, 287–318. Hamburg: Meiner Verlag.

Jack, Clifford R., David A. Bennett, Kaj Blennow, Maria C. Carrillo, Billy Dunn, Samantha Budd Haeberlein, David M. Holtzman, et al. 2018. NIA-AA research framework: Toward a biological definition of Alzheimer's disease. *Alzheimer's & Dementia: The Journal of the Alzheimer's Association* 14 (4): 535–562. https://doi.org/10.1016/j.jalz.2018.02.018.

Keller, Evelyn Fox. 2000. Models of and models for: Theory and practice in contemporary biology. *Philosophy of Science* 67 (September): 72–86. https://doi.org/10.1086/392810.

Kelley, Truman Lee. 1927. *Interpretation of educational measurements*. Yonkers/Chicago: World Book Company.

Kendler, Kenneth S., and Josef Parnas. 2012. *Philosophical issues in psychiatry. II, Nosology*. Oxford: Oxford University Press.

Keuck, Lara. 2020. A window to act: Revisiting the conceptual foundations of Alzheimer's disease in dementia prevention. In *Preventing old age and decline? Critical observations on aging and dementia* ed. A. Leibing, S. Schicktanz. New York, Oxford: Berghahn Publishers, 19–39.

Kitcher, Philip. 2003. *Science, truth, and democracy*. New York: Oxford University Press.

Kutschenko (=Keuck), Lara. 2011a. In quest of 'Good' medical classification systems. *Medicine Studies* 3 (1): 53–70. https://doi.org/10.1007/s12376-011-0065-5.

———. 2011b. How to make sense of broadly applied medical classification systems: Introducing epistemic Hubs. *History and Philosophy of the Life Sciences* 33 (4): 583–602.

Lemoine, Maël. 2011. *La désunité de la médecine: Essai sur les valeurs explicatives de la science médicale*. Paris: Hermann, Impr.

Leonelli, Sabina. 2019. What distinguishes data from models? *European Journal for Philosophy of Science* 9: 22. https://doi.org/10.1007/s13194-018-0246-0.

Longino, Helen. 2002. *The fate of knowledge*. Princeton: Princeton University Press.

Markus, Keith A., and Denny Borsboom. 2013. *Frontiers of test validity theory: Measurement, causation and meaning*. New York: Routledge.

McKinney, William T., and William E. Bunney. 1969. Animal models of depression. *Archives of General Psychiatry* 127: 240–248.

Meunier, Robert. 2012. Stages in the development of a model organism as a platform for mechanistic models in developmental biology: Zebrafish, 1970–2000. *Studies in History and Philosophy of Science Part C: Studies in History and Philosophy of Biological and Biomedical Sciences* 43 (2): 522–531. https://doi.org/10.1016/j.shpsc.2011.11.013.

Mitchell, Sandra D. 2009. *Unsimple truths: Science, complexity, and policy*. Chicago: University of Chicago Press.

Mol, Annemarie. 2002. *The body multiple: Ontology in medical practice*. Durham: Duke University Press.

Mullan, Mike, Fiona Crawford, Karin Axelman, Henry Houlden, Lena Lilius, Bengt Winblad, and Lars Lannfelt. 1992. A pathogenic mutation for probable Alzheimer's disease in the APP gene at the N–Terminus of β–Amyloid. *Nature Genetics* 1 (5): 345–347.

Parker, Wendy S. 2020. Model evaluation: An adequacy-for-purpose view. *Philosophy of Science* 87 (3): 457–477.

Plutynski, Anya. 2018. Review of philosophy of cancer: A dynamic and relational view, by Marta Bertolaso. *History and Philosophy of the Life Sciences* 40: 1. https://doi.org/10.1007/s40656-017-0165-y.

Qizilbash, Nawab, John Gregson, Michelle E. Johnson, Neil Pearce, Ian Douglas, Kevin Wing, Stephen J.W. Evans, and Stuart J. Pocock. 2015. BMI and risk of dementia in two million people over two decades: A retrospective cohort study. *The Lancet Diabetes & Endocrinology* 3 (6): 431–436. https://doi.org/10.1016/s2213-8587(15)00033-9.

Rheinberger, Hans-Jörg. 2006a. Vers la fin des organismes modèles? In *Les organismes modèles dans la recherche médicale*, ed. G. Gachelin, 275–277. Paris: Presses Universitaires de France.

———. 2006b. Gaston Bachelard und der Begriff der ‚Phänomenotechnik'. In *Epistemologie des Konkreten: Studien zur Geschichte der modernen Biologie*, 37–54. Frankfurt am Main: Suhrkamp.

Robins, Eli, and Samuel B. Guze. 1970. Establishment of diagnostic validity in psychiatric illness: Its application to Schizophrenia. *American Journal of Psychiatry* 126 (7): 983–987. https://doi.org/10.1176/ajp.126.7.983.

Rosenberg, Charles E. 2002. The Tyranny of diagnosis: Specific entities and individual experience. *The Milbank Quarterly* 80 (2): 237–260. https://doi.org/10.1111/1468-0009.t01-1-00003.

Schaffner, Kenneth F. 1998. Genes, behavior, and developmental emergentism: One process, indivisible? *Philosophy of Science* 65 (2): 209–252. https://doi.org/10.1086/392635.

———. 2012. A philosophical overview of the problems of validity for psychiatric disorders. In *Philosophical issues in psychiatry II: Nosology*, ed. Kenneth S. Kendler and Josef Parnas, 169–189. Oxford: Oxford University Press.

Schaffner, Kenneth F. Forthcoming. Construct validity in psychology and psychiatry. (Draft version Oct 2018; available from the author on request at kfs@pitt.edu).

Solomon, Miriam. 2015. *Making medical knowledge*. Oxford: Oxford University Press.

———. 2022. On validators for psychiatric categories. *Philosophy of Medicine* 3 (1): 1–23.

Steel, Daniel. 2008. *Across the boundaries: Extrapolation in biology and social science*. Oxford: Oxford University Press.

Strzebonska, Karolina, and Marcin Waligora. 2019. Umbrella and basket trials in oncology: Ethical challenges. *BMC Medical Ethics* 20: 58. https://doi.org/10.1186/s12910-019-0395-5.

Sullivan, Jacqueline A. 2009. The multiplicity of experimental protocols: A challenge to reductionist and non-reductionist models of the unity of neuroscience. *Synthese* 167 (3): 511–539. https://doi.org/10.1007/s11229-008-9389-4.

Tabb, Kathryn. 2017. Philosophy of psychiatry after diagnostic kinds. *Synthese* 196 (6): 2177–2195. https://doi.org/10.1007/s11229-017-1659-6.

Tabb, Kathryn, and Maël Lemoine. 2021. The prospects of precision psychiatry. *Theoretical Medicine and Bioethics* 42: 193–210. https://doi.org/10.1007/s11017-022-09558-3.

Wagenknecht, Susann, Nancy J. Nersessian, and Hanne Andersen. 2015. *Empirical philosophy of science: Introducing qualitative methods into philosophy of science.* Dorndrecht: Springer.

Willner, Paul. 1984. The validity of animal models of depression. *Psychopharmacology* 83 (1): 1–16. https://doi.org/10.1007/bf00427414.

———. 1994. Animal models of depression. In *Handbook of depression and anxiety. A biological approach*, ed. Siegfried Kasper, Johan A. den Boer, and J.M. Ad Sitsen, 291–316. New York: Marcel Dekker.

Willner, Paul, and Paul John Mitchell. 2002. Animal models of depression: A diathesis-stress approach. In *Textbook of Biological Psychiatry*, ed. H. D'Haenen, Johan A. den Boer, and Paul Willner, 703–706. Chichester: Wiley.

Chapter 10
Scope Validity in Medicine: An Asset to the Epidemiologist's Armoury

Frank J. Wolters

In recent years, increasing attention for internal validity has aimed to reduce bias and increase reproducibility of translational biomedical research (Van der Worp et al. 2010; Sena et al. 2010). Yet, translation of findings across research domains remains challenging. In a chapter that is conveniently pragmatic for a medical doctor and pleasantly methodological for an epidemiologist, Lara Keuck (Chap. 9, this volume) argues that the translation of biomedical research from bench to bedside may benefit from a focus on 'scope validity': the extent to which an experimental target matches the intended application. Keuck reasons that more awareness and better articulation of the scope of preclinical investigations could support epistemic reasoning as well as knowledge utilisation in biomedical research.

Aforementioned definition of scope validity positions it amidst construct validity (i.e., the extent to which a measure assesses what it is supposed to) and external validity (i.e., the extent to which findings of a study are generalisable to other contexts). While the difference with the former quickly becomes clear, the distinction between scope validity and external validity warrants closer inspection. Keuck builds notably on examples from the fields of Alzheimer's disease and multiple sclerosis. In the latter example, external validity represents the extent to which the observed effect of a treatment in an animal model can be generalised to the entire population of patients with multiple sclerosis. Scope validity then represents the concordance between multiple sclerosis in the patient population and multiple sclerosis in the animal model. Any such comparison—of the disease between different contexts—relies in part on the definition of the disease of interest. In other words, we cannot fully appreciate the notion of scope validity without a closer inspection of the nosology and ontological aspects of the disease.

F. J. Wolters (✉)
Department of Epidemiology, Erasmus MC – University Medical Centre Rotterdam, Rotterdam, The Netherlands
e-mail: f.j.wolters@erasmusmc.nl

M. Schermer, N. Binney (eds.), *A Pragmatic Approach to Conceptualization of Health and Disease*, Philosophy and Medicine 151,
https://doi.org/10.1007/978-3-031-62241-0_10

The examples of multiple sclerosis and Alzheimer's disease illustrate that any mismatch between an experimental target and its intended application could be due at least in part to uncertainty about the definition of disease. Rare monogenic variants in genes like *PSEN1* and *APP* are definite causes of Alzheimer's disease, presumably through the accumulation of the amyloid-β protein, and these genetic variants have long served as a benchmark for animal models of the disease. If all cases of clinical Alzheimer's disease had a shared underlying mechanism, it would follow that treatment trials in those animals would conceptually apply to all clinical cases in humans, i.e., the animal trials would have scope validity, albeit not necessarily external validity.

Over the years, however, it became clear that the aetiology of clinically defined Alzheimer's disease is multifactorial, as distinct mechanisms give rise to very similar clinical phenotypes. In other words, the more or less typical conglomeration of various clinical symptoms (i.e., a *syndrome*) often has multiple potential underlying causes, occurring either jointly or in isolation. Each of these causes in itself can be seen as a distinct *disease*, i.e., an object that is generally defined by its presumed single pathophysiological pathway. Medicine is rich in clinical syndromes of which the precise causes are uncertain or—equally often—cannot be determined in the individual patient due to the concurrence of several potential contributing objects (Rothman et al. 2008). This is the case for dementia, depression, and many other syndromes alike. Moreover, the definition of a disease often changes with our understanding of the underlying pathophysiology. For example, in the realm of cardiovascular medicine, atherosclerosis (i.e., the thickening or hardening of the arteries) for decades was considered a single entity, from which basic science models were construed. Owing to increasingly fine-grained assessments using novel technology, it now becomes clear that the nature of atherosclerosis in fact is rather heterogeneous (Slenders et al. 2022), reflecting multiple underlying disease processes. Similar developments are on the horizon for Alzheimer's disease (Tijms et al. 2020).

The changes in disease constructs over time call into question the assumed scope validity of many prior studies. As aetiological heterogeneity of a disease or syndrome emerges, the prior scope can no longer be assumed to capture the whole disease. It raises the question how we may reliably determine scope validity *avant la lettre*, in the absence of a clear disease label. Keuck (Chap. 9, this volume) addresses this by saying scope validity "need not mean that we should change the diagnostic criteria of a given disease altogether, because in other contexts than that of drug testing such as, for instance, the assessment of socioeconomic factors that impact on (mental) health, it might be more adequate to sample the patient population in a different way." Although for certain questions, notably related to public health, indeed syndrome diagnoses are the best reflection of disease burden, they are often incapable of dealing with the large heterogeneity in clinical presentations and disease course. In patient care therefore, changes in disease entity seem indispensable in improving diagnosis, prognosis, and treatment. When limitations in knowledge about a disease give rise to scope mismatch, empirical evaluation may be the only solution. Still, progress from empirical science could benefit from revisiting past research through both historical and contemporary lenses.

Inherent to empirical science and the process of scientific discovery is the urge to generalise observations—through inductive reasoning—to more abstract notions of reality. Scope validity calls for more caution when crossing the boundaries of our scope. Is this a problem in the scientific quest for access to a mind-independent world, where truth awaits?

Keuck's introduction suggests she might take a pluralist or relativists stance in answering this question. "Validity is never unmediated, never absolute," she writes in the opening paragraph. This aligns with an 'adequacy-for-purpose' stance towards validity, stating that the truth of claims holds only in the context of the specified framework of assessment. Scope validity fits with the notion that physical theories can be at odds with each other and yet compatible with all possible data; they can be logically incompatible and empirically equivalent (Quine 1970). To come back to the example of Alzheimer's disease: the assertion that environmental factors affect amyloid-β levels may be valid *in the context of the APP and PSEN1 mutations*, but not necessarily for sporadic amyloidosis. Yet, many researchers felt that similarities between monogenetic and sporadic Alzheimer's disease allowed for a broader extrapolation of results. Their disagreements may well have been fed by incorrect perception of the scope of prior research, part of which boiled down to a semantic confusion of tongues. More careful definition of terms may have circumvented long-lasting feuds in the scientific discourse on Alzheimer's disease, and an explicit mention of study scope could have facilitated this conversation. However, the Alzheimer disagreement was fuelled too by different weights people gave to the available scientific and pragmatic arguments (e.g., the need for better treatment) that urged them to either err on the side of caution or sail closer to the wind. It illustrates that even a correct verdict on scope validity provides no guarantee for valid theory choice and policy against the backdrop of other contextual values. It calls to mind the arguments for underdetermination made by Imre Lakatos and Paul Feyerabend that "the difference between empirically successful and empirically unsuccessful theories lies in the talents and resources of their respective advocates" (Laudan 1990). Perhaps here, we should borrow from teachings of epidemiology on external validity, and likewise consider scope validity in the end a matter of consensus.

Moving towards the applicability of scoping methods, Keuck draws parallels with other methodical disciplines that aim to improve the match between experimental target and intended application. In particular, she turns to life-course epidemiology, in which people are followed through life for many years if not decades. This allows for mapping of disease trajectories in relation to preceding risk factors, an approach that bears resemblance to medical practice, which often starts with a clinical observation and then tries to look for explanatory factors in a patient's history. When done prospectively in population cohorts, one can better discern signs and symptoms that would easily go unnoticed and unreported with a more advanced disease presentation in clinical setting. Such approaches have helped, for example, to acknowledge the early importance of cognitive deficits, in addition to motor function impairment, in patients with Parkinson's disease (Darweesh et al. 2017).

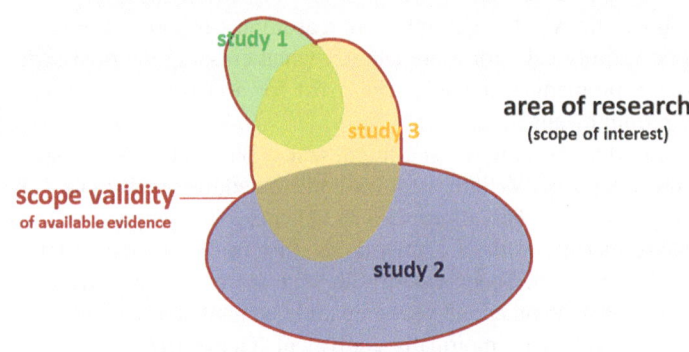

Fig. 10.1 Mapping scope validity across different studies could provide insight in the applicability of results. The coloured areas represent the scope validity of different studies, with the red lining demonstrating the overall scope validity of available evidence. (Inspired by Sjøberg et al. 2008)

Examining a topic of investigation through the lens of scope validity, a two-dimensional model may be construed that depicts the state of translational evidence (Fig. 10.1). The clustering or dispersity of studies within this framework could visualise the extent to which the available evidence is consistent with apparently incompatible theories. For example, the notion that discrepancies between studies 1 and 2 in Fig. 10.1 could be explained by scope differences is further explored by designing and mapping an additional investigation (e.g., study 3) on the intersect of prior evidence. Using such modes of visualisation, one might start to explore the utility of scope validity in identifying knowledge gaps and weighing arguments in favour or against the viability of a scientific theory. This map could be redrawn as knowledge accumulates, like geographical maps after voyages to previously uncharted territory. As the resemblances across species in gene effects and clinical disease phenotypes are unravelled (The Zoonomia Project 2023; Clarkson 2022), assessment of scope validity too may change.

The incorporation of scope validity in the medical sciences can borrow from prior applications in other fields. In software engineering, scope validity has helped to define "the universe of discourse in which a theory is applicable" (Sjøberg et al. 2008). In commercial enterprises, scope verification and validation ensure that products meet the intended requirements and stakeholders' needs. These fields have in common with contemporary medical academia that they are judged largely by their usefulness in achieving societal goals, a rather pragmatic focus on valorisation and knowledge utilisation that paves the way for scope validity in the biomedical arena. The notion of *precision medicine* further encourages approaches to dissolve heterogeneity in disease course (Phillips 2020), and a contextual approach like scope validity promotes the unravelling of scope-specific mechanisms, like in the abovementioned case of atherosclerosis.

Clear communication is at the foundation of scientific discourse, and this involves internal validity as well as the population and scope to which results are generalisable. Attention for the scope validity of empirical studies is therefore a welcome guest to the table. Scope validity advocates a clear notion among researchers about their position within the broader field of study. In an era of increasing superspecialisation, this alone is a task as daunting as it is imperative.

References

Clarkson, Josie. 2022. *Dementia in other animals*. Dementia Platform UK. https://www.dementiasplatform.uk/news-and-media/blog/dementia-in-other-animals. Accessed 1 Aug 2023.

Darweesh, Sirwan K.L., Vincentius J.A. Verlinden, Bruno H. Stricker, Albert Hofman, Peter J. Koudstaal, and M. Arfan Ikram. 2017. Trajectories of prediagnostic functioning in Parkinson's disease. *Brain* 140 (2): 429–441. https://doi.org/10.1093/brain/aww291.

Laudan, Larry. 1990. Demystifying underdetermination. In *Scientific theories*, ed. C. Wade, 267–297. Minneapolis: University of Minnesota Press.

Phillips, Christopher J. 2020. Precision medicine and its imprecise history. *Harvard Data Science Review* 2 (1). https://doi.org/10.1162/99608f92.3e85b56a.

Quine, William V. 1970. On the reasons for indeterminacy of translation. *Journal of Philosophy* 67 (6): 178–183.

Rothman, Kenneth J., Sander Greenland, and Timothy L. Lash. 2008. *Modern epidemiology*. 3rd ed. Philadelphia: Wolters Kluwer Health / Lippincott Williams & Wilkins.

Sena, Emily S., H. Bart van der Worp, Philip M.W. Bath, David W. Howells, and Malcolm R. Macleod. 2010. Publication bias in reports of animal stroke studies leads to major overstatement of efficacy. *PLoS Biology* 8 (3): e1000344. https://doi.org/10.1371/journal.pbio.1000344.

Sjøberg, Dag I.K., Tore Dybå, Bente C.D. Anda, and Jo E. Hannay. 2008. Building theories in software engineering. In *Guide to advanced empirical software engineering*, ed. F. Shull, J. Singer, and D.I.K. Sjøberg, 312–336. New York: Springer. https://doi.org/10.1007/978-1-84800-044-5_12.

Slenders, Lotte, Daniëlle E. Tessels, Sander W. van der Laan, Gerard Pasterkamp, and Michal Mokry. 2022. The applications of single-cell RNA sequencing in atherosclerotic disease. *Frontiers in Cardiovascular Medicine* 9: 826103. https://doi.org/10.3389/fcvm.2022.826103.

The Zoonomia Project. 2023. https://zoonomiaproject.org/. Accessed 1 Aug 2023.

Tijms, Betty M., Johan Gobom, Lianne Reus, Iris Jansen, Shengjun Hong, Valerija Dobricic, Fabian Kilpert, et al. 2020. Pathophysiological subtypes of Alzheimer's disease based on cerebrospinal fluid proteomics. *Brain* 143 (12): 3776–3792. https://doi.org/10.1093/brain/awaa325.

Van der Worp, H., David W. Bart, Emily S. Howells, Michelle J. Sena, Sara Rewell Porritt, Victoria O'Collins, and Malcolm R. Macleod. 2010. Can animal models of disease reliably inform human studies? *PLoS Medicine* 7 (3): e1000245. https://doi.org/10.1371/journal.pmed.1000245.

Chapter 11
The Biomarkerization of Alzheimer's Disease: From (Early) Diagnosis to Anticipation?

Marianne Boenink and Lennart van der Molen

11.1 Introduction

In the last decennia the number of tests to diagnose disease has increased exponentially, thanks to the rise of novel technologies like imaging, and biochemical, genetic and other molecular tests. Both the bodily parameters examined and the information produced by these technologies have come to be referred to as 'biomarkers'. Biomarkers are viewed as crucial vehicles to realise 'personalized' or 'precision medicine' (Hood 2019; Faulkner et al. 2020; see also Boenink 2016a). This 'biomarkerization' (Metzler 2010) of diagnosis has brought along an increasing number of 'asymptomatic diseases'—but also many controversies regarding the desirability of such disease concepts (e.g. Elgart et al. 2023; Mark and Brehmer 2022; Vogt et al. 2016, 2019; Fiala et al. 2019; Tinland 2022).

Practices of diagnosing asymptomatic disease are usually associated with the aim of (secondary) prevention: by identifying pathology before symptoms arise it may be possible to intervene earlier and thus to decrease severity or even fully avoid symptoms later on. 'A-symptomatic disease' is thus interpreted as 'pre-symptomatic disease'. As indicated above, the identification of such disease is enabled by technologies claiming to offer a 'window' into disease and health. Disease without symptoms typically is *technologically constituted disease* (Hofmann 2001).

Author contributions Lennart van der Molen conducted a systematic analysis of the publications on the disease criteria for Alzheimer's disease and reviewed the manuscript. Marianne Boenink performed the conceptual analysis, conceived the argument of this paper, and drafted the manuscript.

M. Boenink (✉) · L. van der Molen
Ethics of Healthcare, Department of IQ Health, Radboud University Medical Centre Nijmegen, Nijmegen, The Netherlands
e-mail: Marianne.Boenink@radboudumc.nl; Lennart.vanderMolen@radboudumc.nl

M. Schermer, N. Binney (eds.), *A Pragmatic Approach to Conceptualization of Health and Disease*, Philosophy and Medicine 151,
https://doi.org/10.1007/978-3-031-62241-0_11

The concept of 'asymptomatic disease', however, has been contested since its emergence. First of all because it challenges the traditional view that the aim of healthcare is to respond to and, where possible, alleviate complaints, symptoms and functional impairment. Moreover, many critics have pointed out that identification of asymptomatic disease is associated with the harms and costs of potential overdiagnosis and overtreatment (Hofmann et al. 2021; Brodersen et al. 2014; Hofmann and Skolbekken 2017; Kreiner and Hunt 2014). Additionally, such diagnoses are said to medicalize people who do not feel ill, which may incite stress, anxiety and stigma, but also relocate responsibility for health and ignore non-medical interventions (Mark and Brehmer 2022; Hofmann 2022). Diagnosing asymptomatic disease creates "patients in waiting" (Timmermans and Buchbinder 2010) or even "previvors": individuals identifying as survivors of breast cancer thanks to predictive DNA diagnostics (Getachew-Smith et al. 2020). In any case, the implied (but contested) normativity of diagnosing and acting upon asymptomatic disease is that the value of health should trump all other values in human life. A diagnosis of asymptomatic disease not only *redescribes* what is going on in one's body, but also tends to re-evaluate what is at stake, what should be done, who should decide and take which action—in brief, of how to pursue a good life.

Recently, controversy has been particularly prevalent in the domain of Alzheimer's disease (AD). In the last 15 years the emergence of AD biomarkers has led to repeated adjustments of two sets of criteria used to define and diagnose AD. Both of these included proposals to introduce the category of 'preclinical Alzheimer's disease', initially for research purposes only (Jack et al. 2011, 2018; NIA-AA 2023a, b; Dubois et al. 2007, 2010, 2014, 2021). Each revision of the two sets was followed by critical commentary from a broad range of authors (including one of the authors of this chapter). These critics pointed out conceptual confusions, but mainly raised concerns about the drawbacks, risks and/or futility of labelling individuals with preclinical AD, in particular when an effective cure for AD is lacking (e.g. George et al. 2011; Boenink 2018; Schermer and Richard 2019; Isaacs and Boenink 2020; Schermer 2023). The impact of such criticism has been negligible. In the recent draft revision of the NIA-AA guidelines the concept of 'preclinical AD' is considered sufficiently robust for clinical use (NIA-AA 2023a).

This chapter offers an analysis of recent developments in the AD biomarker field and their impact on conceptualization of the phenomenon 'AD'. Rather than focus on the emergence of 'preclinical AD' only, however, we aim to show how biomarkers have more broadly impacted the interrelation between symptoms, pathological observations and AD categories. This first of all helps to better understand why there is so much confusion and controversy surrounding biomarker-based diagnosis of AD. Secondly, the analysis suggests that it might be more productive to shift attention from diagnosis to prognosis, or rather: anticipation.

In what follows we first outline developments in the field of AD biomarkers in the last decennia. We then offer a conceptual reflection on biomarkers, arguing that these may be informed by and reinforce two analytically distinct modes of thinking about and dealing with disease, the ontological and the physiological one. In the next sections we use this distinction to make sense of the developments and controversies in the AD domain, pointing out that gradually, but often implicitly, a shift is taking place towards a physiological, rather than an ontological mode of understanding AD. We suggest, moreover, that such a shift may indeed fit with the heterogeneity of disease opened up by biomarker technologies. We continue by pointing out the implications of this shift for the experience of AD (with or without symptoms) and conclude with some reflections on why and how similar trends may be visible (and helpful) in other disease domains.

11.2   Biomarkers for Alzheimer's Disease: Food for Thought

From its inception in the early twentieth century, the golden standard for diagnosing AD were the "plaques and tangles" found at autopsy in the brain of patients who during life had manifested symptoms of dementia (Ballenger 2006). So, AD has always been a diagnosis identifying a specific subset of all patients with dementia symptoms. As Ballenger shows, which symptoms and patients were included actually varied over time. For example, until the 1970s AD was thought to manifest only in patients whose dementia started at a relatively young age (before 65), thus suggesting AD was "pre-senile dementia". By the end of the 1970s the age criterion was dropped for various reasons, leading to a steep rise in the prevalence of AD. On the pathological side things were also complicated, since the 'plaques and tangles' typical for AD could be made visible at autopsy only. This meant that any *in vivo* diagnosis of AD remained uncertain. As a result, AD was diagnosed for decennia by exclusion: if diagnostic examination revealed no other explanations for the dementia symptoms, the patient probably suffered from AD.

In 1984 the first AD diagnostic criteria were published, to facilitate the establishment of the natural history of the disease, incorporation in research protocols, comparison of clinical trials and, importantly, improve the reliability of the diagnosis (McKhann et al. 1984). The committee stresses the importance of the conception of these criteria by reporting that about 20% of individuals clinically diagnosed with AD during life show pathology of other conditions, and not AD, at autopsy (McKhann et al. 1984). The 1984 AD criteria feature a two-step diagnostic approach. First the presence of a dementia syndrome is established (including "a history of decline in performance and by abnormalities noted from clinical examination and neuropsychological tests" (McKhann et al. 1984: 940). Subsequently, the likelihood of AD causing this syndrome is assessed (requiring "a typical insidious onset of dementia with progression and (...) no other systemic or brain diseases that could account for the progressive memory and other cognitive deficits" (ibid.: 940)). Whereas the first is based on reports and measurement of behaviour, the latter can include lab tests of bodily material to exclude alternative causes for the dementia syndrome. The criteria propose three diagnostic categories, referring to the level of certainty about the diagnosis: 'possible AD' (in case of symptoms of progressive dementia but an aberrant onset, presentation or clinical course), 'probable AD' (progressive deficits in two or more cognitive domains, established via neuropsychological testing, and no alternative explanation of these progressive deficits) and 'definite AD' (to be established only if 'probable AD' is confirmed by histopathologic evidence from a biopsy or autopsy). Overall, then, a reliable AD diagnosis according to these criteria requires both symptoms and biological evidence. An AD diagnosis is impossible without symptoms, but full certainty about the diagnosis is gained only post mortem. The possibility that individuals could show pathology post mortem while not having suffered from symptoms during life is not even mentioned.

Although 'disease without symptoms' was not an option in the 1984 criteria, soon afterwards some started asking questions regarding how many symptoms had to be present and how severe these should be to justify an AD diagnosis. After all, since onset of dementia symptoms in case of AD was typically considered 'insidious', one might wonder where to draw the line between minor complaints and possible AD dementia. In the 1990s the concept of 'Mild Cognitive Impairment' or MCI (Petersen et al. 1999) was introduced to refer to individuals who suffered from memory or other cognitive impairments that were more serious than common for their age and education, while not satisfying criteria for dementia syndrome. Controversies ensued whether MCI should or should not be understood as a precursor to dementia or AD. Subsequently, cohort studies indicated that a minority of subjects labelled with MCI deteriorates towards dementia or AD in the following years, that some may return to 'normal' functioning, and that 'turnover' is lower when subjects are recruited from a community setting rather than among visitors of a memory clinic (Petersen et al. 1999; Busse et al. 2006). Some critics questioned whether the MCI concept was useful at all, since mild cognitive complaints might belong to 'normal ageing', and assigning a label might have a medicalising impact (Whitehouse and Juengst 2005). Thus, although it has been generally acknowledged that clinical symptoms of AD (and dementia more generally) develop and deteriorate over time, drawing a clear boundary between health and disease proved challenging.

The emergence in the late twentieth century of technologies to investigate the human brain and the central nervous system (like PET scanning and CSF analysis) led to attempts to observe the presence of plaques and neurofibrillary tangles (now known to consist of the proteins amyloid and tau respectively) in living persons rather than post mortem (Lock 2013). In due time, both amyloid and tau could successfully be visualized and quantified by PET-scanning as well as CSF-analysis. The resulting biomarkers were claimed to decrease the uncertainty of clinical AD diagnosis by providing insight in pathology in vivo. Moreover, basic research into the development of the amyloid plaques and tau tangles had led to the conclusion that these might be developing long before symptoms appeared. Applying the novel biomarker technologies to individuals with no complaints or MCI might be a good way to enable not only a more reliable, but also an *earlier* diagnosis.

As a result of the developments in the biomarker field, AD disease criteria started to be (re-)adjusted; since 2007 a wealth of disease criteria have been published by multiple working groups and regulatory bodies. For reasons of space, we will reconstruct here the development of the criteria published by the American National

Institute of Ageing and the Alzheimer's Association (NIA-AA) only. Rather similar, but sometimes also slightly different, adjustments were made by an international (largely European) working group (Dubois et al. 2007, 2010, 2014, 2021) and in the DSM5 (APA 2013; for comparisons of these sets of criteria see Boenink 2018 and Schermer 2023).

11.2.1   NIA-AA 2011

The first NIA-AA revision of the 1984 criteria was published in 2011, now distinguishing criteria for three different diagnostic categories: 'dementia due to AD', 'MCI due to AD' and 'preclinical AD' (McKhann et al. 2011; Albert et al. 2011; Sperling et al. 2011; accompanied by an overarching introductory paper (Jack et al. 2011)). These phrasings first of all show that the novel biomarkers were considered to increase certainty about the pathology underlying the clinical symptoms: the latter are now said to be "due to" the former. In the case of 'dementia due to AD', however, the authors (led by the same McKhann as in 1984) prefer to speak of "probable/possible AD with evidence of the pathophysiological process" and stress that AD remains a clinical diagnosis, which positive biomarker tests may help corroborate in specific situations. They seem to be more careful than the authors of the other two papers: "In persons who meet the core clinical Criteria for probable AD dementia biomarker evidence may **increase the certainty that the basis of the clinical dementia syndrome is the AD pathophysiological process.**" (McKhann et al. 2011: 266; original emphasis and capital). Another important change is that the criteria show a multiplication of disease labels, covering not only those with minor symptoms ('MCI due to AD') but also individuals lacking any symptoms ('preclinical AD'). The latter category fully depends on evidence of deviant biomarkers and is said to be fit for research purposes only.

These pre-disease categories were considered particularly useful to enable clinical trials with AD drugs that had not, or hardly, been effective in patients with symptoms (Lock 2013). The idea is that prescribing treatment to individuals with preclinical AD might actually be much more effective in removing or reducing (in particular) amyloid load, and thus prevent symptoms later on. This prevention-oriented way of thinking and the focus on amyloid rather than tau biomarkers are supported by the widespread 'amyloid cascade hypothesis', which reconstructs in a rather linear and deterministic way how amyloid plaques gradually develop and subsequently lead to the formation of neurofibrillary tangles, ultimately resulting in cognitive deficits and impaired psycho-behavioral functioning (Hardy and Higgins 1992; Jacobs and Theunissen 2022).

However, research into the correlation between plaques and tangles on the one hand and symptoms on the other had already repeatedly shown that the connection between these two is not as tight as often assumed (Whitehouse et al. 2000; Jacobs and Theunissen 2022). Whereas some individuals with increased amyloid will suffer from symptoms later on, not all of them will. This observation was confirmed in research trying to identify amyloid-based biomarkers. This raises the question, then, what AD biomarkers actually mark: the presence of disease-causing pathology, the presence of a risk factor making emergence of symptoms more likely, or meaningless biological variation (Boenink 2016b)? Many sceptics doubted whether it was wise to introduce the category of preclinical AD and pursue AD prevention in view of such ambiguous scientific underpinning (Schermer and Richard 2019; Isaacs and Boenink 2020; Schermer 2023). As with many other pre-diseases, the harms and risks might outweigh the benefits.

11.2.2 NIA-AA 2018

In 2018, the NIA-AA published a diagnostic framework that was solely intended for use in research settings, thereby proposing another revision of the diagnostic criteria (Jack et al. 2018). Although the authors present the changes as building on the 2011 criteria, the 2018 framework proposes substantial conceptual amendments. The authors now opt for a *biological* definition of AD, turning around the burden of proof from biomarkers to the clinical phenomena:

> [D]ementia is not a 'disease' but rather is a syndrome composed of signs and symptoms that can be caused by multiple diseases, one of which is AD. As we elaborate in the following paragraph, there are two major problems with using a syndrome to define AD; it is neither sensitive nor specific for the neuropathologic changes that define the disease, and it cannot identify individuals who have biological evidence of the disease but do not (yet) manifest signs or symptoms.(Jack et al. 2018: 538)

Hence, the 2018 criteria reserve the term 'AD' for plaques and tangles proven in vivo and refer to the set of clinically ascertained symptoms previously considered typical for AD as 'Alzheimer clinical syndrome'. The authors firmly claim "**A** [amyloid] **and T** [tau] **proteinopathies define AD as a unique disease** among the many that can lead to dementia" (ibid.: 553, emphasis in the original).

Moreover, the 2018 criteria endorse the view that AD is a "patho-physiological continuum" (Jack et al. 2018; Aisen et al. 2017). Rather than identifying separate clinically defined disease states (AD, Mild Cognitive Impairment and Preclinical AD) and looking for biomarkers that identify and differentiate each of these states, the 2018 criteria approach AD pathology as continuously developing over a long period of time. As a result, both clinical manifestations and the biomarker results are now supposed to gradually change over time, with the change of the latter considered to start before the onset of the first, and with amyloid markers changing earlier than tau markers.

Additionally, biomarkers are now categorized as markers related to aggregated Aβ (A), to aggregated tau (T) or to neurodegeneration or neuronal injury (N); this includes both imaging or fluid biomarkers. The function of the groups differs: 'A' biomarkers are said to establish whether an individual is on the AD continuum, 'T' biomarkers confirm that an individual has AD, and 'N' biomarkers help indicate disease severity. The latter are indicated in parentheses because they are not specific to AD. In the limitations section of the paper, the authors state that "None of the biomarkers are as sensitive as direct examination of tissue at autopsy" (ibid.: 544). This does not prevent them from suggesting that much confusion surrounding the use of the label 'AD' might be avoided if we would refer to someone's biomarker profile and clinical manifestations only (e.g. $A^+T^{+(}N)^-$ with dementia). Such a differentiation, they suggest, "will presumably be useful in tailoring treatment to the individual when appropriate specific treatments become available" (ibid.: 542).

In line with the view that AD is a pathophysiological continuum, the 2018 criteria translate the three clinical states defined in 2011 (preclinical AD, MCI and dementia) into stages, stating "these categories have at times been interpreted to indicate three distinct entities. In the research framework, we avoid the notion of separate entities and instead refer to the 'cognitive continuum'" (Jack et al. 2018: 545). Two types of clinical staging schemes are proposed, in which cognitive and behavioural manifestations are linked to biomarker profiles. This results in a variety of categories for those without symptoms: this group includes individuals with 'preclinical AD' (profile $A^+T^+(N)^-$ or $A^+T^+(N)^+$), with 'preclinical Alzheimer's pathologic change' (in case of $A^+T^-(N)^-$), or with 'Alzheimer's and concomitant suspected non Alzheimer's pathologic change, cognitively unimpaired' ($A^+T^-(N)^+$). In case of mild symptoms (MCI) similar differentiations are proposed. Whether the biomarker profiles sufficiently map onto the clinical stages to justify biomarker-based claims about how an individual's situation is likely to develop, will have to be confirmed. However, when comparing their framework to the proposals of another group (Dubois et al. 2014), the authors do explain that asymptomatic individuals with biomarker evidence of AD in the 2018 framework are not "at risk of AD", but *have* AD and "*are at risk of future cognitive decline*" (Jack et al. 2018: 551, our emphasis), implying that the labels they propose have a predictive, as well as diagnostic function.

11.2.3 Draft NIA-AA 2023

NIA-AA is currently preparing another update of their AD criteria; the draft was open for feedback while this chapter was written (NIA-AA 2023a, b). This time the criteria are proposed to be ready for clinical use: the categories are now considered ready to move from research settings to the clinic, although relatively novel biomarkers mentioned in the paper may have to be further validated. The draft is very much in line with the 2018 criteria, extending these into an even more elaborate framework. We will highlight only the changes most salient to the points made in this chapter.

First of all, only one abnormal 'core biomarker' (i.e. either A or T) is now considered sufficient to conclude that Alzheimer pathology is present, further lowering the threshold for the diagnosis of AD (which is still defined biologically). The reasoning for this reduced threshold becomes almost circular: "Our rational for diagnosing AD by the presence of any abnormal core biomarker is that the disease exists when the earliest manifestation of AD pathophysiology can be detected by biomarkers, even though onset of symptoms may be years in the future." (NIA-AA 2023a: 10) The authors explain that this also means that an in vivo biomarker-based diagnosis may be at odds with the gold standard for neuropathological diagnosis, since it allows for lower amounts of pathology (ibid.: 11).

The biological redefinition of AD is more radically implemented by proposing to stop using the term 'Alzheimer's Clinical Syndrome' for those with symptoms but without (known) biomarkers: "The terms Alzheimer's, Alzheimer's disease, or AD dementia should be reserved for situations where the presence of AD has been diagnosed biologically" (NIA-AA 2023b: 5, Text box 5). Symptoms should be referred to in observational and syndromal terms and indications of severity (e.g. 'mild dementia'). In effect, this makes the use of the term Alzheimer fully dependent on biomarker testing or neuropathological examination.

In addition, three new types of biomarkers are added (not all of them considered ready for clinical use), resulting in an extension of the nomenclature with I- (indicating inflammation), V- (vascular brain injury) and S- (α synuclein) markers. Whereas I-markers are claimed to refer to non-specific biomarkers of tissue reactions involved in AD pathophysiology, V- and S-markers are considered helpful to identify non-AD co-pathology. In principle, this could result in an extensive multi-modal (ATNISV) biomarker profile of individuals, although the authors admit that in practice partial profiling is more likely. For each of the earlier A/T/N/markers the criteria now distinguish more clearly between imaging-based and fluid-based markers since their functions appear to differ. Imaging markers point to cumulative pathological effects and are therefore thought to better align with neuropathological findings, whereas fluid markers measure (rates of) production or clearance of pathological substances.

The number of functions biomarkers are supposed to fulfil is further increasing in the draft 2023 criteria. Although diagnosis is still important, much more attention is now paid to disease staging and prognosis, while identifying co-pathology and measuring treatment effects are also discussed. The authors advise to base staging on core biomarkers (A and T) only, and reserve it for those who have been proven to be on the AD continuum (having at least one abnormal core biomarker). In other words, biomarker-based diagnosis should precede staging and prognosis.

With regard to staging and prognosis, the 2023 document attempts to outline (in a formal way, ignoring specific biomarkers) how biological (biomarker-based) stages might be associated with clinical stages. In the typical progression trajectory these two are claimed to be fully aligned. According to the authors, temporal interrelation-ships between amyloid PET, tau PET and clinical symptoms provide evidence of such a typical trajectory. It is acknowledged, however, that biomarker test results can be worse or better than one would expect based on clinical symptoms (and vice versa), which may be due to "morbid pathology" and "exceptional cognitive reserve or resistance" respectively (NIA-AA 2023b: 12, Table 6). In the last section of the document most suggestions for future work are also related to staging and prognos-tication. The authors conclude by saying that they "envision creating a comprehen-sive system to stratify risk of progression by incorporating all biomarkers (core AD, non-core, and biomarkers of non-AD copathology) along with demographics and genetics. However, all these goals will depend first on standardization/harmonization of biofluid assays, standardized quantification of tau PET, and standardization of cutpoints for all fluid and PET biomarkers." (NIA-AA 2023a: 28).

Overall then, the criteria to identify AD have changed substantially since the emergence of AD biomarkers. Before 2011 'AD' was viewed as a clinico-pathological entity and a disease that could be diagnosed with certainty only post mortem. Less than 15 years later AD has evolved into a pathophysiological continuum and a biological condition diagnosed in vivo via biomarker testing. Only one deviant imaging or fluid biomarker test result, regardless of clinical symptoms, suffices for the diagnosis—although the 2023 authors urge readers to only use stringently validated biomarker tests, to interpret results in a conservative manner and always put them in a clinical context (whatever this may mean for individuals subjected to biomarker testing without symptoms). These developments clearly entail that living individuals without symptoms can now be diagnosed with 'AD' and that the threshold for AD diagnosis has been lowered substantially. However, the meaning of such a diagnosis also shifted.

In the next section we aim to provide tools for a better understanding of these developments by reflecting on the concept of 'biomarker' and the various ways in which such markers may relate to 'disease'.

11.3 Biomarkers and Disease

Explanations of what biomarkers are and what they do are often full of metaphors. They are for example called 'windows' into health and disease (e.g. Weston and Hood 2004; Yurkovich and Hood 2019), 'footprints' (e.g. Ullah and Aatif 2009; Daiber et al. 2021), 'signatures' or 'profiles', in particular when multiple biomarkers are combined (e.g. Wilson 2017; Chipi et al. 2019). Each of these metaphors has slightly different implications, but they all imply that biomarkers somehow refer to something else. Windows open up a world 'out there' for our observation. Footprints suggests that traces are left by a perpetrator, who might be identified and caught when following the signs. And signatures and profiles stress the personal, specific character of what is signified. Biomarkers, these metaphors imply, are *signs* signifying a phenomenon that is larger than and/or underlying what is being observed, and which is apparently hard to access fully or directly.

According to the most recent FDA definition a biomarker is "A defined characteristic that is measured as an indicator of normal biological processes, pathogenic processes, or biological responses to an exposure or intervention, including therapeutic interventions. Biomarkers may include molecular, histologic, radiographic, or physiologic characteristics. A biomarker is not a measure of how an individual feels, functions, or survives." (FDA 2016, lemma 'biomarker'). This definition, like the metaphors above, shows that biomarkers are *indicators:* signs of something else. It adds two more things. First, biomarkers look at *biological* processes and not at subjective feeling or functioning. Secondly, they infer something about the *charac-ter* of these biological processes, which may be normal, pathogenic, or a response to a preceding event or intervention.

Neither the FDA definition, nor the metaphors used are very precise about *what* is signified. If biomarkers are supposed to indicate whether biological processes are 'normal' or 'pathological', the ultimate phenomena referred to could be identified as 'health' and 'disease'. However, as the FDA glossary shows, this relation between biomarkers and disease/health is manifold. Markers can indicate (and often quantify)

(a) susceptibility/risk of disease,
(b) presence of disease,
(c) disease state,
(d) the likelihood of certain future events, recurrence or progression,
(e) predicted response to an intervention,
(f) actual response, or
(g) adverse effects of an intervention (FDA 2016, lemma 'biomarker').

At least the first four functions may play a role in establishing disease without symptoms: biomarkers can show that an individual is at increased risk of disease (a) or that one is already (latently) diseased (b), and if so, in what stage this latent disease is (c), and/or how this disease is likely to evolve. With regard to the distinction between increased risk and disease, according to the FDA glossary *risk biomarkers* estimate the likelihood of future disease or a condition in an individual "who does not currently have clinically apparent disease or the medical condition" (ibid. lemma 'susceptibility/risk biomarker'), whereas a *diagnostic biomarker* detects or confirms the actual presence of a disease or condition (or a subtype) (cf. lemma 'diagnostic biomarker').

The FDA's insistence that biomarkers refer to biological processes and not to subjective feeling or functioning places them firmly in the realm of 'disease' as a bodily phenomenon, as contrasted with 'illness' (the subjective experience) or 'sick-ness' (diminished social functioning). However, this still leaves open the question how 'disease' or a 'condition' is actually conceptualized. As the reconstruction of the AD criteria shows, the meaning of 'disease' in the context of biomarkers should not be taken for granted. What kind of phenomenon is this disease that biomarkers point to?

Several authors distinguish between an 'ontological' and a 'physiological' mode of understanding of disease (Temkin 1963; Rosenberg 2003). Without claiming to do justice to the historical work underlying this distinction, we suggest it is useful to grasp contemporary developments in biomarker research in and possibly beyond the AD domain. We use the distinction here as an *analytical* tool. The terms 'ontolog-ical' and 'physiological' in this chapter do not refer to separate types of diseases, nor to subsequent historical understandings of disease, but to *different modes of thinking about and approaching disease*. These cover various aspects of approaching 'dis-ease' that often tend to go together, even though these aspects are not associated by logical or causal necessity. The two modes of thinking can be, and often are, used in relation to the same disease, but this may lead to confusion about what a test is actually diagnosing or what a treatment is supposed to achieve. Distinguishing the two modes helps to pinpoint (and where necessary contest) the different 'logics' implied for thinking about and dealing with disease. Moreover, the distinction helps to recognize how specific technologies may be driven by and reinforce specific modes of understanding and dealing with disease.

In the *ontological* mode disease is understood as an *object in space*, a three-dimensional entity absent from healthy bodies. This entity is seen as independent from the individual body, and each 'disease' is understood as having a specific identity of its own, regardless of the body in which it manifests itself. This implies that the disease is caused by the same pathology, and that this pathology expresses itself and evolves in the same way in any individual (if not, the patient is suffering from 'a-typical' disease). In this ontological mode of understanding the boundary between disease and health is relatively clearcut: the disease entity is either present or absent. Tumours are often viewed in the ontological mode, but also atherosclerosis, or a hernia are examples of diseases that tend to be understood in this way. Diagnostic technologies inviting this way of thinking are often extensions of sight and touch, like microscopy, X-rays or MRI-scanners. In view of the assumed specific character of each disease, diagnostic progress in this mode means that these specificities are increasingly recognised. This mode also assumes that prognosis is implied by diagnosis; if the specific disease is known, the prognosis follows (e.g. the specific type of tumour is typically progressing fast or slow, or responding well to treatment A but not to B). As for therapeutic remedies, the ontological mode suggests that the disease entity should be removed or destroyed, for example by surgery, radiation therapy or drug treatment. Ideally, treatment targets the specific character of the disease entity, fitting the 'disease lock' like a key. A specific diagnosis is required, then, for both prognosis and treatment.

In the *physiological* mode disease is understood as a dynamic *process in time*. In contrast with the ontological mode, which focuses on presence or absence of deviant entities, the physiological mode focuses on deviations in bodily processes. These processes are understood to be the result of (often complex) interactions within and between a particular body and its evolving, particular environmental circumstances (both material and social). As a result, a disease manifests itself in a unique way in each individual. Moreover, in this mode of thinking health and disease form a continuum, and it is more or less arbitrary where exactly to draw the line between the two. Diabetes is a paradigmatic example of a disease that is usually understood in a physiological mode, and so is rheumatoid arthritis. Diagnosis in this mode requires repeated measuring of bodily functioning, as one deviant observation may be insufficient to correctly characterise what is going on. Hence, technologies facilitating regular measurement of bodily functioning, like thermometers or biochemical analyses of urine and blood, tend to invite a physiological mode of thinking about disease. In this approach diagnosis is not very informative about prognosis, since the same disease may evolve very differently in different individuals. As for ways to counter disease, the physiological mode of thinking invites therapies that somehow restore (normalize) the bodily balance. These may include a variety of options (drug treatment, but also dieting or exercising for example) and need not be specific to the disease at hand. Finally, rather than providing a 'one off' treatment, such treatments may have to be administered repeatedly or even permanently to prevent the process of going astray again. Whereas according to the ontological logic disease evolves in a predictable ('typical') way, in one direction only, the physiological mode of thinking allows for a variety of routes, including remission, relapse and other complexities.

To understand how biomarkers are both shaped by and shape our understanding of disease, it is important to be aware that biomarker technologies can construct disease in an ontological or a physiological mode. In the first case, biomarker test results indicate the presence/absence of disease and/or of specific disease characteristics. This mode of thinking is visible, among others, in discourse on personalised and precision medicine when promoting the vision that biomarkers help to better distinguish subtypes of disease thus far lumped together. The physiological mode of thinking is at play, for example, in visions of data-intensive healthcare when repeated measuring of biomarkers is promoted to monitor the development of a person's disease (or risk of disease).

Which mode of thinking is dominant may depend on the technology used. Bell (2013) observed, for example, that technological evolution in the context of cancer biomarkers has led to a shift in understanding disease. She argues that the emergence of molecular, quantitative technologies in the biomarker field shifted the relation between sign (biomarker test result) and signified (health/disease). Traditional imaging markers were understood as direct representations of the disease, due to the similarity of source and image (e.g. mammography showing a suspicious light clump looking like a tumour). They ascertained the absence or presence of disease. Molecular technologies (including quantitative imaging modalities), in contrast, construct the relation between biomarkers and disease in what Bell (referring to Peirce) calls an "indexical" way. They measure one element of an ongoing causal process (as smoke refers to a fire). In doing so they also position an individual on a continuum of disease and health. The shift Bell observes in the field of tumour biomarker technologies can be rephrased as a shift from an ontological to a physiological understanding of disease. Interestingly, she shows how such a shift also affects the functions ascribed to biomarkers. Whereas the ontological thinking stimulated by traditional imaging tends to stress biomarkers' potential for identification of asymptomatic disease, the physiological thinking invited by molecular technologies stresses their potential for measuring risk and predicting future development of disease, *both* for subjects without complaints *and* for those already diagnosed with disease. In our view, a similar shift is visible in the field of AD.

11.4 AD Biomarkers: Promising Homogeneity and Certainty, Producing Heterogeneity and Probabilities

Looking at the history of Alzheimer's disease with these conceptual tools in mind, the term 'AD' when coined at the start of the twentieth century was first and foremost defined in an *ontological* mode. 'Disease' in AD referred to the presence of specific pathological entities in the brain at autopsy, now known as amyloid plaques and tau tangles, which (together) were considered specific for AD, distinguishing AD from other dementias. Since then, however, it has remained challenging to align this ontological mode of thinking about AD with the clinical observations and experiences the plaques and tangles were supposed to explain. From a clinical perspective the diagnosis of a dementia syndrome that might be AD-related still covered a wide variety of presentations, qua type and severity of symptoms, their order of appearance and speed of deterioration. Moreover, while the symptoms thought to be associated with 'AD' actually varied in different periods (Ballenger 2006), all were progressive. From a clinical perspective, therefore, it was not obvious that the heterogeneous manifestation and the varied evolution in symptom presentations warranted delineation of AD as one specific disease entity.

Since the inception of 'AD', then, the mode of understanding 'disease' in AD pathology was at odds with clinical experiences and observations of AD. Not only was the gold standard for diagnosing AD based on post-mortem pathology, while during life only symptoms could be accessed; the ontological mode of thinking about disease in AD pathology was hard to reconcile with experiences in the AD clinic.

The 1984 disease criteria can be understood as an attempt to align clinical and pathological perspectives, mainly driven by a desire to standardize the meaning of 'AD' in research settings and hence produce more comparable scientific evidence about the natural history of the disease and the effectiveness of treatment. Although the criteria are meant to inform clinical diagnosis, these stress that such a diagnosis is inherently uncertain, as for a definite diagnosis both symptoms and pathology are required. Here we see again the ontological mode of thinking at work, in particular in the rule that if AD specific pathology is lacking in individuals diagnosed with possible or probable AD, the AD label is to be discarded. Histopathological examination is the ultimate test to distinguish the specific AD-type of dementia from other dementia syndromes. The introduction of the paper also suggests an ontological mode of thinking when describing AD as a "brain disorder characterized by progressive dementia" (McKhann et al. 1984: 939) and hinting at specific disease mechanism at play. However, the paper also leaves open the possibility that there may be multiple pathological pathways leading to plaques and tangles. The authors also state that there is variety in the order and speed of dementia progression, thus acknowledging pathological and clinical elements that fit better with a physiological understanding of disease.

When biomarkers entered the AD-scene, this initially seems to have been driven by the ontological mode of thinking; it also seems to have strengthened such thinking to some extent. In view of the perceived uncertainty of clinical diagnosis it need not surprise that these tests were welcomed for their potential to improve *in vivo AD-diagnostics*: these tests promised confirmation or rejection of the probable AD diagnosis during life. Images showing (for example) atrophy or high amyloid load in the brain of individuals suffering from dementia seemed to directly prove 'AD'. Implicitly this boosted the ontological mode of thinking about AD, because such proof not only enabled the delineation of a specific group of living dementia patients; it might even help to develop specific, targeted treatment to remove or stop development of the AD pathology. If proteins in the brain are the problem, we should remove them or prevent that they build up.

This ontological mode of thinking also fed emerging ideas about 'AD without symptoms'. Studying individuals with MCI showed that they also could test positive on biomarkers for AD pathology, and so did some people without any cognitive complaints. Biomarker research thus confirmed earlier histopathological findings that the pathology considered specific for AD could also be present in individuals with minor or no symptoms (a possibility not yet considered in the 1984 criteria because the first publications regarding this mismatch date from the 1990s). Such findings in the brains of living asymptomatic individuals were often interpreted as proof of *latent* disease. Like mammography enabling the identification of smaller breast cancer lesions, biomarker technologies were thought to identify the presence of 'minor AD pathology', which in turn seemed to justify extension of the AD label to individuals without symptoms. The amyloid cascade hypothesis, which gained popularity in the same period, outlined a mechanism to explain how the pathological hallmarks of AD might gradually develop, thus providing a biological underpinning for the possibility of latent disease. Interpreting both in vivo signs of AD pathology and the mechanism producing it in an ontological mode—as a disease characterised by a specific biological pathway leading to specific pathology resulting in dementia—the function ascribed to biomarker tests for asymptomatic and MCI individuals was *diagnostic*: biomarkers seemed to facilitate the early identification of previously invisible cases of disease. In this vein it makes sense to assume that individuals suffering from dementia, MCI or without symptoms who test positive on the same biomarkers *have* the same disease, warranting the 2011 extension of the guidelines with criteria for 'MCI due to AD' and 'preclinical AD'.

As described above, the 2018 and 2023 criteria reframe this merger of dementia, MCI and preclinical AD as the "AD continuum" (although the term is already mentioned in the 2011 criteria for preclinical AD: Sperling et al. 2011: 282). In many respects these criteria continue the ontological mode of thinking about disease as a specific, unified entity brought about by a specific pathway. The more recent criteria consider what is now called 'A and T proteinopathies' as the unifying element of AD, distinguishing it from other dementias. Positive A and T biomarkers (both in 2018, one of them in 2023) provide evidence, and hence certainty about the presence of such specific proteinopathies, meaning one can get rid of the terms 'possible' or 'probable'. Moreover, assuming a unified and deterministic disease pathway, biomarker evidence of proteinopathies in individuals with minor or no complaints is explicitly understood as early diagnosis of AD and presented as an important advantage of diagnosing AD with biomarker tests (Jack et al. 2018). Biomarkers thus are said to enable a more precise, a more certain and an earlier diagnosis.

However, the possibility to observe AD pathology during life did not always deliver the certainty and clarity hoped for. It also meant that the *bodily process* leading to AD pathology, which was previously obscure, could now be observed. The increasing amount of measurements of the increasing number of bodily parameters becoming available turned out to put into question the ontological assumption that AD is characterized by a unified specific pathological mechanism (the amyloid cascade). Although biomarker results might help to distinguish individuals with evidence of AD pathology as a specific subset among the heterogeneous group of those with dementia syndrome, these individuals (now identified as 'being on the AD continuum') still displayed rather heterogeneous disease dynamics. Although some did embody the expectations inspired by the amyloid cascade, many did not. Thus, a wide variety in pathological dynamics was added to the variety in clinical manifestations, making it harder rather than easier to align pathological and clinical observations. Thus, the emergence and increase of biomarker testing actually opened a Pandora's box: it multiplied, rather than unified the meaning of 'AD', even when defined biologically.

This (re)invites a physiological mode of thinking about the disease, in which each individual may manifest a disease in a highly specific manner. This is visible already in the 2018 criteria talk about 'profiling' of individuals with the A, T and N parameters all of which may be measured in multiple ways (Jack et al. 2018: 540). Interestingly, the authors of the 2023 draft criteria not only expand this scheme with 3 more parameters (I, V, and S). They also shift the function of the biomarker profiles into a scheme for biological *staging*, which is conceptualised as independent from clinical staging (NIA-AA 2023a: 15 and 2023b: 9, Table 4). Also striking is the 2023 observation that fluid markers, in contrast to imaging markers, can not only measure production or clearance of certain products, but also the *rates* with which these processes take place. The authors of the 2023 draft criteria even advise to interpret PET in a quantitative way to facilitate functions related to disease dynamics, e.g. to identify pathologic changes in cases where amyloid-β levels are still considered 'normal' (NIA-AA 2023a: 13). As described above, the 2023 draft criteria are very much focused on staging, and although the authors state that in most individuals pathology and symptoms will emerge in a standard combination, they acknowledge that both biological and social circumstances may lead to different patterns. In their view the future is about stratified risk of progression, for which one will need non-AD biomarkers, genetic testing and demographic data in addition to AD biomarker results.

The shift towards a physiological mode of thinking about AD is also consequential for those without symptoms. The multiplication of measurements and parameters implies that showing evidence of A and/or T biomarkers (and thus 'being on the AD continuum') in itself is not very telling about what to expect. Confirmed positive biomarkers do not guarantee that an individual with MCI will 'convert' to dementia, let alone when and with what symptoms. Population research shows that positive test results indicate increased risk and thus have predictive, rather than diagnostic value. It would be more fitting to say that a person with positive biomarkers is at risk for dementia. This shows, among others, in the decision of the 2023 committee to use 'asymptomatic disease', rather than preclinical disease.

11.5   Implications for AD, With or Without Symptoms

To be clear, we do not claim that the physiological mode of thinking about AD has fully replaced the ontological mode. We do contend that much confusion and controversy in the AD biomarker domain, especially regarding the meaning and desirability of identifying AD without symptoms, is due to these conflicting approaches of disease. Moreover, with the emergence of in vivo biomarkers the balance between the two seems to be gradually shifting in favour of the physiological mode. If this analysis makes sense, this has several implications for discussions on the desirability of diagnosing 'disease without symptoms' and beyond.

First of all: yes, the emergence of biomarkers has facilitated the identification of AD pathology in people without symptoms and led to the delineation of 'preclinical' and 'asymptomatic AD'. This could potentially lead to the negative impacts often linked with diagnosing pre-disease summarized in the introduction. However, as the shifting terminology suggests, the meaning of the category 'preclinical/asymptomatic AD' is shifting. Although the initial assumption was that having preclinical AD came with specific characteristics and a clear prognosis (i.e., that the individual would end up with dementia syndrome), the wealth of biomarker measurements now shows that this is not necessarily the case.

Secondly, the meaning of the term 'AD' itself not only changed substantially over the years, from a clinico-pathological entity to a biological continuum, but in the process has actually become less meaningful. Being on the AD continuum implies that there is evidence of A and/or T proteinopathy, but what this observation entails for the individual is neither clear nor certain.

Thirdly, the lack of meaning has brought about an increasing interest in staging and prognostication, rather than (early) diagnosis. To be useful, biological parameters should be monitored repeatedly, indicating the temporal development of an individual's pathology, and should be put in relation to the development of their complaints, which may then allow for predictions about future developments of a person's disease. Such predictions thus require individual monitoring, as well as population studies.

Fourthly, and most importantly, the observations above not only apply to preclinical/asymptomatic AD, but also to clinically manifest disease. Because the meaning of 'AD' is not as specific as many assumed it to be, offering an AD diagnosis to a person suffering from dementia is also less meaningful than often suggested. The level of certainty about the pathology underlying the symptoms increased, but a biomarker-based diagnosis at present does not give many more practical clues for what to expect or how to treat than the previous clinical diagnosis. As in the case of asymptomatic AD, this currently seems to be bringing about a shift in biomarker research, from diagnosis to prediction and prognosis. If future AD biomarkers will indeed enable sorting out how an individual's disease trajectory is likely to evolve, not only the asymptomatic stage, but also the symptomatic stages will become pre-stages. The meaning of biomarker testing may then lie in prognostication of what is likely ahead for this particular individual, rather than in confirming one's current state.

11.6 Towards Anticipatory Healthcare

This last point links with Bell's observation, mentioned earlier, that cancer bio-markers increasingly measure risk for future events, not only in healthy individuals, but also in those with symptoms (Bell 2013). When biomarkers are repeatedly used to measure one's present bodily state, they may infer projections of future risk, or, with a term Bell borrows from Gillespie, of "measured vulnerability" (see also Tinland 2022). A similar point has been made by Aronowitz in the context of chronic disease more generally (Aronowitz 2009). In his view, not only are those at increased risk of disease often treated as if they were diseased; with increased monitoring the experience of disease has become more risk-like. Whereas the boundary between risk and disease is blurred, the experiences *on either side of the line* become more similar. Both, one could say, become literally more 'chronic'.

The earlier observations by Aronowitz and Bell suggest that the trends observed in the field of AD biomarkers are neither unique to AD, nor dependent on biomarkers only. Nonetheless, biomarkers contribute to the increased recruitment of individuals into disease categories, as well as to the intensification of disease monitoring—two factors listed by Aronowitz as potential explanations for the increased convergence of risk and disease. Those studying other disease domains would do well, then, to pay attention to the way biomarkers shift our understanding of both risk and disease, both for asymptomatic and symptomatic individuals. In doing so, the distinction between ontological and a physiological mode of understanding may prove helpful.

Another reason to hypothesize that similar trends may already occur in other disease domains is the widespread promotion of visions of personalised and preci-sion medicine. The core claim of these approaches is that the use of biomarkers will better account for and help to tackle disease heterogeneity Prainsack (2017). Such visions are heir to what Rosenberg (2002) called the "specificity revolution", assuming that by sorting out which specific type of disease you have, physicians will be able to select or develop tailored treatment. This disease specificity in visions of personalized or precision medicine tends to be understood in an ontological mode, claiming that biomarker-based diagnosis will suffice to distinguish the relevant disease types. To the extent that such an understanding neglects or simplifies understanding of the temporal dimension of disease, pursuing these visions may very well lead to more 'Pandora's box experiences': adding to, rather than grasping complexity of disease Hofmann (2023).

Finally, if shifts and trends like the ones reconstructed here for AD are also at play in other disease domains, we may actually be witnessing a 'demotion' of the role of diagnosis in healthcare. This demotion seems driven by a pragmatist approach to healthcare. Rather than aiming to identify and causally explain the specific disease an individual is suffering from, clinical professionals aim for information that helps to advise an individual and/or decide what to do. Useful knowledge in this case anticipates what is likely to happen in the future, and what can be done to achieve a (more) desirable outcome. This point was made earlier by Armstrong (2019), who discusses why many biomarker tests do not move from the lab to the clinic. Although such tests may offer better insight in the specific molecular pathology underlying a disease, he argues, biomarkers identifying a disease subtype are viable only if they are clinically useful. This means that the proposed stratification of individuals should help to *predict a relevant endpoint* in a clinical population. In practice biomarkers can do so in two ways, both related to the temporal dimension of disease. First, biomarkers can indicate the probable future trajectory of the individual—whether or not the individual already experiences symptoms does not make much of a difference here. Second, biomarkers can assess, or indicate the probability of, treatment-responsiveness. In both cases, the series of all biomarker test results up to the present is used to predict future events, and repeated biomarker monitoring helps to finetune such predictions and interventions. In this context, Armstrong argues, diagnosis would no longer be the stable foundation of medicine; prediction/prognosis would be guiding treatment. Such healthcare, we would say, is anticipatory.

11.7 In Conclusion

We hope to have shown that shifting from an ontological to a physiological mode of understanding helps to make sense of the trends in and perplexities of AD biomarker research. Such a shift has several advantages, which may very well extend beyond the AD domain. First of all, it helps understand how the emergence of biomarker testing affects the conceptualization of both risk and disease and changes the experiences of those with and without symptoms. This is a crucial first step to study the implications of such changes and evaluate their desirability. Secondly, it provides a warning that attempts to specify diagnosis—whether or not inspired by personalized/precision medicine –, while trying to sort out disease heterogeneity, are likely faced with a further explosion of such heterogeneity. Finally, it reminds us that establishing a specific (precise) and certain disease label in itself has limited practical value for the person concerned as well as healthcare professionals. It may be more fruitful to address the heterogeneity of disease by a shift towards anticipatory healthcare, in which biomarkers help to estimate and respond (in a variety of ways) to an individual's vulnerability for future events. Taking the promise of personalized medicine literally, such an approach would put the person, rather than the disease, at the center.

Acknowledgements This chapter profited from the input given on the first draft by the participants of the workshop 'Health and Disease as Practical Concepts', in particular from the meticulous comments made by Bjorn Hofmann. MB is also grateful to Nicholas Binney, Maartje Schermer, Timo Bolt and Rik van der Linden for taking the effort to comment on a second draft.

References

Aisen, Paul S., Jeffrey Cummings, Clifford R. Jack, John C. Morris, Reisa Sperling, Lutz Fröhlich, Roy W. Jones, et al. 2017. On the path to 2025: Understanding the Alzheimer's disease continuum. *Alzheimer's Research and Therapy* 9: 1–10.

Albert, Marilyn S., Steven T. DeKosky, Dennis Dickson, Bruno Dubois, Howard H. Feldman, Nick C. Fox, Anthony Gamst, et al. 2011. The diagnosis of mild cognitive impairment due to Alzheimer's disease: Recommendations from the National Institute on Aging-Alzheimer's Association workgroups on diagnostic guidelines for Alzheimer's disease. *Alzheimer's & Dementia* 7 (3): 270–279.

APA – American Psychiatric Association. 2013. *The diagnostic and statistical manual of mental disorders.* 5th ed. Washington, DC: American Psychiatric Publishing.

Armstrong, David. 2019. Diagnosis: From classification to prediction. *Social Science & Medicine* 237: 112444.

Aronowitz, Robert A. 2009. The converged experience of risk and disease. *The Milbank Quarterly* 87 (2): 417–442.

Ballenger, Jesse F. 2006. *Self, senility, and Alzheimer's disease in modern America. A History.* Baltimore: The Johns Hopkins University Press.

Bell, Kirsten. 2013. Biomarkers, the molecular gaze and the transformation of cancer survivorship. *BioSocieties* 8: 124–143.

Boenink, Marianne. 2016a. Disease in the era of genomic and molecular medicine. In *The Bloomsbury companion to philosophy of medicine*, ed. James Marcum, 65–92. London: Bloomsbury.

———. 2016b. Chapter 4: Biomarkers for Alzheimer's disease: Searching for the missing link between biology and clinic. In *Emerging technologies for diagnosing Alzheimer's disease: Innovating with care*, ed. Marianne Boenink, Harro van Lente, and Ellen Moors. Basingstoke: Palgrave.

———. 2018. Gatekeeping and trailblazing: The role of biomarkers in novel guidelines for diagnosing Alzheimer's disease. *BioSocieties* 13 (1): 213–231.

Brodersen, John, Lisa M. Schwartz, and Steven Woloshin. 2014. Overdiagnosis: How cancer screening can turn indolent pathology into illness. *APMIS* 122 (8): 683–689.

Busse, Anja, Anke Hensel, Uta Guhne, Matthias C. Angermeyer, and Steffi G. Riedel-Heller. 2006. Mild cognitive impairment: Long-term course of four clinical subtypes. *Neurology* 67 (12): 2176–2185.

Chipi, Eelena, Nicola Salvadori, Lucia Farotti, and Lucilla Parnetti. 2019. Biomarker-based signature of Alzheimer's disease in pre-MCI individuals. *Brain Sciences* 9 (9): 213.

Daiber, Andreas, Omar Hahad, Ioanna Andreadou, Sebastian Steven, Steffen Daub, and Thomas Münzel. 2021. Redox-related biomarkers in human cardiovascular disease-classical footprints and beyond. *Redox Biology* 42: 101875.

Dubois, Bruno, Howard H. Feldman, Claudia Jacova, Steven T. Dekosky, Pascale Barberger-Gateau, Jeffrey Cummings, André Delacourte, et al. 2007. Research criteria for the diagnosis of Alzheimer's disease: Revising the NINCDS–ADRDA criteria. *The Lancet Neurology* 6 (8): 734–746.

———. 2010. Revising the definition of Alzheimer's disease: A new lexicon. *The Lancet Neurology* 9 (11): 1118–1127.

Dubois, Bruno, Howard H. Feldman, C. Claudia Jacova, Harald Hampel, José Luis Molinuevo, Kaj Blennow, Steven T. Dekosky, et al. 2014. Advancing research diagnostic criteria for Alzheimer's disease: The IWG-2 criteria. *The Lancet Neurology* 13 (6): 614–629.

Dubois, Bruno, Nicolas Villain, Giovanni B. Frisoni, Gil D. Rabinovich, Marwan Sabbagh, Stefano Cappa, Alexandre Bejanin, et al. 2021. Clinical diagnosis of Alzheimer's disease: Recommendations of the International Working Group. *The Lancet Neurology* 20 (6): 484–496.

Elgart, Jorge F., Rocío Torrieri, Matias Ré, Martin R. Salazar, Walter G. Espeche, Julieta Angelini, Carmen Martínez, et al. 2023. Prediabetes is more than a pre-disease: Additional evidences supporting the importance of its early diagnosis and appropriate treatment. *Endocrine* 79: 80–85.

Faulkner, Eric, Anke-Peggy Holtorf, Surrey Walton, Christine Y. Liu, Hwee Lin, Eman Biltaj, Diana Brixner, et al. 2020. Being precise about precision medicine: What should value frameworks incorporate to address precision medicine? A report of the personalized precision medicine special interest group. *Value in Health* 23 (5): 529–539.

FDA-NIH Biomarker Working Group. 2016. *BEST (Biomarkers, EndpointS, and other Tools) Resource*. Silver Spring: Food and Drug Administration (US); 2016-. Glossary 2016 Jan 28 [Updated 2021 Nov 29]. Available from: https://www.ncbi.nlm.nih.gov/books/NBK338448/. Co-published by National Institutes of Health (US), Bethesda (MD). Accessed 25 Aug 2023.

Fiala, Clare, Jennifer Taher, and Eleftherios P. Diamandis. 2019. P4 medicine or O4 medicine? Hippocrates provides the answer. *The Journal of Applied Laboratory Medicine* 4 (1): 108–119.

George, Daniel R., Peter J. Whitehouse, and Jesse Ballenger. 2011. The evolving classification of dementia: Placing the DSM-V in a meaningful historical and cultural context and pondering the future of "Alzheimer's". *Culture, Medicine, and Psychiatry* 35: 417–435.

Getachew-Smith, Hannah, Amy A. Ross, Courtney L. Scherr, Marleah Dean, and Meredith L. Clements. 2020. Previving: How unaffected women with a BRCA1/2 mutation navigate previvor identity. *Health Communication* 35 (10): 1256–1265.

Hardy, John A., and Gerald A. Higgins. 1992. Alzheimer's disease: The amyloid cascade hypothesis. *Science* 256: 184–185.

Hofmann, Bjørn. 2001. The technological invention of disease. *Medical Humanities* 27 (1): 10–19.
———. 2022. Too much, too mild, too early: Diagnosing the excessive expansion of diagnoses. *International Journal of General Medicine* 15: 6441–6450.
———. 2023. Temporal uncertainty in disease diagnosis. *Medicine, Health Care and Philosophy* 26: 1–11.

Hofmann, Bjørn, and John-Arne Skolbekken. 2017. Surge in publications on early detection. *BMJ* 2017 (357): j2102.

Hofmann, Bjørn, Lynette Reid, Stacy Carter, and Wendy Rogers. 2021. Overdiagnosis: One concept, three perspectives, and a model. *European Journal of Epidemiology* 36: 361–366.

Hood, Lee. 2019. How technology, big data, and systems approaches are transforming medicine. *Research-Technology Management* 62 (6): 24–30.

Isaacs, Jeremy D., and Marianne Boenink. 2020. Biomarkers for dementia: Too soon for routine clinical use. *The Lancet Neurology* 19 (11): 884–885.

Jack Jr, Clifford R., Marilyn S. Albert, David S. Knopman, Guy M. McKhann, Reisa A. Sperling, Maria C. Carillo, Bill Thies, and Creighton H. Phelps. 2011. Introduction to the recommendations from the National Institute on Aging-Alzheimer's Association workgroups on diagnostic guidelines for Alzheimer's disease. *Alzheimer's & Dementia* 7 (3): 257–262.

Jack Jr, Clifford R., David A. Bennett, Kaj Blennow, Maria C. Carillo, Billy Dunn, Samantha B. Haeberleinn, David M. Holtzman, et al. 2018. NIA-AA research framework: Toward a biological definition of Alzheimer's disease. *Alzheimer's & Dementia* 14 (4): 535–562.

Jacobs, Noortje, and Bert Theunissen. 2022. It's groundhog day! What can the history of science say about the crisis in Alzheimer's disease research? *Journal of Alzheimer's Disease* 90 (Preprint): 1–15.

Kreiner, Meta J., and Linda M. Hunt. 2014. The pursuit of preventive care for chronic illness: Turning healthy people into chronic patients. *Sociology of Health and Illness* 36 (6): 870–884.

Lock, Margaret. 2013. *The Alzheimer conundrum: Entanglements of dementia and aging*. Princeton: Princeton University Press.

Mark, Rurth E., and Yvonne Brehmer. 2022. Preclinical Alzheimer's dementia: A useful concept or another dead end? *European Journal of Ageing* 19: 1–8.

McKhann, Guy M., David Drachman, Marshall Folstein, Robert Katzman, Donald Price, and Emanuel M. Stadlan. 1984. Clinical diagnosis of Alzheimer's disease: Report of the NINCDS-ADRDA Work Group* under the auspices of Department of Health and Human Services Task Force on Alzheimer's disease. *Neurology* 34 (7): 939–939.

McKhann, Guy M., David S. Knopman, Howard Chertkow, Bradley T. Hyman, Clifford R. Jack Jr, Claudia H. Kawas, William E. Klunk, et al. 2011. The diagnosis of dementia due to Alzheimer's disease: Recommendations from the National Institute on Aging-Alzheimer's Association workgroups on diagnostic guidelines for Alzheimer's disease. *Alzheimer's & Dementia* 7 (3): 263–269.

Metzler, Ingrid. 2010. Biomarkers and their consequences for the biomedical profession: A social science perspective. *Personalized Medicine* 7 (4): 407–420.

NIA-AA – National Institute of Ageing – Alzheimer's Association. 2023a. NIA-AA revised clinical criteria for Alzheimer's disease (main text), draft as of July 15, 2023. https://aaic.alz.org/nia-aa. asp. Accessed 25 Aug 2023.

———. 2023b. NIA-AA revised clinical criteria for Alzheimer's disease (text boxes, tables and figures), draft as of July 15, 2023. https://aaic.alz.org/nia-aa.asp. Accessed 25 Aug 2023.

Petersen, Ronald C., Glenn E. Smith, Stephen C. Waring, Robert J. Ivnik, Eric G. Tangalos, and Emre Kokmen. 1999. Mild cognitive impairment: Clinical characterization and outcome. *Archives of Neurology* 56: 303–308.

Prainsack, Barbara. 2017. *Personalized medicine.* New York: New York University Press.

Rosenberg, Charles E. 2002. The tyranny of diagnosis: Specific entities and individual experience. *The Milbank Quarterly* 80 (2): 237–260.

———. 2003. What is disease? In memory of Owsei Temkin. *Bulletin of the History of Medicine* 77: 491–505.

Schermer, Maartje H.N. 2023. Preclinical disease or risk factor? Alzheimer's disease as a case study of changing conceptualizations of disease. *The Journal of Medicine and Philosophy: A Forum for Bioethics and Philosophy of Medicine* 48 (4): 322–334.

Schermer, Maartje H.N., and Edo Richard. 2019. On the reconceptualization of Alzheimer's disease. *Bioethics* 33: 138–145.

Sperling, Reisa A., Paul S. Aisen, Laurel A. Beckett, David A. Bennett, Suzanne Craft, Anne M. Fagan, Takeshi Iwatsubo, et al. 2011. Toward defining the preclinical stages of Alzheimer's disease: Recommendations from the National Institute on Aging-Alzheimer's Association workgroups on diagnostic guidelines for Alzheimer's disease. *Alzheimer's & Dementia* 7 (3): 280–292.

Temkin, Owsei. 1963/1977. The scientific approach to disease: Specific entity and individual sickness. In *The double face of Janus and other essays in the history of medicine*, 441–455. Baltimore: Johns Hopkins University Press.

Timmermans, Stefan, and Mara Buchbinder. 2010. Patients-in-waiting: Living between sickness and health in the genomics era. *Journal of Health and Social Behavior* 51 (4): 408–423.

Tinland, Julia. (2022). Personalised prevention: Increasing or decreasing over-medicalisation, overdiagnosis and overtreatment? In *Personalized medicine in the making: Philosophical perspectives from biology to healthcare*, ed. C. Beneduce and M. Bertolaso, 87–111. Cham: Springer.

Ullah, Mohd F., and Mohammed Aatif. 2009. The footprints of cancer development: Cancer biomarkers. *Cancer Treatment Reviews* 35 (3): 193–200.

Vogt, Henrik, Bjørn Hofmann, and Linn Getz. 2016. The new holism: P4 systems medicine and the medicalization of health and life itself. *Medicine, Health Care and Philosophy* 19: 307–323.

Vogt, Henrik, Sara Green, Claus T. Ekstrøm, and John Brodersen. 2019. How precision medicine and screening with big data could increase overdiagnosis. *BMJ* 366: l5270.

Weston, Andrea D., and Leroy Hood. 2004. Systems biology, proteomics, and the future of health care: Toward predictive, preventative, and personalized medicine. *Journal of Proteome Research* 3 (2): 179–196.

Whitehouse, Peter J., and Eric T. Juengst. 2005. Antiaging medicine and mild cognitive impairment: Practice and policy issues for geriatrics. *Journal of the American Geriatrics Society* 53 (8): 1417–1422.

Whitehouse, Peter J., Konrad Maurer, and Jesse F. Ballenger, eds. 2000. *Concepts of Alzheimer disease: Biological, clinical, and cultural perspectives*. Baltimore: The Johns Hopkins University Press.

Wilson, Alphus D. 2017. Biomarker metabolite signatures pave the way for electronic-nose applications in early clinical disease diagnoses. *Current Metabolomics* 5 (2): 90–101.

Yurkovich, James T., and Leroy Hood. 2019. Blood is a window into health and disease. *Clinical Chemistry* 65 (10): 1204–1206.

Chapter 12
Biomarking Life

Bjørn Hofmann

In their insightful chapter, Marianne Boenink and Lennart van der Molen argue that biomarkers contribute to shift the focus of medicine from action-guiding diagnosis towards characterizing and prediction of future events. Accordingly, healthcare is becoming more anticipatory, and as such can "put the person, rather than the disease, at the center" (Boenink and Molen, Chap. 11, this volume). To do so, they argue that biomarkers have shifted the concept of disease from an ontological to a physiological one.

12.1 Do Biomarkers Promote a Shift from Ontological Concepts to Physiological Concepts?

It is worth to note that the battle between ontological and physiological conceptions of disease goes back to conceptual disputes between Knidians and Coans in ancient Greece (Hofmann 2001). In particular the physiological conception of disease has been called "Hippocratic" "biographical", "historical", "Coan", and "empirical" (Hofmann 2001).

However, technology has promoted both ontological and physiological conceptions of disease. As biomarkers identify ontological entities, they tend to promote ontological conceptions of disease. However, when the markers are not diagnostically decisive, e.g., when there are no clear cut-off limits, they have to be monitored (for change) and balanced against each other (and other signs). In such cases they

B. Hofmann (✉)
Centre for Medical Ethics, University of Oslo, Oslo, Norway

Norwegian University for Science and Technology, Gjøvik, Norway
e-mail: b.m.hofmann@medisin.uio.no

© The Author(s) 2024
M. Schermer, N. Binney (eds.), *A Pragmatic Approach to Conceptualization of Health and Disease*, Philosophy and Medicine 151,
https://doi.org/10.1007/978-3-031-62241-0_12

may come to take on and promote physiological conceptions of disease, as Boenink and Molen show. However, one could still argue that biomarkers promote both ontological and a physiological conceptions of disease. Hence, their argumentation may need more elaboration.

12.2   Are Biomarkers Responsible for the Shift to Anticipation?

Boenink and Molen raise the important question of whether the increasing importance biomarkers in the detection and management of Alzheimer's Disease (AD) have changed the strategy from early diagnosis to anticipation—from detection to monitoring—and whether this makes medicine put the person more than the disease at center.

Clearly more biomarkers are included in the assessment of AD, and as the authors eloquently show, they have become less decisive with respect to deciding a diagnosis. However, whether this results in a more temporal or anticipatory approach may need more scrutiny. First, diagnosis is in itself a temporal process and is closely connected to both anamnesis and prognosis (Hofmann 2023). Hence, diagnosis includes both temporal elements and anticipatory aspects.

Second, if "biomarkers help to estimate and respond (in a variety of ways) to an individual's vulnerability for future events" (Boenink and Molen, Chap. 11, in this volume) there is a danger that the estimates and responses may increase the vulnerability of the persons. Having disadvantageous constellations of biomarkers may not always empower persons, and anticipating dementia is not necessary a good thing. Hence, putting the persons (and their biomarkers) at the center may have stigmatizing side effects worth taking into consideration.

Third, diagnosis is imbued with temporal uncertainty (Hofmann 2023). Using biomarkers less closely connected to what matters to people, i.e., pain, suffering, and dysfunction, will increase this uncertainty. Despite substantial hype in precision medicine, P4 (or P7) medicine (Vogt et al. 2016), and Deep Medicine (Topol 2019) the anticipation of individual persons' development based on biomarkers can be anticipated to increase uncertainty (Fortmann-Roe 2012). If this is so, it may make uncertainty more biomarker-centered, but less personally relevant.

One could argue that early detection as such is anticipatory, or more precisely imbued with temporal uncertainty. Biomarkers are not constitutive to this. They are only contributory. The prognostic uncertainty of all types of indicators (biomarkers included) render what the indicators indicate indicative—or anticipatory, if you prefer.

12.3 Do Biomarkers Decouple Disease from Suffering and Put the Person at the Center?

As pointed out by Boenink and Molen, there are very many biomarkers connected to AD, but none of them are inevitably connected to severe suffering from dementia. Even more, the biomarkerization has created unexperienced AD: "the emergence of biomarkers has facilitated the identification of AD pathology in people without symptoms and led to the delineation of 'preclinical' and 'asymptomatic AD'" (Boenink and Molen, Chap. 11, in this volume). However, the biomarkerization of AD have not identified pre-symptomatic markers, i.e., biomarkers having a clear prognosis.

This points to a general challenge with the biomarkerization of human disease, i.e., the decoupling of biomarkers and what matters to people, i.e., pain, suffering, and loss of function (Hofmann 2018, 2019, 2021). Indicators, such as biomarkers, are identified because they in some way can be connected to manifest disease and suffering in individuals. However, while the connection between the identification of a cancerous lump on the patient's body and his suffering traditionally was strong, the connection between a biomarker and manifest disease and suffering has become very weak.

Kristin Zeiler has provided a normative framework to address various aspects of the decoupling between biomarkers and people's suffering from disease in the context of AD (Zeiler 2020). The reason why this decoupling is so interesting is of course because it alters the moral impetus of medicine. There seems to be a stronger moral appeal from people's experienced suffering than from their biomarker makeup (unless they are closely connected) (Hofmann 2024). Hence, the biomarkerization of disease connects to the goal of medicine as such.

An important implication is that if the biomarkerization of disease decouples the tasks of medicine from individuals' suffering, then it is not clear that it will "put the person, rather than the disease, at the center" (Boenink and Molen, Chap. 11, in this volume). Clearly, the person's individual biomarkers will be in focus, but if they are not experienced by the person, it is not clear that the person will be at the center.

12.4 From Marking (What Is) Bad to Defining What Is Good

The meaning of biomarkers depends on their purpose. While some biomarkers may be useful in research, others are useful clinically. Moreover, some are helpful for (early) detection of disease (including presymptomatic detection), others for predicting disease or disease development (prognosis), for assessing the suitability of treatment options, for monitoring progression (staging) of specific conditions, or for assessing treatment outcomes (as surrogate endpoints). The significance of biomarkers stems from their usefulness.

The case of AD indicates that this usefulness is shifted from detecting and treating pain and suffering to monitoring and anticipating what may be bad in the future. Even more, biomarkers can come to monitor our wellbeing. Our conception of how well we are may be informed and framed by biomarkers.

Hence, biomarkers can themselves influence what we take to be good. Our attention on biomarkers can influence how we feel, how we assess our health, and how we think about ourselves (Stites et al. 2018; Karceski and Antonopoulos 2023). This illustrates a general tendency where (biomarker) technology not only alters our (biomedical) conceptions of disease but also our experiential experience of illness (Bergen and Verbeek 2021; Hofmann and Svenaeus 2018). Although a "biomarker is not a measure of how an individual feels, functions, or survives" (FDA-NIH Biomarker Working Group 2016) it can come to influence how we do, resulting in what Ian Hacking has called the "looping effect" (Hacking 1995). Health information can be relieving, but also detrimental to people's health. For example, diagnostic labelling influences people's self-rated health (Jørgensen et al. 2015). Moreover, diagnostic labels alter persons' relations (Undeland and Malterud 2007) and bodily experiences (Reventlow et al. 2006).

Hence, biomarkers do not only influence our conception of disease (biomarkerization of disease). There is a biomarkerization of illness and sickness as well. This is because there is an extensive interaction and dynamics between the three dimensions of human malady, i.e., disease, illness, and sickness (Hofmann 2011). Accordingly, one could argue that there is a biomarkerization of human malady.

12.5   Conclusion

Boenink and Molen help us reflect on how biomarkers come to change our conceptions of disease and patient care in the future. Whether biomarkers make disease more physiological and anticipatory and medicine more person centred may need more research. However, biomarkers can decouple medicine and healthcare from what matters to people (such as pain, dysfunction, and suffering), reducing the moral relevance of medicine.

Hence, putting biomarkers at the center of medicine may not mean that we set persons at the center of medicine. On the contrary, a biomarkerization of medicine may make us all diseased, as there are no healthy persons, only persons that have not been sufficiently biomarkerized.

Biomarkers may do more than detecting or anticipating disease. They may come to define "the good life" and how we feel and fare. Therefore, there is not only a biomarkerization of *disease*, but also of *illness* and *sickness*. Biomarkers come to influence how we experience disease and how we behave and our social role (i.e., our sick role). Hence, *there is a biomarkerization of human malady*. When asked, "how are you?" we may well come to answer "my biomarkers are very fine, thank you."

References

Bergen, Jan Peter, and Peter Paul Verbeek. 2021. To-do is to be: Foucault, Levinas, and techno-logically mediated subjectivation. *Philosophy & Technology* 34 (June): 325–348. https://doi.org/10.1007/s13347-019-00390-7.

FDA-NIH Biomarker Working Group. 2016. *BEST (Biomarkers, EndpointS, and Other Tools) resource.* Silver Spring: Food and Drug Administration (US). http://www.ncbi.nlm.nih.gov/books/NBK326791/.

Fortmann-Roe, Scott. 2012. *Understanding the bias-variance tradeoff.* Available at https://scott.fortmann-roe.com/docs/BiasVariance.html. Accessed 26 Sept 2023.

Hacking, Ian. 1995. The looping effects of human kinds. In *Causal cognition: A multidisciplinary debate*, ed. Dan Sperber, David Premack, and Ann James Premack. Oxford University Press. https://doi.org/10.1093/acprof:oso/9780198524021.003.0012.

Hofmann, Bjørn. 2001. Complexity of the concept of disease as shown through rival theoretical frameworks. *Theoretical Medicine and Bioethics* 22 (3): 211–236. https://doi.org/10.1023/A:1011416302494.

———. 2011. On the dynamics of sickness in work absence. In *Social aspects of illness, disease and sickness absence*, ed. Halvor Nordby, Rolf Rønning, and Gunnar Tellnes, 47–62. Oslo: Unipub forlag.

———. 2018. Looking for trouble? Diagnostics expanding disease and producing patients. *Journal of Evaluation in Clinical Practice* 24 (5): 978–982. https://doi.org/10.1111/jep.12941.

———. 2019. Expanding disease and undermining the ethos of medicine. *European Journal of Epidemiology* 34 (7): 613–619. https://doi.org/10.1007/s10654-019-00496-4.

———. 2021. How to draw the line between health and disease? Start with suffering. *Health Care Analysis* 29 (2): 127–143. https://doi.org/10.1007/s10728-021-00434-0.

———. 2023. Temporal uncertainty in disease diagnosis. *Medicine, Health Care, and Philosophy* 26 (3): 401–411. https://doi.org/10.1007/s11019-023-10154-y.

———. 2024. Moral obligations towards human persons' wellbeing versus their suffering: an analysis of perspectives of moral philosophy. *Health Policy* 142: 105031.

Hofmann, Bjørn, and Fredrik Svenaeus. 2018. How medical technologies shape the experience of illness. *Life Sciences, Society and Policy* 14 (1): 3. https://doi.org/10.1186/s40504-018-0069-y.

Jørgensen, Pål, Arnulf Langhammer, Steinar Krokstad, and Siri Forsmo. 2015. Diagnostic labelling influences self-rated health. A prospective Cohort study: The HUNT study, Norway. *Family Practice* 32 (5): 492–499. https://doi.org/10.1093/fampra/cmv065.

Karceski, Steven, and Milan Antonopoulos. 2023. Biomarkers in Alzheimer disease: A review. *Neurology* 101 (4): e461–e463. https://doi.org/10.1212/WNL.0000000000207630.

Reventlow, Susanne Dalsgaard, Lotte Hvas, and Kirsti Malterud. 2006. Making the invisible body visible. Bone scans, osteoporosis and women's bodily experiences. *Social Science & Medicine (1982)* 62 (11): 2720–2731. https://doi.org/10.1016/j.socscimed.2005.11.009.

Stites, Shana D., Richard Milne, and Jason Karlawish. 2018. Advances in Alzheimer's imaging are changing the experience of Alzheimer's disease. *Alzheimer's & Dementia (Amsterdam, Netherlands)* 10: 285–300. https://doi.org/10.1016/j.dadm.2018.02.006.

Topol, Eric J. 2019. *Deep medicine: How artificial intelligence can make healthcare human again.* Basic Books.

Undeland, Merete, and Kirsti Malterud. 2007. The Fibromyalgia diagnosis: Hardly helpful for the patients? A qualitative focus group study. *Scandinavian Journal of Primary Health Care* 25 (4): 250–255. https://doi.org/10.1080/02813430701706568.

Vogt, Henrik, Bjørn Hofmann, and Linn Getz. 2016. The New Holism: P4 systems medicine and the medicalization of health and life itself. *Medicine, Health Care, and Philosophy* 19 (2): 307–323. https://doi.org/10.1007/s11019-016-9683-8.

Zeiler, Kristin. 2020. An analytic framework for conceptualisations of disease: Nine structuring questions and how some conceptualisations of Alzheimer's disease can lead to "Diseasisation". *Medicine, Health Care, and Philosophy* 23 (4): 677–693. https://doi.org/10.1007/s11019-020-09963-2.

Chapter 13
Risk and Disease: Two Alternative Ways of Modelling Health Phenomena

Élodie Giroux

13.1 Introduction

Since the 1970s, the concepts of 'risk' and 'risk factor' for disease have become a key element in contemporary medical discourse, thinking, and practice. Prevention at the individual level has been renewed by means of drug treatments. The risk of cardiovascular disease, for example, can be reduced by medically lowering the relevant parameter such as hypertension or hypercholesterolemia. As a result, people who felt 'normal' are classified as 'at risk' and are led to take lifelong medication in order to reduce their risk of potential and future ill health outcomes. New categories of asymptomatic people emerge: "partial patients", "perpetual patients" or "patients-in-waiting" (Timmerman and Buchbinder 2010). This 'at risk' status puts into question the traditional normal-pathological demarcation which is at the basis of pathophysiology, as well as the categorical and binary diagnostic approach central to traditional clinical practice and decision making. Moreover, it is not clear whether a risk factor such as hypertension is only a probable and partial cause of a future disease or the marker of a slow process that is already pathological, or even a process for which there is no clear threshold demarcating the normal from the pathological. Some risk factors, in particular those that Peter Schwartz (2008) calls "risk-based diseases" (hypertension and hypercholesterolemia, but also obesity and

This chapter is a development of an earlier chapter published in French in 2022 under the title "Risque et maladie: confusion ou alternative?" in Mathieu Arminjon et al., *Le normal et le pathologique: des categories périmées*?, Editions Matériologiques, 2022, https://doi.org/10.3917/edmat.armin.2022.01.0057.

É. Giroux (✉)
Faculty of philosophy, Lyon 3 Jean Moulin University and Institute for Philosophical Research of Lyon (UR 4187), Lyon, France
e-mail: elodie.giroux@univ-lyon3.fr

osteoporosis), tend to be understood, treated and experienced by patients and clinicians as current pathological conditions (Greene 2007; Timmerman and Buchbinder 2010). We are witnessing an extension of the pathological, and not only of medicalisation.[1]

This reminds us of the problem of iatrogeny that Ivan Illich (1976), a virulent critic of medicine, had already denounced in the 1970s, and which is partly reformulated today in criticisms of medicalisation, pathologisation, overdiagnosis, and overtreatment (Welch et al. 2011; Hofmann 2018; see also Kukla, Chap. 21, this volume). Indeed, while early detection and the preventive treatment of specific diseases may save some patients, it is not so for a large number of other people and conditions. Early diagnosis and the categorisation of at-risk people as 'patients' can lead to harm that is likely to outweigh any benefit these persons may receive. Overdiagnosis is inefficient in its creation of unnecessary healthcare, but also actively harmful with regards to the effects that such investigations and treatments can have on patients (Croft et al. 2015). Moreover, the role of the pharmaceutical industry in this expansion cannot be ignored, as its economic interests are obviously at stake since the market for drugs increases in proportion to the number of 'at risk' people who are candidates for treatment (Greene 2007; Dumit 2012).

There is thus a need to consider how to distinguish the categories of risk and disease, which is the main objective of this paper. At first glance, the risk-disease distinction seems clear. These categories refer to two different ontologies: a disposition or tendency and a current state. This is the difference between having a current pathological condition and tending towards getting one with a greater or lesser probability, but without necessity. This is what separates the possible or probable from the actual and real, the present and the future, but also the certain and the uncertain. In practice, however, this distinction is not so clearcut: a continuum and/or an amalgam seems to exist between the possible and the real. Could the conceptual analysis of disease and, in particular, of naturalist theories which rely on the dysfunction criterion to delimit the category of disease, provide some clarification?

First, *via* an analysis of the concepts of risk and risk factor, I examine the sources and reasons for the confusion between these two categories of risk and disease, despite the apparent evidence of their conceptual and ontological differences. Second, I focus on the concept of disease and analyse whether its definition, in the tradition of conceptual analysis, provides a solution for the disease-risk distinction. I show that the function-requiring definition of the naturalist camp, which has gone furthest in precisely delimiting the category of disease (and in excluding 'at risk' conditions from this concept), leads to a paradoxical result: the concept of disease

---

[1] Sholl (2017) proposed a distinction between pathologisation (defining and diagnosing a condition as a disease) and medicalisation (proposing an intervention or treatment based on medical knowledge) by showing their relative independence even though these two phenomena are often very much linked. Such a distinction makes it possible to better understand the phenomenon of medicalisation.

itself, and more particularly the function-dysfunction demarcation, must integrate a measure of risk or prognosis into its definition. Finally, I defend the idea that a more relevant approach to the risk-disease distinction emerges if these two concepts are understood as referring to two different—albeit complementary—ways of modelling health phenomena.[2] Indeed, in adopting a historical epistemological approach, it is enlightening to think of the context of the emergence of the concepts of risk and risk factor within epidemiology. This enables us to consider the concept of risk as part of an alternative approach to health phenomena which relies on a gradualist and comparative theory of health and which differs as such from the pathophysiological approach based on a binary normal-pathological distinction and a non-comparative theory of health.[3] This different way is dimensional, gradualist, and prognostic. It accords central importance to the question of health outcomes, the level of severity of health processes or states, and underlines the necessity of contextualisation for each individual. This way of understanding the disease-risk distinction does not solve all issues concerning overdiagnosis and overmedicalisation, but it shifts our understanding of them in such a way that it can help untangle some of them. Consequently, the confusion between the categories of risk and disease is explained by the ambiguity surrounding how the risk approach is considered: a *modulation or adaptation* of the traditional and dominant binary normal-pathological approach or a genuine *alternative* modelling of health phenomena.

13.2   Explaining the Blurring of the Disease-Risk Distinction

While we are dealing *a priori* with two quite distinct categories, separated by the same line as that which demarcates the possible and the real or the probable and the certain, many situations lead us to confuse the risk of disease and the disease itself. This first section aims to explain this confusion and its multiple origins.

---

[2] By 'health phenomena' I mean both those states that physiology characterises as healthy and functional and that pathology characterises as pathological. 'Health' then has a broad and inclusive meaning. I use it as a generic expression to be distinguished from the meaning retained in the normal-pathological dichotomy.

[3] Claude Bernard had pointed out the 'quantitative' continuity of the normal and the pathological in pathophysiology, leaning on his discovery of the continuity of the normal and pathological glycaemia level in diabetics. I discussed elsewhere the difference between this continuity for which Bernard argues and that which is observed in epidemiology and its risk approach (Giroux 2010). Even if there could be a quantitative continuity, the normal-pathological binarity is still central in Bernard's work and more generally in the pathophysiological way of modelling health phenomena, as rightly shown by Boorse's conceptual analysis of the pathophysiological concept of disease. Indeed, to Boorse, the normal-pathological distinction is the core foundation of modern medicine, which is based on physiology (Boorse 1977).

13.2.1 The Plasticity of the Concepts of Risk and Risk Factor

A first source of confusion can be found in a form of indeterminacy characterising the concepts of 'risk' and 'risk factor'. Their extension is potentially very broad and, in any case, broader than that of the concept of disease.[4]

To begin with, let's have a look at the definition of 'risk'. The term commonly refers to situations in which it is possible but not certain that some adverse or unwanted event will occur. But is there a more precise definition that we might agree upon? As soon as we refine the definition, a certain equivocation appears. Sven Ove Hansson (2022) proposes a series of different definitions, including the following three:

1. risk = an unwanted *event* that may or may not occur
2. risk = the *cause* of an unwanted event that may or may not occur.
3. risk = the *probability* of an unwanted event that may or may not occur.

In the first two definitions, there is an ambiguity regarding whether a risk is an *event* or the *cause* of an event, i.e., for our purposes, an unwanted *disease-event* or the *cause* of such an event. The following Sect. 13.2.2. will return to this ambiguity in exploring the more general blurring of the demarcation between the cause of a disease and the disease itself. Despite important nuances, these definitions have in common that they all refer to a certain category of events (*'unwanted'*) and to the *possibility* of this type of event. Definitions (3) make this *possibility* a defining characteristic of risk itself. It should be said here that a *disease* is precisely, and in an emblematic way, an unwanted event, but that disease is itself also the cause of another unwanted event often used in the calculus of the statistical expectation value of an unwanted event: *death*. Thus, the concept of risk can include disease as an unwanted event but also as a constitutive cause of another unwanted event, such as death.

As for 'risk factor', the concept is just as ambiguous and, in any case, very broad and heterogeneous in its extension. It is seemingly sufficient to observe a statistical correlation with a poor health outcome. Its link to the notion of causality is not always clear. It can refer to variables that are simple predictive markers of the effect (but correlation is not causation) or to variables that have a causal role in the effect (hypertension in cardiovascular disease for example). The notion is generally used to refer to any attribute, characteristic or exposure of a subject that increases the likelihood of a decrease in health or of suffering a disease. Among important recognised health risk factors one finds for example, being under- and over-weight, having unprotected sex, high blood pressure, tobacco or alcohol consumption, unsafe drinking water, and inadequate hygiene or sanitation. The notion is then used in relation to very different kinds of things, such as a behaviour, a pathology, a lifestyle, a specific environment (work conditions and exposure, etc.), a

---

[4]The concept of disease is also plurivocal and its definition is highly debated, but the plasticity of the concept of risk seems stronger and, above all, its extension much wider.

physiological variable (hypertension, hypercholesterolemia, etc.), or a sociodemographic characteristic (profession, marital status, income, etc.).

Thus, and this adds to the confusion, a *disease* can be *a risk factor* for *another disease*—this is the case for diabetes and cardiovascular diseases—and it is very often a risk factor for *death*. Hypertension (HTA), which is emblematic of the emergence of the concept of 'risk factor' itself, notably in the context of the pioneering Framingham Heart Study (see Giroux 2006, 2013) is, as we shall see later, a particular source of ambiguity, because it can also be considered a pathology in its own right, at least in certain contexts and at certain levels of severity. This ambiguity is present in the first cohort studies of cardiovascular epidemiology: there is a certain hesitation about whether to study HTA as an exposure (explicative variable) or as an outcome (dependent variable), in other words, as a risk factor for cardiovascular disease or as a cardiovascular disease whose risk factors should be studied.

We may conclude that, on a purely conceptual level, the categories of risk and risk factor and their extension remain largely underdetermined. François Ewald wrote in his book on the welfare state: "Nothing is in itself a risk, there is no risk in reality; conversely, everything can be a risk; it all depends on how one analyses the danger, considers the event" (my translation, Ewald 2014: 173).

13.2.2   About the Distinction Between Cause(s) of Disease and Disease

A second source of confusion between risk and disease concerns the definition of risk as the cause of the unwanted event *and* the unwanted event itself, and, more generally, the relative porosity between a particular (and selected) cause which serves to define the disease *and* the disease itself. This porosity stems from the fact that the (or a) cause of disease can, and does, ideally serve to define the disease when a specific and necessary cause can be identified and thereby allows for an *aetiological definition* of the disease. Indeed, a prevalent ideal—even if far from being universally accepted—in the definition and classification of diseases is that causation must serve as the main or privileged criterion. The anatomical criterion is often insufficient because several organs can be affected and the phenomenological criterion consisting in describing the symptoms lacks specificity. The cause has the advantage of being potentially more specific and of guiding prognosis and treatment more efficiently (Clarke 2011). In the case of tuberculosis (TB), for example, the disease is identified from its necessary cause: Koch's bacillus. The presence of this bacillus is not sufficient for TB to exist—there are healthy carriers—, but it is necessary for its identification and diagnosis. It is this bacillus that *defines* the disease. This is what is usually called the 'monocausal model' of disease.

A first objection to this alleged difficult distinction between the *cause* of the disease and the *disease* itself could be that identifying the (or a) particular cause that *defines* the disease does not necessarily imply that the cause of the disease and the disease are one and the same thing. The disease is the manifestation of the effects of the cause and not solely the particular cause chosen for its definition. Koch's bacillus is a necessary but not sufficient condition for TB. As we will see in the next Sect. 13.3.1., the philosopher Christopher Boorse insists on the difference between the *pathologic* and the *pathogenic* to argue for the risk-disease distinction. However, besides the fact that this distinction is not so easy to draw (Giroux 2017), it is sufficient for our purposes here to point out a certain porosity and conceptual slippage in the importance given to the identification of a necessary cause of the disease and the characterisation or definition of the disease itself.

A second, more important objection is that this type of slippage or confusion does not concern a great many (even most) diseases, for which there are no necessary causes. Indeed, the monocausal model and the aetiological—or pathogenic—definition of disease are valid within a deterministic causal framework if at least one necessary cause of disease can be identified. However, many chronic diseases are complex and multifactorial in their aetiology and it is rarely possible to identify a particular cause that is more determinant and specific than others, such that it could have a special status in defining the disease entity. In this multifactorial context, the statistical and probabilistic methodology of the modern analytical epidemiology that emerged during the 1950–1960s has proved to be particularly useful to identify what was then called 'risk factors' for disease. The 'risk approach to disease' (Aronowitz 1998; Giroux 2006, 2015) has renewed and enriched ways of understanding disease causation in opening the path to a probabilistic interpretation. It has proved to be particularly well suited to studying the complex aetiology of chronic diseases such as cancers and cardiovascular diseases, allowing for a multiplicity of causes of the same disease to be taken into account in a multifactorial model. But as a consequence, the definition of disease on the basis of its aetiology became particularly elusive, if not impossible; this is one of the reasons for the criticism of the 'risk factor approach': its inherent multifactorialism (Broadbent 2013: chapter 10). It no longer seems possible to classify and define diseases by their aetiology insofar as a risk factor is only a partial and probable cause and often intervenes in different types of pathologies. A risk factor thus lacks specificity to contribute to the definition of a disease.

Does this particularity of risk factors lead to a clarification and reinforcement of the difference between the *cause* of disease and the *disease* and thus constitute a relevant and sufficient argument for the distinction between risk factor and disease?[5] This does not seem to be the case for two main reasons. Firstly, the difference between the monocausal and the multifactorial model of disease should be nuanced (Stegenga 2018). On the one hand, any cause of disease is only a partial cause. Even in the context of the monocausal model, we are dealing with the necessary presence of conditions for the necessary cause to produce the expected effect. This means that, in a way, any disease is in fact multifactorial. For example, exposure to someone

---

[5]This is what I have first defended in Giroux 2017.

who has TB does not necessarily result in infection with TB; other conditions, such as proximity to an infected person, frequency and duration of exposure, ventilation, nutritional status as well as immune status also play a role in determining whether active TB will develop. On the other hand, it is not obvious that the multifactorial model of disease aetiology is incompatible with a deterministic framework which is inherent to the monocausal model. Indeed, a deterministic interpretation is still possible, like in the "sufficient-component cause model" defended in epidemiology by Kenneth Rothman (1976) and inspired by the model of the Insufficient, but Necessary part of an Unnecessary but Sufficient (INUS) condition elaborated by the philosopher John Mackie (1965). In this model of causation, a risk factor is neither sufficient nor necessary on its own, but it becomes a necessary component cause if a certain set of factors are combined that, all together, are sufficient for the effect. A 'sufficient cause' is defined here as a complete causal mechanism that inevitably produces disease. It is not a single factor, but a minimum set of factors and circumstances that, if all present for a given individual, will necessarily produce the disease. This modelling maintains necessity and sufficiency and a relative specificity within a multifactorial scheme, and a close link between a *set of causes* of disease and *disease* definition.[6] Such a model could then explain why a person who possesses a large number of the known risk factors for the same disease is classified as being at 'high risk' of disease and tends to be considered and to consider herself/himself as having a disease.

Secondly, even if one were to admit that there is a clear difference between these two models of disease causation, arguing for a difference between risk factor and disease—based on the indeterministic and probabilistic nature of causal risk factors in the aetiology of chronic diseases—can backfire. Indeed, by disassociating the definition of disease from its causes and assuming the non-specificity of risk factors, one is led to make the risk factor a *clinical problem* in itself, independently of the multiple diseases to which it is linked, and it is thus easy to assimilate it to a disease. This non-specificity ultimately becomes another source of confusion between risk factor and disease. Indeed, it is for risk factors with very low specificity that confusion with the concept of disease seems to be strongest. Obesity, for example, is identified as a risk factor for a very large number of diseases: diabetes, metabolic diseases, cardiovascular diseases, hypertension, respiratory problems, joint diseases and problems, cancers, etc. This probably explains why it is increasingly considered as a disease in its own right, although essentially on the basis of the epidemiological criterion of excess risk of morbidity and mortality.[7]

---

[6]Broadbent (2013) argues for a similar but different solution to the problem of 'multifactorialism' discussed above and the question of how causal risk factors can still contribute to the definition of disease. His solution does not necessarily require the adoption of a deterministic model as in the INUS model; he proposes to focus the attention on the risk factors that enter into a contrastive model of disease, i.e. this model amounts to considering that the risk factors relevant to defining a disease will be those (the particular combination of them) whose effect is greatest.

[7]For a critical analysis of the pathological status of obesity, see Hofmann (2016a, b).

13.2.3   Probabilistic Modelling of Chronic Diseases

A third source of confusion between risk and disease is a certain convergence of the experience of being at risk of disease and the experience of disease itself for patients with chronic diseases (Aronowitz 2009). Indeed, the probabilistic approach in terms of risk has gained significant weight in the field of chronic diseases care. The evolution of knowledge and interventions, in particular at the molecular level, has modified the understanding of the natural history of diseases such as cancers, which are increasingly modelled probabilistically as 'at risk conditions' and very long processes (see Boenink and van der Molen, Chap. 11, this volume). It becomes difficult to discern the precise onset of the disease in the continuous process involved. The blurring of the demarcation is not only at its onset but also at its end. There is also an indeterminacy of the prognosis. For patients who have contracted cancer and have been treated, the risk of recurrence is sometimes, and even often, like a sword of Damocles hanging over their heads. Rather than talking about a 'cure', we adopt the probabilistic notion of 'remission'. The risk approach is no longer limited to prevention and screening in the early and asymptomatic stages. It is also used to assess the severity of the health situation and the prognosis along a continuous spectrum of health levels.

The importance taken by the risk approach can be explained by the fact that its gradualist nature is more suitable than a categorical approach for chronic diseases that develop slowly and incrementally. For diseases that are ontologically better considered as *processes* rather than *states or events*, identifying their onset, the transition from risk to disease, and also their termination, is often particularly difficult. The threshold between an abnormal cell architecture and a cancerous stage is hard to determine. What we choose to define as 'cancer' is generally a judgment on the stage of the carcinogenic process that we think is severe enough. Because of this processual nature and the difficulty of delimiting the beginning and the end, cancer should be considered as a "probabilistic process" (Reid 2017).

In this section, I have analysed and made more explicit the sources of confusion between the categories of risk and disease, despite a seemingly simple and obvious distinction between the real and the probable, a state and a disposition, a cause and its effect. A blur or confusion exists that can be partly explained on semantic, epistemological, and ontological grounds, but this partial explanation does not help us counter the problem of pathologisation and overdiagnosis.

13.3   Limits of the Functionalist Conceptual Analysis of Disease for the Risk-Disease Distinction

Conceptual analysis in its classical style aims at identifying essentialist (necessary *and* sufficient) characteristics for the definition of a concept, thus regulating and clarifying its use. Could the definition of the concept of disease determine whether 'risk factors' are to be included or excluded from this category? In what follows, I discuss whether the naturalist conceptual analysis, with its dysfunction-required account of disease, succeeds in delivering a criterion for the risk-disease distinction.

13.3.1 The Functionalist Criterion for the Risk-Disease Distinction

In the philosophical debate on health concepts that opposes naturalists and normativists, it is mainly the naturalist camp that defends the importance and possibility of a distinction between risk and disease.[8] Christopher Boorse is the main advocate of the naturalist camp. His Bio-Statistical Theory (BST) is based on the criteria of biological dysfunction and statistical subnormality. He defines disease as a "type of internal state which is either an impairment of normal functional ability, i.e. reduction of one or more functional abilities below typical efficiency, or a limitation on functional ability caused by environmental agents" (Boorse 1997: 7–8). In his seminal articles, Boorse argued for the importance of distinguishing the *pathological* from that which tends to produce it, the *pathogenic* (1977). This leads him to introduce the notion of "instrumental health"—concerned with the pathogenic—and to distinguish it from "intrinsic health". There are variations in instrumental health and different levels of risk, but the demarcation between intrinsic health and disease remains categorical and the normal-pathological dichotomy, according to him, is the fundamental concept of occidental physiological medicine. However, in this presentation of his theory, Boorse does not take a clear and explicit position on clinical risk factors such as hypertension, or on what he has more recently called "protodiseases" (Boorse 2011: 18).

In his article "Risk and Disease", Peter Schwartz (2008) directly addresses the problem of this distinction. His definition of disease is inspired by Boorse's BST. For him, it is a strength of the naturalist functionalist definition that it allows for a clear distinction between risk and disease, and thus between prevention and treatment. He shows that risk factors or what he calls "risk-based diseases" such as hypertension, hypercholesterolaemia, diabetes, osteoporosis and obesity should not be considered diseases because these conditions are not biological dysfunctions when they are not associated with complications.

Thus, for example, *stage 1 hypertension*, diagnosed for an asymptomatic but slightly raised blood pressure of 140–159 mmHg (systolic) and of 90–99 mmHg (diastolic)—Schwartz is relying here on the definition of the Recommendations of 2003 by the JNC 7 (Chobanian et al. 2003) –, is not a dysfunction because there are no identifiable pathological changes in the system that regulates blood pressure.

---

[8]Indeed, if there is any mention of the notions of disease risk or risk factor in the normativist literature, it is more in the direction of including it in the category of the pathological. See Giroux (2017).

Furthermore, at this level of blood pressure, the organs perform their typical functions with adequate efficiency: blood flow is ensured without the pressure impeding blood distribution or inducing immediate damage. Schwartz (2008) argues that stage 1 hypertension should then be considered a *risk factor* and not a disease, but should still be treated, because such preventive treatment lowers the risk of future cardiovascular and other diseases. Nevertheless, this should count as the prevention of disease, not its treatment. Higher levels of blood pressure such as levels ≥ 160/100 mmHg, i.e. the *stage 2 hypertension* and more elevated levels of blood pressure for the JNC 7, might produce symptoms and functional complications—for example headache and blurred vision, or damage the brain or the kidneys—which would count as dysfunction and then as current pathological conditions.

However, the distinction is not as clear as Schwartz seems to think. Indeed, firstly, *stage 2 hypertension* need not necessarily damage organs or produce symptoms either. Secondly, the threshold for and then the definition of stage 2 hypertension has changed since the JNC 7. In the new guidelines of the American College of Cardiology and American Heart Association of 2017 (Whelton et al. 2018), this threshold is lower: blood pressure ≥ 150/90 mmHg. This means that, in order to distinguish between hypertension *as a risk factor* or hypertension *as a disease* we cannot any more rely on the categories defined by current medical guidelines. The demarcation should be drawn inside the category of *stage 2 hypertension*. A third and deeper objection could be raised to Schwartz's approach of the distinction: is it so obvious that *stage 1 hypertension* is not dysfunctional? Can we not consider that the function of the blood pressure system is also to maintain blood pressure at a level that minimises the probability or risk of heart attack or stroke? As Schwartz himself points out, to consider that a system with stage 1 hypertension is functional is to exclude from the outset the idea that protecting against cardiovascular risk may be a function of that system. This may seem rather costly. In fact, both Schwartz (2008) and, recently, Boorse (2023) recognize that reducing the risk of cardiovascular mortality may be a function of the blood pressure system.

13.3.2   Risk Level Is Used to Determine the Threshold for Disease

In a later paper (2014), Schwartz deals with the same problem of the risk-disease (and normal-pathological) distinction, but in the context of increasingly early detection of pre-diseases, such as small asymptomatic tumours and in particular, ductal carcinoma in situ (DCIS), the most frequent non-invasive form of breast cancer in women.[9] He acknowledges that it is very difficult to determine physiologically the

---

[9]In DCIS, the cancer cells are only present in the lining of the breast duct. They do not spread outside the duct to neighbouring breast tissue or to other organs. DCIS is too small to be felt, and is often detected only during screening mammography.

dysfunction threshold, and thus to distinguish disease from risk-based conditions. But what interests us most here is that in order to defend the demarcation between the functional-normal and the dysfunctional-pathological, Schwartz appeals more specifically to the criterion of the severity of unwanted effects, or what he calls "negative consequences". This criterion introduces an assessment of the status of a condition based on its effects, which is characteristic of the statistical and probabilistic approach of modern analytical epidemiology (Giroux 2015). According to Schwartz, these negative consequences do not necessarily have to be already present but can occur later and thus are very close to what is embraced by the notion of risk. The term 'risk' is explicitly used here to specify the criterion of negative consequences which "have to be understood in terms of the risk that the lesion will progress, spread, and cause morbidity and mortality" (Schwartz 2014: 994). If there are no negative consequences, a level of functioning will not be deemed pathological, even if it is atypical. Thus, it is better to consider for the risk-disease distinction that stage 1 hypertension is not pathological *because* it is quite typical *and because* the risk of negative consequences is low. DCIS is not pathological *because* the risk of negative consequences is low, even if it is atypical. But it appears here that there is a paradox involved in including the notion of risk as a defining criterion of the pathological, while the objective is to distinguish risk and disease as two exclusive categories. This enables us to see the limits of binary (normal-pathological) and functionalist models of health phenomena when we apply them to chronic and probabilistic processual conditions.

Before closing this third section, I wish to mention briefly a completely different way for conceptual analysis to solve this problem of the risk-disease distinction: renouncing the essentialist way of defining a concept. Following Kazem Sadegh-Zadeh (2008), the concept of disease can be considered non-classical and non-essentialist. Category membership thus becomes gradual in nature. There is room for vagueness at the boundaries of the concept of disease, and of some specific diseases (see also the "vague cluster approach" defended by Mary Walker and Wendy Rogers, Chap. 15, in this volume). The risk factor could then be considered as simply part of this blurred area between the normal and the pathological (see also for example De Vreese 2017). But it seems that this other way of resolving the problem remains based on a framework that prevents us from accurately apprehending what the risk approach brings to our way of modelling health phenomena. We remain stuck in the pathophysiological normal-pathological binary model into which we try to integrate the gradualist risk approach. In the next and final section, I propose another solution for the risk-disease distinction which considers the risk approach as an alternative and complementary way of modelling health phenomena.

13.4 Risk Beyond the Normal-Pathological Dichotomy: An Alternative Gradualist Approach of Health

13.4.1 The Epidemiological Risk Approach as an Alternative Way of Modelling Health Phenomena

The terminological and conceptual analysis of "disease" have failed to take into account the historical context of the emergence of the risk factor concept and the epistemological particularities of the probabilistic approach to health phenomena associated with it. This approach was developed within modern analytic epidemiology in order to gain a better understanding of the natural history of certain chronic diseases and to overcome the limits that were then facing the pathophysiological and monocausal model of disease. Indeed, in the 1930s and 1940s, while cardiovascular diseases and cancers were becoming prevalent, very little was known in pathophysiology about their pathogenesis and there were no therapeutic solutions. Above all, as noted above, these chronic diseases were slow, silent, and progressive in their development and their aetiology was complex and multifactorial. The case of myocardial infarction is emblematic. Doctors did not see patients until it was too late to act. It was in this context that the risk approach, based on aetiological research using the probabilistic and statistical tools of the then emergent analytic epidemiology and its multifactorial model, appeared as ways of breaking this deadlock, whilst research into the pathophysiology of those diseases continued.

In the 1950–1960s, new designs of analytical studies, now called 'cohort studies' and 'case-control studies', were developed and gave epidemiology and the probabilistic approach a new role in research on disease causation. Risk correlations of exposure with health outcomes such as negative consequences (e.g. morbidity and mortality rates) at the population level have led to the identification of difference-making factors that have since been called 'risk factors'. The Framingham Heart Study, a cardiovascular U.S.A. cohort study, is considered to have played a major role in framing the 'risk factor approach' of the emergent analytical epidemiology (Aronowitz 1998; Oppenheimer 2005; Giroux 2006, 2008a). It is notable that epidemiology relied and still relies on existing knowledge from pathophysiology and nosology. However, its way of knowing is based on a very distinct framework and leads to a different way of modelling health phenomena.

First, and as mentioned in the Sect. 13.2, the 'risk factor approach' deploys a different way of considering causality than the experimental, deterministic and mechanistic framework of pathophysiology and bacteriology, which mainly relies on a monocausal and proximal model of disease causation. Epidemiological causal analysis is based on a relative and comparative method which allows for the identification of difference-making factors (see Giroux 2008b; Russo 2009). It bases its generalisations on the observation and comparison of populations. While pathophysiology is based on a binary and experimental type of thinking, relying on the definition of a referential or absolute normal state of functioning for the organism, epidemiology is, on the other hand, based on a statistical and probabilistic type of thinking, intrinsically comparativist and gradualist.[10] This alternative way of dealing with disease causation is generally seen as complementary to the mechanistic approach.[11]

A second main difference appears here: while pathophysiology starts from the description of the *normal* functioning as absolute category of the human organism in a reference class, so as to then identify the pathological as a deviation from this norm—the so-called "broken-normal view" (Moghaddam-Taaheri 2011)—, the epidemiological approach starts historically and logically from an observation of *negative consequences* or contrasting levels of health outcome (not necessarily "pathologies") in order to identify, through the use of statistical and probabilistic tools, the factors that make a difference in the frequency of those health outcomes. Further, by starting with health outcomes as a means for establishing correlations and then identifying contrasting levels of these health outcomes, the epidemiological risk approach puts central emphasis on prognosis in its modelling of health phenomena. It does not necessarily require a normal-pathological distinction or a diagnosis as a point of departure; and, as is particularly visible in the "epidemiology of health", it can establish a gradation in the level of health outcomes which will then inform the clinical judgment and the treatment decision.[12] The following table proposes a synthesis on the differences between the two main ways of modelling health phenomena (Table 13.1).

13.4.2   Neither Normal Nor Pathological

In this context of a comparativist and gradualist approach to health, what sense does it make to maintain a binary normal-pathological and risk-disease framework? Its

---

[10] In another paper I proposed to characterise this difference between pathophysiology and epidemiology using Ernst Mayr's distinction between *typological* (or essentialist) and *population* types of thinking. This distinction was introduced by Mayr to characterise the evolution of knowledge in biology due to the passage from, respectively, an Aristotelian biology to a Darwinian evolutionist biology (Giroux 2008b).

[11] Risk correlations (difference-making factors) allow for the identification of the *existence* of a causal link without necessarily determining the *nature* (mechanism) of the link. In the so-called Russo and Williamson Thesis (2007), these *two aspects of causation* are considered to be necessary if we are to infer causation in medicine. They are associated with *two types of evidence*, the establishment of a continuous chain of necessary events and the establishment of a statistical difference in frequency, which are themselves more or less reciprocally produced by the two disciplines of pathophysiology and epidemiology.

[12] The importance assumed by diseases and medical nosology in epidemiology could be seen as a consequence of the weight that the medical and pathophysiological model has in Western medicine, as well as on the pragmatic advantages of relying on measurements of mortality and morbidity, rather than of levels of health. But in the context of the emergence of modern epidemiology in the U.S.A. there was a place for an "epidemiology of health" in which a gradualist vision of health in all its range was central (see Galdston 1953).

Table 13.1 Two alternative ways of modelling health phenomena

	Pathophysiological framework	Epidemiological risk framework
Health concept	Fundamentally binary (the normal-pathological distinction is central)	Fundamentally gradualist
Theory of health	Fundamentally non-comparative	Fundamentally comparative
Causal approach and type of explanation	Mechanistic	Difference-making and probabilistic
Type of thinking	Typological thinking	Populational thinking
Clinical goal	Improving the current symptoms (narrow focus on disease as the determinant of the outcome)	Improving outcomes for patients in their total biological, social and psychological environment
Clinical approach	Diagnostic	Prognostic

continued existence could be explained by the need for a dichotomy to ground clinical decision-making, but also, more generally, by the human psychological preference for categorical and binary thinking. But it could also be asked whether sticking to this binary framework, in the context of some chronic conditions, is not a misguided attempt to integrate two different ways of modelling health phenomena that ought rather to be kept separate and seen as complementary.

The reflection on the pathological status of hypertension by a pioneering cardiologist in hypertensiology, Sir George Pickering, in the 1960–70s, is eloquent. Faced with the problem of the status of essential hypertension, a type of blood pressure with no specific cause, Pickering defended the idea that the distinction between normal and pathological blood pressure is as fallacious as the phlogiston theory. Part of Pickering's argument is that there is no sufficiently unified causal pattern for essential hypertension to be called a disease. But it is not normal either, because it correlates with an increased risk of cardiovascular morbidity and mortality. For Pickering, such a condition constitutes what he initially called a "quantitative disease" (Pickering 1961). But in the end he decided to move away from the disease-health or normal-pathological dichotomy and vocabulary. According to him, classifying blood pressure into one or the other is harmful. In a paper entitled "Normotension and hypertension: the mysterious viability of the false" (1978: 561–562), he wrote:

> They (the average doctor) persist in treating arterial pressure as a quality and dividing it into two normotension and hypertension, physiologic and pathologic, healthy and diseased, good and bad. (. . .) It seems that unless these terms are used, new facts cannot find a place in the mind of the contemporary doctor. (. . .) That the current practice of treating a quantity as a quality should have arisen in the first place is not difficult to understand. What never ceases to astonish me is that it should persist in the face of developments in science and the growth of established fact. (. . .) The doctor will tell you that he has to decide whether or not his patient has hypertension before he can diagnose and treat him, and he must have a dividing line to assist him in this. I have never had any doubt that this practice does untold harm to the patient. In the first place, the patient becomes identified with the "grim label" hypertension. It has been shown (. . .) that the moment when this label is attached often marks the beginning

of symptoms, such as headache, that were absent before. (. . .) The second danger to the patient that arises from this practice is the initiation of treatment whose beneficial effects are conjectural and in which the drugs used may prove harmful in the long term.

In 1961, Pickering had also already written: "I often think that the greatest contribution to the sanitation of the mind would be the abolition of the terms normal and physiological and their opposites" (1961: 131). It is by taking into account another type of information and many other individual variables that one should determine the severity of the prognosis for the individual and the interest of treating or not treating them, without necessarily going through a previously determined diagnostic category.

The notion of 'risk factor' has meanwhile become established in the medical literature to designate conditions which are continuously distributed in the population, such as hypertension, and which should be integrated within a multivariable approach. In this alternative epidemiological risk approach, the question of whether or not a given risk factor is normal or pathological becomes nonsensical. And it is no longer relevant even for medical decision-making. The relevant question becomes: to what extent does this level of variable, or rather this level of this *set* of variables in *this* individual, increase the risk of negative consequences, and at what level of risk would the patient actually benefit from treatment? In this multifactorial and probabilistic context, the level of an isolated risk variable has no relevant clinical meaning. For certain conditions, some authors go so far as to defend, not without echoing George Pickering's words, that we ought to "move beyond the binary, diagnostic thinking that has dominated medicine for so long and embrace a quantitative approach" to patient management (Vickers et al. 2008: 203). What I have just characterised as the epidemiological modelling of health phenomena, an alternative to the pathophysiological modelling, is in adequation with the "risk prediction approach" (Vickers et al. 2008) and the "patient prognostic approach" (Croft et al. 2015) which are considered preferable, for certain conditions, to the traditional "diagnostic approach" as a framework for clinical decision-making. Vickers and his co-writers (2008: 201) described the two approaches in the following way: "The diagnostic approach to blood pressure is to divide the population into 2 groups, those with hypertension and those without hypertension, and then to treat one group but not the other. The prediction alternative is to use a statistical model to estimate the probability that a patient will have a clinically important event, such as a myocardial infarction, within a certain period, such as 10 years".

13.4.3   Advantages of the Risk Approach and a Gradualist Concept of Health

I now address the advantages and limits of this alternative way of modelling health phenomena that implies the adoption of a gradualist and comparativist theory of health.

First, and as noted earlier in the Sect. 13.2.3, the continuous model of health appears to be more adapted to probabilistic and processual conditions whose beginning and end are uncertain and hard to establish. According to Vickers et al. (2008: 200), a large set of diseases of particular concern for industrialised countries are a matter of degree, reflecting a range of severity. They mention: cardiovascular disease, type 2 diabetes, obesity, depression, developmental disorders (such as autism and hyperactivity), back pain, arthritis, and cancer. To them "even arteriosclerosis is a matter of degree, because most adults have at least some level of endothelial dysfunction". Categorising patients as either having or not having the condition depends on choosing a somewhat arbitrary cut-point of severity. According to Croft et al. (2015: 4–5), "the underlying 'disease' is often a continuous distribution of probability for future health states. Diagnosis is then not 'have you got it?' but 'how much of it have you got?'" The point here is to consider those variables like blood pressure, blood glucose, etc. as sources of information about the probability of future events, which provide a quantitative estimate of individual risk for particular outcomes, rather than to artificially dichotomise them and then treat only those states characterised as diseases (Vickers et al. 2008).

Second, this approach has the potential for being more precise and more individualised and thus for improving clinical judgment. In avoiding arbitrary and artificial dichotomisation of continuous variables like blood pressure, it avoids the consequent loss of relevant information for prognosis. Moreover, as said above, in the context of these especially complex and multifactorial diseases, it is always a set of variables which has to be taken into account. It is with the presence and level of other variables (cholesterol, diabetes, age, sex, and smoking history) that a specific variable (blood pressure) takes on clinical meaning. Both of these aspects were precisely addressed by the multivariate equation introduced by the statistician of the National Institutes of Health, Jerome Cornfield, to deal with the analysis of data from the Framingham Heart Study. The multivariate analysis has been crucial for the development of the risk factor concept and approach (Giroux 2013). The equation was first used for causal analysis and was then rapidly extended to clinical prediction. Then, by combining different risk factors for a single individual into a statistical model, the relevant threshold for management of risk factors is further individualised for each person (Giroux 2012). Furthermore, according to Vickers et al. (2008), in the prognostic approach more information relevant for patient outcomes could be integrated into the model: "Modelling an individual's prognosis can draw on the full range of relevant and available information, both clinical and non-clinical." Patient preference, for example, could also be better taken into account. Indeed, a prediction model provides probabilities of events and a patient can weigh these according to his or her preferences.

Thirdly, in giving priority to the optimisation of outcomes, the risk/prognostic approach could help identify overdiagnosis. Overdiagnosis arises when a pathological lesion or state is identified, and the patient is defined as having a disease, in the absence of any evidence that this state leads to a difference in health outcome, and without necessarily making it possible to undertake an investigation or administer a treatment that clearly advantages the patient. As stated by Croft et al. (2015: 3):

Overdiagnosis flourishes in the vacuum created by a culture of 'underprognosis', i.e., lack of critical enquiry, information, or evidence about the likely future benefits or harms of identifying a condition as an abnormal disease state. A prognostic framework for clinical practice would help to resist evidence-free diagnostic novelty. Prognostic evidence highlights when overenthusiastic search for pathology leads to irrelevant treatments and needless anxiety, such as disc anomalies on MRI of the spine, but can reassure people who need neither active intervention nor a diagnosis, and identify those in whom diagnosis does guide decisions that improve outcomes.

The conviction here is that the risk/prognostic approach, in integrating wider and more diverse information in the evaluation of outcomes for a particular individual, allows for greater precision and may avoid false positives and the negative impact of overdiagnosis and overtreatment.

Fourthly, and lastly, the gradualist and comparativist concept of health at the core of the risk/prognostic approach has several conceptual and practical advantages. Andrew Schroeder (2013) has highlighted these advantages in his discussion about the relative superiority of this type of theory of health with respect to traditional fundamentally non-comparative theories like the BST. *Fundamentally non-comparative* theories, in which the comparative form of a concept is possible but considered less basic than the non-comparative form, define health in such a way that a significant number of people living today are, in fact, healthy. This is realistic. In such theories of health, we "should begin by defining a state that a single organism might be in" (Schroeder 2013: 134). In *fundamentally comparative* theories, as presupposed by the epidemiological approach, to be healthy is to be healthier than a sufficient number of people in some comparison class. The "theorist should begin by defining a relation, presumably between organisms or states of organisms" (2013: 134). According to Schroeder, comparativist theories of health can better deal with the problem of arbitrariness regarding the precise demarcation line between the subnormal and normal statistical levels of functional efficiency in the BST, and it also avoids the "problem of common disease", since ill health and differences in health levels in general are always relative to particular populations (see also Giroux 2015). Another important conceptual advantage is that, while in both the broken-normal view and the diagnosis approach, it is essential to give an absolute and decontextualised definition of health at the beginning of any investigation into health phenomena, in the risk comparative approach such a definition—which remains elusive in the philosophical debates—is not necessary.

The choice between a fundamentally comparative and gradualist concept of health at one side and a fundamentally non-comparative and binary one at the other is of practical significance: it has important consequences for clinical and public health strategies. In individual care, adopting a continuous framework for thinking about health care, in the case of a person having a risk factor such as hypertension, avoid the potential harm involved in labelling them as diseased, while drawing attention to multiple other variables that could be useful for delivering a precise prognostic. It also points towards the possibility of delivering more information on the risk-benefit balance of an intervention and on individual context to decide with the patient whether an intervention is required and, if so, what form it might take. In conflating this framework with the binary one, the danger arises of assimilating hypertension or a high-risk status to a disease and then automatically opting for medical treatment without taking prognostic information into account. Indeed, even if pathologisation should not be assimilated to medicalisation, as rightly shown by Boorse (1977) and Sholl (2017), designating a condition as a pathology leads—at least in our ordinary representations of a pathology—to consider it as requiring medical care and as legitimising such care. In a continuous framework, the decision should always be made in an individual and particular context.

Concerning public health and population health more generally, Schroeder (2013) thinks that one of the main practical benefits of fundamentally comparative theories is that they facilitate intergenerational assessments of health that are difficult to achieve in the context of fundamentally non-comparative theories, but also, and most importantly, that they improve health metrics and have a positive impact on the distribution of health resources. Indeed, traditional health metrics rely on non-comparativist theories of health, and then mainly capture the bottom-range of health differences. Yet such a range is too narrow. Measuring the full range of possible health states is very useful and relevant for resource allocation. Indeed, differences in middle-range health can be of great moral importance. A treatment 'B' can do better than a treatment 'A' if we have a better understanding of the range of improvements it allows in terms of health. Moreover, malnutrition and parasitic infections cause small but significant intelligence loss in whole populations, but these are not picked up by traditional metrics that rely on non-comparative approaches (Schroeder 2013: 151). It should also be noted that comparativist and gradualist theories of health are favoured in the recent public health approach called "population health science (PHS)" (Keyes and Galea 2016) but also in the population strategy promoted by the cardiovascular epidemiologist Geoffrey Rose (1992), in which the PHS has its roots. Only such an approach can correct the limitations of what he called the "high risk strategy" of prevention which focuses on high-risk people and still relies on the dichotomous normal-pathological model. The "population strategy", which relies on a continuous and comparative concept of health, is different: the aim is to move the whole distribution curve of a trait or risk prognosis for an entire population.

13.4.4 Challenges for the Risk Approach and a Gradualist Concept of Health

The risk approach, along with the gradualist and comparativist concept of health, do however raise important challenges, especially when one considers that most of our modern healthcare systems and practices are based on a binary disease-diagnostic approach, which is itself deeply anchored in our representations of health phenomena.

A first important series of challenges concerns the validation and translation of prognostic and risk models into practice. A notable difficulty is to obtain clinical evidence of the relevance of certain prognostic markers and prediction models. These models should be evaluated using decision analysis methods to determine their effects in clinical terms (Vickers et al. 2008: 202). And even if good prediction models are available, there remain the problems of how to translate them into practice, as well as the tendency to still rely on categorisation (high-medium-low risk of a particular outcome) to drive stratified care: "prognostic classification for its own sake should not replace diagnosis for its own sake. Nor should individual patient prognosis be a static classification in time – it needs to be updated" (Croft et al. 2015: 5).

Secondly, if the risk prediction approach has the potential to help with the overdiagnosis challenge, it is still subject to issues associated with overmedicalisation. "The potential for a prognostic model of care to solve these problems has to be weighed against the possibility that such a model may create its own version of overmedicalisation" (Croft et al. 2015). Since pathologisation is not necessary for medicalisation, there is no certitude about the fact that escaping the vocabulary of disease will solve all the problems associated with the risk of "too much medicine" (Moynihan and Smith 2002). It could even be argued that having a spectrum of health makes all potential limits to medicine disappear.

Thirdly, another important issue concerns the public, clinical, and patient understanding of risk, probability, and non-categorical, or at least non-binary, thinking. The notion of risk is complex both to understand and communicate and, if its measurement, analysis, and presentations are not well understood by doctors and the public, it is easily open to misinterpretations (Gigerenzer 2003). To Croft et al. (2015) research about how best to inform and justify a prognostic model of clinical practice is crucial. Moreover, its proper understanding is made difficult by the weight of the binary normal-pathological approach that has long been dominant in, and structuring for, the development of modern scientific medicine, rooted in diagnosis and pathophysiology. Adding to this difficulty is the fact that the comparativist risk approach tends to be integrated into the fundamentally non-comparative normal-pathological one. Indeed, the pathophysiological and functionalist approach can incorporate risk information, as is illustrated by Schwartz's definition of disease that integrates the criterion of "negative consequences". Rothman's deterministic "sufficient-component cause model", Boorse's distinction between "intrinsic" and "instrumental" health and the "high-risk strategy" are other examples of the tentative integration of risk into the fundamentally non-comparative pathophysiological approach, without great awareness that two very different epistemological frameworks are actually in play. However, it is precisely by maintaining these approaches as distinct that potentially harmful amalgams can be avoided. It is only when we are aware of the existence of two different logics and frameworks that we will be able to maintain distinct meaning and use for the concepts of disease and risk factor.

It should be reminded here that, in any case, defending the relevance of the risk/prognostic approach does not imply that this must always be considered as the most useful approach in clinical practice. The non-comparativist binary (normal-pathological) approach is complementary and still considered relevant for acute health problems that are limited in time. The challenge then becomes to identify when each of those two models of health phenomena is more appropriate.

Thus, advocating for the distinction between risk and disease means defending the distinction between, and complementarity of, two models of health phenomena. This distinction has the advantage of highlighting the fact that the values and implications of a clinical decision are not the same, nor are the determinants of a decision. In a risk approach, value judgments and patient preferences are much more important and should therefore be made more explicit, precisely because of the absence of prior normal-pathology categories and also because the decision concerns a future instead a current harm. Likewise, it is also of greater importance to recognize the uncertainty and underdetermination of a condition as well as the corresponding course of treatment.

13.5   Conclusion

I began by making explicit various sources of confusion between the concepts of disease and risk. I then showed that the essentialist conceptual analysis of disease, especially the naturalistic dysfunction-requiring account, does not provide a satisfactory solution to this problem of the disease-risk demarcation and to the related issues concerning overpathologisation, overdiagnosis and overtreatment. Finally, by considering that analytical epidemiology participates in the construction of knowledge about health in an alternative manner to pathophysiology, I have shown that risk factors belong to a way of modelling health phenomena that relies on a different framework than that of the disease concept inherited from pathophysiology and based on a binary and fundamentally non-comparativist theory of health centred on diagnosis. The risk approach is anchored instead in a gradualist and fundamentally comparativist theory of health centred on prognosis. These two frameworks could be considered as two complementary medical approaches to dealing with health phenomena.

As it stands, however, the fundamentally non-comparativist approach remains dominant, even when trying to integrate the comparativist one. I have argued that we should insist instead on their distinction and consider that we have two different approaches to health phenomena. The fundamentally comparativist and gradualist approach is better suited to health problems of a chronic and progressive nature. From this perspective, the distinction between risk factor and disease should be based on the distinction between these two ways of modelling health phenomena. The confusion or slippage between the two may then be linked to the fact that, rather than seeing the gradualist risk approach as an alternative approach, it is seen as a kind of adaptation or appendix to the classical binary pathophysiological model, used in order to address its limitations in dealing with certain chronic diseases.

The distinction between the risk and disease frameworks is important because neither the criteria they focus on nor the issues they raise regarding clinical decision-making are the same. The risk/prognostic approach has the advantage of forcing us to be more explicit about the criteria leading to a decision about whether or not to treat (being more precise about the benefit of a treatment), while also lessening the risks of overdiagnosis and overpathologisation. However, it does not solve all the issues related to the extension of prevention or surveillance in healthcare or to overmedicalisation. A gradualist concept of health has its limits, just as the diagnostic and binary approach also has limits. The main challenge then becomes for the public, the patient, and clinicians to be well aware of the differences between these two frameworks and to identify when—or for which clinical condition—one or the other approach is most appropriate.

References

Aronowitz, Robert A. 1998. *Making sense of illness: Studies in twentieth century medical thought.* Cambridge: Cambridge University Press.
———. 2009. The converged experience of risk and disease. *The Milbank Quarterly* 87: 417–442. https://doi.org/10.1111/j.1468-0009.2009.00563.x.
Boorse, Christopher. 1977. Health as a theoretical concept. *Philosophy of Science* 44: 542–573.
———. 1997. A rebuttal on health. In *What is disease?* Biomedical Ethics Reviews, ed. James M. Humber and Robert F. Almeder, 1–134. Humana Press.
———. 2011. Concepts of health and disease. In *Philosophy of medicine*, ed. Fred Gifford, Dov Gabbay, Paul Thagard, and John Woods, vol. 16, 13–64. North Holland: Elsevier.
———. 2023. Boundaries of disease: Disease and risk. *Medical Research Archives*, (online) 11: 4. https://doi.org/10.18103/mra.v11i4.3599.
Broadbent, Alex. 2013. *Philosophy of epidemiology*. Palgrave Macmillan.
Chobanian, Aram V., George L. Bakris, Henry R. Black, William C. Cushman, Lee A. Green, Joseph L. Izzo Jr, Daniel W. Jones, Barry J. Materson, Suzanne Oparil, and Jackson T. Wright Jr. 2003. The seventh report of the joint national committee on prevention, detection, evaluation, and treatment of high blood pressure: The JNC 7 report. *Journal of the American Medical Association* 289: 2560–2571.
Clarke, Brendan. 2011. Causation and melanoma classification. *Theoretical Medicine and Bioethics* 32: 19–32.
Croft, Peter, Douglas G. Altman, Jonathan J. Deeks, Kate M. Dunn, Alastair D. Hay, Harry Hemingway, Linda LeResche, George Peat, Pablo Perel, and Steffen E. Petersen. 2015. The science of clinical practice: Disease diagnosis or patient prognosis? Evidence about "what is likely to happen" should shape clinical practice. *BMC Medicine* 13: 20.
De Vreese, Leen. 2017. How to proceed in the disease concept debate? A pragmatic approach. *The Journal of Medicine and Philosophy: A Forum for Bioethics and Philosophy of Medicine* 42: 424–446. https://doi.org/10.1093/jmp/jhx011.
Dumit, Joseph. 2012. *Drugs for life: How pharmaceutical companies define our health*. Durham: Duke University Press.
Ewald, François. 2014. *L'Etat providence*. Paris: Grasset.
Galdston, Iago. 1953. The epidemiology of health. In *The epidemiology of health*. New York Academy of Medicine.
Gigerenzer, Gerd. 2003. *Reckoning with risk: Learning to live with uncertainty*. London: Penguin.
Giroux, Élodie. 2006. *Épidémiologie des facteurs de risque: genèse d'une nouvelle approche de la maladie. Thèse de philosophie*. Paris: Université Paris 1 Panthéon Sorbonne.

———. 2008a. Enquête de cohorte et analyse multivariée: une analyse épistémologique et historique du rôle fondateur de l'étude de Framingham. *Revue d'Epidémiologie et de Santé Publique* 56: 177–188.

———. 2008b. L'épidémiologie entre population et individu: quelques clarifications à partir de la notion de la pensée populationnelle'. *Bulletin d'Histoire et d'Épistémologie des Sciences de la Vie* 1: 37–52.

———. 2010. Les facteurs de risque et le problème de la démarcation entre le normal et le pathologique: une analyse épistémologique. *La Revue de Médecine Interne* 31: 651–654.

———. 2012. Les modèles de risque en médecine. Quelles conséquences pour la définition des normes et pour le jugement clinique ? exemple du calcul du risque cardiovasculaire global. In *L'émergence de la médecine scientifique*, dir. Anne Fagot-Largeault, 199–215. Paris: Editions Matériologiques. https://doi.org/10.3917/edmat.fagot.2012.01.0199.

———. 2013. The Framingham study and the constitution of a restrictive concept of risk factor. *Social History of Medicine* 26: 94–112.

———. 2015. Epidemiology and the bio-statistical theory of disease: A challenging perspective. *Theoretical Medicine and Bioethics* 36: 175–195.

———. 2017. Risque de maladie: normal ou pathologique? In *Science, philosophie, société*, ed. Alexandre Guay and Stéphanie Ruphy. Presses Universitaires de Franche-Comté.

Greene, Jeremy. 2007. *Prescribing by numbers. Drugs and the definition of disease*. Baltimore: Johns Hopkins University Press.

Hansson, Sven Ove. 2022. Risk. In *The Standford Encyclopedia of Philosophy*, ed. Edward N. Zalta and Uri Nodelman. Metaphysics Research Lab, Standford University.

Hofmann, Bjørn. 2016a. Obesity: Its status as a disease. In *eLS*, 1–8. American Cancer Society. https://doi.org/10.1002/9780470015902.a0027022.

———. 2016b. Obesity as a socially defined disease: Philosophical considerations and implications for policy and care. *Health Care Analysis* 24: 86–100.

———. 2018. Looking for trouble? Diagnostics expanding disease and producing patients. *Journal of Evaluation in Clinical Practice* 24: 978–982. https://doi.org/10.1111/jep.12941.

Illich, Ivan. 1976. *Medical nemesis: The expropriation of health*. New York: Pantheon Books, Random House.

Keyes, Katherine M., and Sandro Galea. 2016. *Population health science*. Oxford: Oxford University Press.

Mackie, J.L. 1965. Causes and conditions. *American Philosophical Quarterly* 2: 245–264.

Moghaddam-Taaheri, Sara. 2011. Understanding pathology in the context of physiological mechanisms: The practicality of a broken-normal view. *Biology and Philosophy* 26: 603–611.

Moynihan, Ray, and Richard Smith. 2002. Too much medicine?: Almost certainly. *British Medical Journal* 324: 859–860.

Oppenheimer, Gerald M. 2005. Becoming the Framingham study 1947–1950. *American Journal of Public Health* 95: 602–610.

Pickering, George. 1961. *The nature of essential hypertension*. Londres: J. & A. Churchill.

———. 1978. Normotension and hypertension: The mysterious viability of the false. *The American Journal of Medicine* 65: 561–563.

Reid, Lynette. 2017. Truth or spin? Disease definition in cancer screening. *The Journal of Medicine and Philosophy: A Forum for Bioethics and Philosophy of Medicine* 42: 385–404. https://doi.org/10.1093/jmp/jhx006.

Rose, Geoffrey. 1992. *The strategy of preventive medicine*. Oxford: Oxford University Press.

Rothman, Kenneth J. 1976. Causes. *American Journal of Epidemiology* 104: 587–592.

Russo, Federica. 2009. Variational causal claims in epidemiology. *Perspectives in Biology and Medicine* 52: 540–554.

Russo, Federica, and Jon Williamson. 2007. Interpreting causality in the health sciences. *International Studies in the Philosophy of Science* 21: 157–170.

Sadegh-Zadeh, Kazem. 2008. A prototype resemblance theory of disease. *Journal of Medicine and Philosophy* 33: 106–139.

Schroeder, S. Andrew. 2013. Rethinking health: Healthy or healthier than? *The British Journal for the Philosophy of Science* 64: 131–159.

Schwartz, Peter H. 2008. Risk and disease. *Perspectives in Biology and Medicine* 51: 320–334. https://doi.org/10.1353/pbm.0.0027.

———. 2014. Small tumors as risk factors not disease. *Philosophy of Science* 81: 986–998. https://doi.org/10.1086/678280.

Sholl, Jonathan. 2017. The muddle of medicalization: Pathologizing or medicalizing? *Theoretical Medicine and Bioethics* 38: 265–278.

Stegenga, Jacob. 2018. *Care and cure: An introduction to philosophy of medicine.* University of Chicago Press.

Timmerman, Stefan, and Mara Buchbinder. 2010. Patients-in-waiting: Living between sickness and health in the genomics area. *Journal of Health and Social Behavior* 51: 408–423.

Vickers, Andrew J., Ethan Basch, and Michael W. Kattan. 2008. Against diagnosis. *Annals of Internal Medicine* 149: 200–203.

Welch, H. Gilbert, Lisa M. Schwartz, and Steve Woloshin. 2011. *Overdiagnosed: Making people sick in the pursuit of health.* Beacon Press.

Whelton, Paul K., Robert M. Carey, Wilbert S. Aronow, Donald E. Casey, Karen J. Collins, Cheryl Dennison Himmelfarb, Sondra M. DePalma, Samuel Gidding, Kenneth A. Jamerson, and Daniel W. Jones. 2018. 2017 ACC/AHA/AAPA/ABC/ACPM/AGS/APhA/ASH/ASPC/NMA/PCNA guideline for the prevention, detection, evaluation, and management of high blood pressure in adults: A report of the American College of Cardiology/American Heart Association Task Force on Clinical Practice Guidelines. *Journal of the American College of Cardiology* 71: e127–e248.

Chapter 14
Fundamental Concepts in Medicine: Why Risk and Disease Are Likely to Stay on Board

Olaf Dekkers

In everyday medical language the words *risk* and *disease* are easily used. Patients and doctors discuss the cause and prognosis of a disease while at the same time consider what the risks associated different treatment options are. As an example, the conversation between a doctor and a patient is about the cause(s) of a newly diagnosed diabetes (obesity, unhealthy lifestyle), with a subsequent discussion about the risks of side effects of the glucose lowering drug metformin (nausea).

Arguably, both risks and diseases belong to the basic stuff in medicine. And medical thinking in terms of risk has clearly emerged in the last century. The textbook example is the smoking-lung cancer case where Doll and Hill used large cohorts to show that smoking was a strong risk factor for lung cancer (Doll and Hill 1950). However, these epidemiological data in itself were not considered a definitive proof and the search for the causative agent emerged, ie what is the pathophysiological mechanism by which smoking was causing lung cancer (Cornfield et al. 2009)? This example shows that a risk-based and a pathophysiological approach are complementary as they approach a specific disease such as lung cancer from a different angle. And obviously the two approaches are linked: only if smoking is a true risk factor for lung cancer (and not a proxy for another true cause), the search for a pathophysiological mechanism would have been meaningful.

On closer philosophical examination, the concepts risk and disease (and their demarcation) are however not unproblematic, and in her chapter *Risk and Disease: two alternative ways of modelling health phenomena*, Giroux outlines several of these conceptual problems, and ultimately argues that the concepts risks and diseases can be seen as different approaches to the same set of health-related phenomena. The risk-based approach has its roots in epidemiological population studies, the

O. Dekkers (✉)
Departments Clinical Epidemiology and Internal Medicine, Leiden University Medical Center, Leiden, The Netherlands
e-mail: o.m.dekkers@lumc.nl

© The Author(s) 2024
M. Schermer, N. Binney (eds.), *A Pragmatic Approach to Conceptualization of Health and Disease*, Philosophy and Medicine 151,
https://doi.org/10.1007/978-3-031-62241-0_14

disease-based approach is rooted in pathophysiological thinking. This two-angled approach was indeed the cornerstone in the lung-cancer smoking debate; interestingly the two approaches were needed for a full proof of the causal relation, but they were also at the root of some of the controversies in the debate (Cornfield et al. 2009).

It is an intriguing phenomenon that everything that classifies as disease can, from another perspective, be seen as risk factor. Diabetes, generally considered a disease, is a risk factor for cardiovascular disease; cancer is a risk factor for death. That we can consider a similar health phenomenon a risk factor but at the same time a disease, underlines the point Giroux makes, i.e. that risks and disease are different ways to look at health phenomena. Of note, the opposite is obviously not true, as not every risk factor is a disease; for example, smoking itself is not a disease (although addictive behaviour may classify as such).

As was shown by the example of smoking and lung, the risk-based approach to medicine starts from population thinking and uses epidemiological techniques to disentangle the causal structure of risk factor—disease associations. The Framingham study is another famous example showing the strength of the epidemiological approach, a study which contributed importantly to the understanding of risk factors for cardiovascular disease (Mahmood et al. 2014). And from a public health perspective, these epidemiological studies can inform health politics (how) to modify risk factors and risk behaviour thereby aiming to improve health for the population (of course this assumes the epidemiological studies to be valid, a discussion that would open a can of worms). Cigarette use has decreased since public health measures were undertaken to discourage their use. Of note, a study assessing risk factors cannot replace the evidence needed to assess the effect of an intervention. Even if we're certain based on epidemiological data that obesity is a risk factor for cardiovascular disease, further studies should be undertaken to assess which interventions aiming to reduce obesity will optimally reduce cardiovascular risks (Hernán and Taubman 2008).

However, accepting risks and disease as two different angles, does not solve all the problems related to these concepts (to be fair, Giroux nowhere in the paper makes the claim that everything is solved). Consider the example of hypertension. From a risk-based perspective, it is a factor that influences the risk for many diseases such a myocardial infarction or renal failure. At the same time researchers can consider the pathophysiology that underlies these associations. But still, this does not solve the question whether we should classify hypertension itself as a disease (and if yes from which blood pressure onwards?). And this is not merely a semantic issue, as disease classification and categorization hugely impact individuals and populations. As Giroux argues in her paper, many attempts to come up with an overarching framework to define or classify diseases are easily refuted by counterexamples.

The interesting question is whether the risk and disease-based approaches to health phenomena will ever merge. It can be argued that personalized medicine just tries that as it aims to link risks and diseases for individuals, not through risk-based epidemiological data, but through detailed, clinical data at the individual level. Personalized medicine holds the promise that detailed geno- and phenotyping will perfectly predict the effect of treatments or will perfectly predict the risk for a future cardiovascular event for individuals. (Mind, that this is not an empirical claim, as personalized medicine in this very strict sense has currently not been proven.)

Suppose Joe is provided with a 10% risk to get a myocardial infarction, and that this prediction is based on epidemiological data such as the Framingham study, meaning that Joe is ultimately compared to the 'average Joe' from the data. Will personalized finally medicine overcome the need for the epidemiological approach? The first question then is, whether at the individual level all risks are ultimately 0 or 1, meaning that it is determined whether someone (Joe) will actually get a myocardial infarction (or not). Let us, for sake of the argument, assume that everything that contributes to Joe's risk is perfectly measured (a pretty hard assumption). Joe's prediction relates to to what will (not) happen in the future, and claiming that Joe's risk is either 0 or 1 at a certain time-point, is arguing that future events will not influence this risk prediction. That this form of determinism is hard to defend (even without taking quantum mechanics into account) is easy to see, as system external disruptors may well interfere with the prediction (Dekkers and Mulder 2020). Joe might die to a car accident, or the politicians might decide to effectively reduce air pollution, factors that will influence Joe's risk in a way that could have not been taken into account at time of the initial prediction. It might even be that Joe changes his lifestyle radically hoping to prevent the prediction to happen (in case of a doomed prediction). That leads to the question whether the 10% risk prediction is as close as we can get even from a personalized medicine perspective. But here the main point is epistemological, as currently the only way to find out is by comparing Joe to other persons who resemble Joe as much as possible with regard to his risk profile. And yes, this means that we are back in the realm of epidemiology, even if populations used to compare are based on more data and more detailed information (Ahlbom 2020).

To conclude, Giroux' plea to consider the concepts risks and diseases being different approaches to the same set of health-related phenomena is valuable. This holds true for daily clinical practice but also for science. Even in the era of personalized medicine.

References

Ahlbom, Anders. 2020. Epidemiology is about disease in populations. *European Journal of Epidemiology* 35 (12): 1111–1113. https://doi.org/10.1007/s10654-020-00701-9.

Cornfield, Jerome, Haenszel William, E. Cuyler Hammond, Abraham M. Lilienfeld, Michael B. Shimkin, and Ernst L. Wynder. 2009. Smoking and lung cancer: Recent evidence and a discussion of some questions. 1959. *International Journal of Epidemiology* 38 (5): 1175–1191. https://doi.org/10.1093/ije/dyp289.

Dekkers, Olaf M., and Jesse M. Mulder. 2020. When will individuals meet their personalized probabilities? A philosophical note on risk prediction. *European Journal of Epidemiology* 35 (12): 1115–1121. https://doi.org/10.1007/s10654-020-00700-w.

Doll, Richard, and A. Bradford Hill. 1950. Smoking and carcinoma of the lung; preliminary report. *British Medical Journal* 2 (4682): 739–748. https://doi.org/10.1136/bmj.2.4682.739.

Hernán, Miguel A., and Sarah L. Taubman. 2008. Does obesity shorten life? The importance of well-defined interventions to answer causal questions. *International Journal of Obesity* 32 (3). Nature Publishing Group: S8–S14. https://doi.org/10.1038/ijo.2008.82.

Mahmood, Syed S., Daniel Levy, Ramachandran S. Vasan, and Thomas J. Wang. 2014. The Framingham heart study and the epidemiology of cardiovascular diseases: A historical perspective. *Lancet* 383 (9921): 999–1008. https://doi.org/10.1016/S0140-6736(13)61752-3.

Chapter 15
A Pragmatic Approach to Understanding the Disease Status of Addiction

Mary Jean Walker and Wendy A. Rogers

15.1 Introduction

In this paper we examine the application of our view of disease as a vague cluster concept (Rogers and Walker 2017, 2018; Walker and Rogers 2017, 2018) to the example of addiction. Addiction is a paradigmatic, but complex case of a borderline condition, whose disease status is contested. Our approach to understanding the concept of disease was partly motivated by seeking to understand borderline disease states, though we focused primarily on conditions with straightforwardly observable physiological dysfunction. In applying our view to addiction we hope to further clarify the view, identify potential challenges, and propose a partial resolution of them in the case of addiction.

In Sect. 2 we summarise the vague cluster account. The account incorporates two main claims—that the concept 'disease' is best treated as a cluster concept, and that it is a vague concept that can be precisified for specific purposes. These claims were developed separately, so we briefly clarify how they fit together. In Sect. 3 we apply the account to illuminate why addiction's disease status is unclear, showing it has the typical features of disease in vague and ambiguous senses. In Sect. 4 we examine pragmatic considerations that might help us precisify our concept 'disease' to decide the disease status of addiction for particular purposes or in particular contexts.

M. J. Walker (✉)
Politics, Media and Philosophy Department, La Trobe University,
Bundoora, Melbourne, VIC, Australia
e-mail: mary.walker@latrobe.edu.au

W. A. Rogers
Philosophy Department and School of Medicine, Macquarie University,
Sydney, NSW, Australia
e-mail: wendy.rogers@mq.edu.au

M. Schermer, N. Binney (eds.), *A Pragmatic Approach to Conceptualization of Health and Disease*, Philosophy and Medicine 151,
https://doi.org/10.1007/978-3-031-62241-0_15

Drawing on this discussion, in Sect. 5 we draw out two issues relevant to making these precisifications: that pathologising addiction brings both positive and negative effects simultaneously; and the practical difficulty of maintaining that addiction's disease status might differ by context. We argue that the pragmatic concerns indicate, at the least, that addiction should not be regarded as mere physiological dysfunction, but may be counted a disease in some broader sense. This suggests a way of reading arguments that addiction is a 'biopsychosocial disease' as a pragmatic compromise that partly resolves the two issues.

This analysis yields general points about the disease status of addiction, and about pragmatic accounts of disease. In relation to addiction, it suggests that while addiction might be considered disease in some contexts and not others, a pragmatic approach does rule out some possible precisifications in this case. It further suggests that the debate about the disease status of addiction is not a substantive disagreement over facts, but rather is partly semantic (hinging on different underlying views of the meaning of 'disease') and normative (hinging on different views about how we should respond to addiction). In relation to pragmatic accounts of disease, our analysis shows their utility in understanding borderline disease states, and their potential to mediate between viewing disease status as a discoverable or a decidable attribute.

15.2 The Vague Cluster Account

We previously argued that the concept 'disease':

1. is vague but may be stipulatively precisified for various purposes (Rogers and Walker 2017); and
2. is not structured around necessary and sufficient conditions, such that there is no feature essential to disease but rather a number of features that are typical of diseases (Walker and Rogers 2018; for other arguments that disease does not have a classical structure see Schwartz 2004, 2007a; Nordby 2006; Sadegh-Zadeh 2015).

15.2.1 Vagueness

A concept is vague when there are borderline cases that are not clearly included within or excluded from the concept. Classic examples are the concepts 'tall', 'bald', or 'heap'. There is no bright line determining the exact height at which a person can be said to be tall, the precise amount of hair loss at which a person becomes bald, or the number of grains of sand required to have a 'heap'.

We argued that the concept 'disease' is vague (Rogers and Walker 2017). Health and disease are not binary, mutually exclusive states. There are many conditions that are not clearly either healthy or diseased, or where the difference between health and disease is a matter of degree. Our argument focused on degrees of physiological dysfunction, drawing on cases where dysfunction increases by degrees and there is no clear point at which function alters from normal to abnormal, from healthy to diseased. As such, accounts that require dysfunction for disease face the question of precisely what level of dysfunction indicates the presence of disease (the "line-drawing problem" as Schwartz (2007b) dubs it). Though there are cases where a person either clearly has, or lacks a physiological dysfunction—a bone may be clearly broken or whole, measles virus may be present or not present—for many diseases there is a continuum, with some level of functional variation labelled 'normal', and disease status attributed at higher levels. Thus insofar as disease requires dysfunction, the application of the concept 'disease' can be unclear where there are variations in function that may or may not be considered to amount to pathology.[1] Examples include microscopic cancers that never grow or metastasise, latent TB, and some 'risk factor' or 'pre-disease' states such as mild hypertension, osteopenia, or (arguably) osteoporosis (Rogers and Walker 2017).

While this way of framing the issue may be questioned, the central idea that the concept 'disease' is vague is not contentious—especially when applied to the everyday concept of disease used by laypeople, rather than the specialised or 'theoretical' notion used by doctors and medical scientists. Recognising this vagueness can be helpful in understanding why the disease status of some conditions is unclear: the concept 'disease' is not precise enough to fully determine what states fall under it. Just as it would be misguided to insist that there must be some determinate answer as to whether or not one is really tall, or whether 5 grains of sand is truly a heap, it is misguided to think that we can always resolve questions about the disease status of borderline conditions. This is not due to some lack of knowledge, but to conceptual indeterminacy.

Philosophers have developed different ways of thinking about vagueness.[2] One influential approach uses many-valued logic to allow that statements of the form 'X is a Y' can be true to a degree between 0 and 1, rather than being either false (0) or true (1). It could allow that 'microscopic, non-progressive cancer is a disease' is true to a degree of 0.2 or that 'latent TB is a disease' may be true to a degree of 0.6. The degree to which some condition can be counted a disease might be determined in

---

[1] While we interpret this as evidence of vagueness in our concept of physiological dysfunction it might alternatively be taken to show a conflation of the logic of risk with that of disease. See Giroux (Chap. 13, this volume).

[2] We will not here consider an epistemicist view that would claim that we just do not know enough about the concept 'disease' or particular states to know how to classify the borderline states, as we deny the concept has some yet-to-be-discovered essence that could resolve borderline cases. In Sorensen's (2001) terms, we take it that at least some borderline cases of disease are absolute, and not merely relative borderline cases, for reasons discussed below. However, a version of epistemicism will re-arise on a pragmatic approach given lack of knowledge about the practical consequences of decisions about disease status, as we show below.

different ways, such as focusing on degrees of similarity to prototype diseases with respect to particular features or dimensions (Sadegh-Zadeh 2000, 2015).

Another approach is supervaluationism, which draws on nonclassical logics to say that a claim of the form 'X is a Y' can be neither true nor false. While the many-valued logic approach implies that states of 'health' may gradually fade into states of 'disease', supervaluationism implies a borderline zone between 'clear health' and 'clear disease' that includes states of indeterminate disease/health status. Which approach to understanding vagueness best captures features of vague concepts is a matter of debate that we leave aside here. Below we will motivate adopting the supervaluationist approach in relation to addiction by showing a number of senses in which it is indeterminate whether addiction has the typical features of diseases.

15.2.2   Cluster Concept Structure

While our argument about vagueness focuses on dysfunction, we do not think dysfunction is the only relevant feature of disease. Drawing on arguments that disease cannot be defined in terms of necessary and sufficient conditions (as per the classical structure of concepts) without revising the everyday concept (Schwartz 2004; Nordby 2006), we developed an approach that treats disease as a cluster concept with a number of typical features (Walker and Rogers 2018), several of which are matters of degree.

In rejecting the classical necessary and sufficient concept structure, we suggested several desiderata for assessing the adequacy of a definition of disease (following Kingsbury and McKeown-Green (2009)). One is *extensional adequacy*: whether the definition follows our intuitive classifications of states as disease or non-disease, capturing those cases we usually think fall under it, and excluding cases we usually think do not. This is the desideratum that much of the debate about the concept 'disease' focuses upon. A second desideratum is *criterial adequacy*: the definition should group together conditions in a way that reflects the reasons we group them together, such as similarities between them. This is why a disjunctive definition like 'X is a disease iff it is a case of P_1, or P_2, or ... P_n' listing every known disease would not be a good definition, even if it were extensionally adequate. Third, a definition should be *motivationally adequate*, that is, there needs to be some reason that we want to be able refer to the class of things as a category, some purpose or context motivating having such a concept.

We argued that regarding disease as having a cluster structure could meet these three desiderata. By cluster structure, we mean taking the concept to be connected to a set of features that are typical of cases that fall under it, but allowing that no particular feature is necessary, and no set of conditions is jointly sufficient, for disease (Gasking 1960).[3] We proposed four features typical of disease: (1) dysfunction, (2) harmfulness in causing a person suffering or incapacitation, (3) being in principle explainable in terms of facts about biology and/or psychology, and (4) being beyond the direct conscious control of an individual (Walker and Rogers 2017).

15.2.3   Disease as a Vague Cluster Concept

Of the four typical features we proposed for the concept 'disease', at least two involve vagueness as they are matters of degree and it is not clear where to draw the line, resulting in a threshold problem. We argued previously that this is true of dysfunction, and it would seem to be true of our second typical feature, harmfulness, understood as suffering and/or incapacitation. The harmfulness of a condition can range from no suffering at all, as with asymptomatic diseases such as occult cancers, through minor irritations, to different levels of pain and finally to death. A disease can cause degrees of incapacitation from nil, to having to give up a sport or hobby, to being unable to work, parent, or engage with the world in other important ways.

Alston (1967) distinguished degree vagueness from 'combinatorial' vagueness, arising from lack of clarity as to which of the usual features of a concept are necessary, in what combination(s). Alston demonstrates using the concept 'religion': among the systems generally recognised to be religions, there are shared features such as belief in god/s, sacred objects, moral codes, and ritual behaviours. But some religions do not have all of these features, for instance atheistic Buddhism. This sense of vagueness fits well with the cluster approach, implying that there may be no final list of typical features, or of what combinations of features are necessary and sufficient for the concept's application.

We also suggest that some of the typical features of disease are ambiguous. Ambiguity is a different form of semantic indeterminacy, involving a term's having two or more possible meanings. Some borderline cases of disease may be so because it is not clear whether they have the typical features in the appropriate sense. For instance, some types of harm or incapacitation might not seem clearly the right 'sort' for disease status, such as a case where congenital limb difference causes a person suffering and limits their capacities, but only in combination with features of the person's physical or social environment (e.g. living in a country where supports are not provided, or where they face discrimination). We demonstrate ambiguity in whether addiction has several of the typical features of disease in the following section.

---

[3] Keil and Stoeker argue for a cluster concept approach to defining disease in psychiatry (Keil and Stoecker 2017). Where we organise our cluster around the typical features of diseases discussed below, theirs is organised around disease (in the sense of dysfunction), illness (experience), and sickness (social inability).

We propose then to picture disease as a fuzzily-bordered space mapped along *n* dimensions. Each dimension represents one typical feature. For each dimension, there are conditions that are clearly within the 'space' of disease, conditions clearly without it, and conditions that lie in the fuzzy or indeterminate 'borderline' zone with respect to that dimension. This allows for a number of ways that conditions can be classified as diseases: particular conditions might have different typical features of disease, to different degrees, and some particular diseases might lack some of the typical features entirely. It also implies that there are different ways a condition might count as a borderline state: some may be borderline because they have some of the typical features of disease, but not all of them (combinatorial vagueness); others will be borderline due to degree-vagueness; some may be borderline because of ambiguity.

The vague cluster approach helps explain why some judgements of disease status are unstable and contested. Speakers may have different assumptions about what features are sufficient for disease, or different views about the degree or sense of a feature required for disease. Or they may be focusing on one typical feature rather than another.

This approach is compatible with stipulatively defining more precise definitions to make decisions about borderline conditions, within the indeterminate or fuzzy boundaries set by the concept. These stipulative decisions are constrained by the typical features of the concept 'disease', but might appeal to pragmatic considerations including the practical implications of disease status. In this sense it is a pragmatic approach. The inclusion of motivational adequacy as a criterion for assessing definitions further implies that, as we precisify our vague concept, it is apt to think about what similarities we are attempting to pick out with it, and what work we want our concept to do.[4] Further, the approach licenses conceptual pluralism: we might adopt different stipulative precisifications of the concept in different contexts, or for different purposes, if doing so is useful. For example, if it were to turn out that clinical labelling of microscopic cancers that have a low chance of progressing as 'diseases' causes net harm (e.g., by contributing to overtreatment), but it is useful to consider them diseases for theoretical or research purposes, the vague cluster account allows us to stipulate the condition is a disease in the second context but not in the first. Like other pragmatic approaches then, the account can do justice to the fact that a concept can have more than one function and play different roles in different contexts (Van der Linden and Schermer 2022; Haverkamp et al. 2018; Keil et al. 2017).

---

[4]This may also be consistent with the view that vagueness itself is interest-relative. See Graff (2000).

15.3 Addiction as a Borderline Disease

Addiction is a paradigmatic borderline case, its disease status having been contested for at least several centuries. A current debate focuses on whether or not addiction is a 'brain disease'. Other positions regard addiction as a biopsychosocial disease; as non-pathological maladaptive learning; or as a rational choice under certain conditions.[5] Most authors also make a distinction between a medical model that takes addiction to be a disease, and a moral model regarding addiction as a sin, vice, or character flaw.[6] In this section we consider addiction in relation to our four 'typical features' to show that it is a borderline condition due to vagueness and/or ambiguity in these features. It is neither clearly disease nor non-disease, but partly or ambiguously has features typical of diseases.

15.3.1 Dysfunction

It is possible to hold that addiction does, or does not, involve dysfunction. Dysfunction might be understood as psychological or physiological. Addiction seems to involve psychological dysfunction, in the sense that the person's psychological states or behaviours have altered in ways that depart from usual psychological functions. There are many possible psychological descriptions of addiction, but it commonly involves experiencing recurrent cravings for a particular substance or behaviour, and reduced pleasure in other activities. These might count as dysfunctions in the sense that they differ from the mental functions that are meant to, or normally do, support survival and/or reproduction (cf Papineau 1994; Boorse 1976). Regarding physiological dysfunction, the brain disease model of addiction holds that addiction involves changes to brain structure and functioning that count as neurological dysfunction. These include structural and/or physiological changes to the brain's reward systems, resulting in a person with an addiction experiencing ongoing cravings with loss of pleasure in other activities. There may also be damage to parts of the brain associated with executive decision-making (Volkow et al. 2016). Some take the physiological dysfunction to underlie the psychological, so that these are deemed descriptions of the same phenomenon at different levels, but typically

---

[5] Some of these debates are reflected in the current diagnostic criteria for addiction ("substance use disorder") in the DSM-5, which include pharmacological or biological criteria (e.g. tolerance, withdrawal symptoms), and normative and social criteria (e.g., social and interpersonal problems, neglecting social roles). The DSM specifies that someone who meets 5 of the 12 conditions has the disorder, implying that different people with addiction may not overlap at all in terms of what criteria they meet, reflecting the breadth and heterogeneity of the concept (Epstein 2020; Kennett et al. 2013; Fisher 2022).

[6] Addiction itself, we should acknowledge, is a vague concept that might be stipulatively defined for different purposes (as argued by Sinnott-Armstrong and Summers 2018). We leave this complication aside for brevity, assuming that there are at least some cases that are clear cases of addiction.

addiction is diagnosed from psychological and behavioural, rather than neurological criteria.

One reason why it is unclear whether addiction involves dysfunction is that dysfunction—on either level of description—will be a matter of degree (Pickard 2022). Psychologically, cravings and reduced pleasure in other activities could range from very strong to very weak. Physiologically, the changes to the brain's reward system, involving changes to the strength of synaptic connections due to changing numbers of receptors (Volkow et al. 2016), could occur to differing degrees.

A recent debate shows that it is also unclear whether these changes count as dysfunction in the right sense; that it is ambiguous whether it is apt to describe them as dysfunction. Lewis (2017) argues that the neurological changes should not be considered 'dysfunction' because they are consistent with the way the brain normally changes when we learn. He suggests that while our brains are designed to change in response to feedback in the form of rewards and costs, feedback from some of the substances and behaviours that are now available to us is stronger than the human brain evolved to cope with. Some argue that humans may have developed a reward response to alcohol, for example, because of survival benefits from eating fermented fruit from the forest floor, a rich source of nutrients and calories. The brain mechanisms that functioned well in evolutionary time, however, become problematic in a context where distilled spirits are readily available (Slingerland 2021). Thus, although it is true that addiction involves brain changes, according to Lewis this is not dysfunction; quite the opposite, it is an instance of a brain functioning normally, albeit in an environment different from that in which it was 'designed' to function.[7]

However, it is possible to agree that the neurological changes are consistent with normal brain function, but think addiction still counts as dysfunction. Wakefield (2017b) argues that dysfunction at the psychological level may not imply dysfunction at a biological level of description, as neurological changes consistent with normal learning processes can lead to dysfunction at the psychological level. As others have put it, there can be malfunctions at the level of software that are not describable on the level of hardware (Papineau 1994; Segal 2022).

To demonstrate, Wakefield gives an example of a gosling that has a neurological disposition to imprint on the first creature it sees upon hatching. This normal process of learning is functional for the gosling (and selected over evolutionary time) because in most cases it leads to goslings sticking close to their mothers and obtaining the survival benefits of doing so. Consider a case where a gosling's first sight is a fox. The gosling's brain functions perfectly at the physiological level, by imprinting on the fox. This leads to behaviours detrimental to its survival. We might consider this a dysfunction, Wakefield argues, despite its being a normal brain response, as it does not fulfil the relevant function of this learning mechanism. Similarly, one might think that addiction involves the normal brain function of

---

[7] We adopt an aetiological way of discussing dysfunction for simplicity, following many in the addiction literature, though we think an analogous argument could be made on a non-aetiological account.

learning, but because it occurs in a context with strong feedback that human brains did not evolve to deal with, it leads to psychological states that don't fulfil any function selected for in evolutionary time—and thus counts as dysfunction (Wakefield 2017b: 59–60).

These authors make conflicting claims about whether addiction counts as dysfunction on the basis of the same facts, indicating that our concept dysfunction is not determinate enough to tell us whether or not the changes observed in addiction are dysfunction. We propose then that it is unclear whether addiction counts as disease or not, in part, because 'dysfunction' is both a matter of degree and ambiguous.

15.3.2   Harm

Our second typical feature is harmfulness, involving either suffering or incapacity. The occurrence of harm in addiction is a matter of degree. It seems possible that a person could be addicted and not experience many of the harms ordinarily associated with addiction, for instance if they have access to a pure version of a drug that is not (in that form) medically harmful, are able to take the substance in a way that does not interfere with their work or relationships, are happy to be dependent on the substance, and so on. Many likely have such an addiction to caffeine. At the other end of the spectrum, an addiction might lead to extreme health harms, cause great suffering and social incapacities, or lead to death. Many cases will be somewhere in between these extremes; what level of harm supports a condition's claim for being a disease will thus involve threshold problems.

It can also be ambiguous whether the harmfulness of addiction is of the right kind to indicate disease status. Some might think being addicted itself is a harm, but this does not seem clearly true, as indicated in the possibility of the happy caffeine addict above. Some addictive substances cause medical or psychological harms, or social incapacities. Some of these follow from the addictive 'dysfunction'—for instance, a person's cravings for a drug leading them to neglect family and work, and so to social harms and incapacities. Other harms, such as overdose, result from drug-taking itself, that is arguably a symptom of the condition. However, many of the harms commonly experienced in addiction are attributable, not to the addiction itself, but to social arrangements surrounding a substance. Harms such as legal troubles arise from the illicit status of particular drugs. Criminalising drug use (and sale) can lead to direct physical harms as it tends to drive down drug purity, leading to supply being 'cut' with contaminants that are significantly more dangerous than the drug itself, and to potentially fatal misjudgements about the strength of substances. Criminalisation can also lead to more dangerous forms of use; for instance injecting instead of smoking heroin in order to obtain the greatest 'high' from a small amount, leading to harms such as vein damage or infection from shared needles. Further, some harms result from social stigma around addiction, such as job loss, discrimination leading to reduced medical care, or subjective suffering linked to feelings of guilt and shame. It seems unclear whether these harms are of the right kind to count as being 'caused by the condition', since they are caused by the condition in combination with surrounding social arrangements and attitudes.

15.3.3 Explanation in Biological/Psychological Terms

Being explainable in the terms of biology or psychology seems to be clearly true of addiction in one sense. Addiction is often explained by drawing on (neuro-) biological or psychological sciences.

However, it would be too quick to conclude that addiction has this typical feature unproblematically. A number of authors argue that the biomedical or 'brain disease' model of addiction is too narrow. Part of this criticism is that addiction often has psychological aspects and is typically diagnosed, and often treated, at a psychological and behavioural, rather than a physiological level. But it further claims that addiction crucially involves social aspects, and that in order to explain, and best treat the condition these need to be acknowledged. For instance, addiction is more likely under conditions of social disadvantage and poverty, and rates vary depending on substance availability (Pickard 2016; Alexander 2022; Hammersley 2022; Fisher 2022). Thus, part of the disagreement about addiction's disease status involves different views on whether or not addiction is *fully* explainable in medical or psychological terms, or whether explanations must also refer to social and environmental factors.

15.3.4 Lack of Direct Conscious Control

Our final 'typical feature' is that disease is beyond the direct conscious control of an affected person. This feature involves lacking control over the symptoms and signs of a disease, rather than lack of control over becoming diseased. A person with measles cannot help having a fever, and a person with cancer cannot prevent its metastasizing by an act of will. Claims about addiction's disease status often emphasise that a person with an addiction cannot help having the cravings they experience and have diminished control in relation to the relevant substance or behaviour.[8]

The lack of control over these symptoms is vague in terms of degree, and some of the debate about addiction's disease status appears to hinge on the threshold question of how much loss of control constitutes addiction. Some argue that addiction is not a

---

[8]It might be argued there is a disanalogy in relation to control over getting a condition: some suggest that people with addictions had at least some control over the actions leading to the development of the addiction (e.g., a series of decisions to take a drug) (Kennett et al. 2014). However, it could be similarly argued that one can have some control over the risk of getting other diseases, e.g. by getting vaccinations, avoiding exposure locations, or changing diet or lifestyle.

disease because people with addictions are able to exert *some* control: some overcome addictions, and some can respond to incentives to change the addictive behaviour.[9] Others argue this view too strong, since most diseases do not require total loss of a function but only degrees of hypo- or hyperactive functioning (Wakefield 2017a: 43). The extent of loss of control involved in diagnosing or defining addiction will involve a threshold decision.

15.3.5   Conclusion on the disease status of addiction

On our account then, addiction is a borderline case, its disease status not determined by our concept. Rather than lack of knowledge, this is due to conceptual indeterminacy. Addiction has typical features of diseases, but cases of addiction may have some of them to a limited degree; and/or have some of them ambiguously, in a sense that is not clearly the relevant one for disease.

This implies that to some extent the debate about whether or not addiction is a disease is a semantic disagreement, with those on different sides of the debate operating with different views of what 'disease' means. But thinking that we will one day be able to resolve these differences and thus know whether or not addiction is a disease is wrong-footed—the same mistake as thinking that at some future point we will understand the essence of tallness in a way that enables a final answer to the question of whether 5 foot 7 is, or is not, really 'tall'. Instead, we may try to precisify our vague concept of disease to decide on addiction's disease status, including by appealing to pragmatic considerations.

15.4   Pragmatic Considerations in Specifying the Disease Status of Addiction

What pragmatic considerations, then, could we draw on to precisify our concept and decide addiction's status? Although to our knowledge no one has explicitly applied a pragmatic conception of disease to understanding the disease status of addiction, much of the debate about addiction's disease status already focuses on practical considerations, rather than engaging with theories of what 'disease' means (Pickard 2022). Many authors seem to be also making normative claims about how we should respond to addiction by arguing it is or is not a disease. That is, while there is vigorous debate about the pros and cons of categorising addiction as a (brain) disease, there is little conceptual work investigating the implications of using a precising definition for specific purposes to mobilise particular consequences.

---

[9]For an overview of empirical evidence supporting this see Pickard (2016); for an argument that it implies addiction is not a disease see Foddy and Savulescu (2007).

Before turning to that task, in this section we summarise the main points canvassed in favour of, and against, using the classification of disease for addiction. This will help to identify what role pragmatic concerns might play in moving towards precising definitions for deciding addiction's disease status.[10] We show that the debate assumes a conception of disease as physiological dysfunction, and so conflates pragmatic arguments for precisifying in favour of addiction being a disease, with a particular view of what disease means.

15.4.1 Pragmatic Reasons for Considering Addiction a Disease

Proponents of considering addiction a disease argue that framing addiction as disease negates the 'moral model' on which addiction is sinful, a character flaw, or some other kind of moral defect. The moral model is thought to be assumed by many laypeople, regarding addiction as ongoing bad or selfish choices for which the addicted person can rightfully be blamed, leading to moral judgement and stigma. Reframing addiction as disease aims to remove such moral judgement and destigmatise the condition (Leshner 1997; Frank and Nagel 2017).

This is linked to a number of further effects. Stigma has a range of negative effects that have been empirically demonstrated, including adversely affecting help-seeking, social connections, employment, and housing stability, and encouraging people to view themselves as defective (Kvaale et al. 2013). In contrast, the disease model may undercut or lessen such effects. From the perspective of an addicted person, regarding addiction as a disease can be therapeutic, encouraging the idea that addiction is treatable, and so empowering people to make different choices. It could also help to reduce the guilt or shame people may feel, which could contribute to denial of any problems, or to continued using (Fingarette 1985; Walker 2010). A study by Fraser and colleagues found that clinicians sometimes use the disease model to teach clients to re-interpret their cravings as "effects of their brain", encouraging a distance between "brain and self" (cited in Keane et al. 2022). Similarly, another study by Barnett and colleagues found that clinicians report that some clients gain insight into their behaviour from the disease model (Barnett et al. 2018).

Further, the disease label is claimed to influence how the person's family and friends are likely to respond to them, encouraging compassion rather than blame and judgement. Sometimes comparisons are drawn to other diseases, such as cancer or the flu, to emphasise that people with addictions do not choose their actions, so that it is not fair to respond to them with blame or approbation. Blaming people for psychological conditions has been shown to potentially contribute to their ongoing nature; and blaming addicted people is a predictor of relapse (Kvaale et al. 2013). On the policy level, claims arise that regarding addiction as a disease will encourage

---

[10]This overview is of course by no means exhaustive, given the breadth of this debate.

policies that respond to addiction as a health condition, such as providing treatment and funding medical research, and discourage policies that treat addiction behaviours as crimes (Pickard 2022).

The various claimed positive effects of the disease designation seem to draw on several of our four typical features of disease. Special treatment of addicted people by others is justified because their condition, and resulting behaviours, are out of their control. Addiction is framed as beyond their control because it involves a dysfunction: the behaviour deemed problematic arises from their condition rather than from their choice, blocking stigma and moral condemnation. Many proponents of pathologising addiction understand the relevant dysfunction as physiological, reflecting the influence of the brain disease model. However, these positive effects may also arise from conceptions of disease that do not have an exclusive physiological emphasis, such as the 'biopsychosocial disease' model of addiction.

15.4.2   Reasons Against Taking Addiction to Be a Disease

The claimed negatives of considering addiction a disease are more closely linked to a conception of disease as physiological dysfunction. They follow from taking disease to involve specific versions of two of our typical features; namely dysfunction (understood as physiological) and explainability (in biological terms). A major criticism of claims that addiction is a disease is that it oversimplifies, presenting a condition that often has complex social and psychological aspects as though it were only a physiological problem (Fisher 2022; Keane et al. 2022).

At the personal level, this may hinder explaining, or helping people understand, their experience. Barnett and colleagues report that according to clinicians, while some addicted individuals find it helpful to view their experience through a 'disease' lens, others find it irrelevant (Barnett et al. 2018). Others note that some people with addictions might take the pathologisation of their condition to mean it is unchangeable, as it is instantiated physiologically. This could lead to disempowerment, reduced hope for change, or decreased treatment uptake (Richter et al. 2019). At the policy level, a disease designation risks focusing our attention on physiological treatment, to the neglect of treatments that focus on social or psychological factors. Locating the problem of addiction in a person's body could divert attention away from the complex social determinants linked to addiction, including poverty, social isolation, or unemployment; or factors such as drug supply or urban design (Barnett et al. 2018).

A further negative consequence is that of implying people with addictions are disordered, non-autonomous, or irrational agents (Pickard 2022). Some of the claimed benefits above seem to rely on this, to excuse people from blame. However agency and responsibility are also status concepts (Mackenzie 2018), so regarding people with addictions as impaired agents implies they are not full members of the moral community, and risks turning them into objects rather than agents. The disease label may also reduce self-esteem and prevent a person from developing self-control

(Frank and Nagel 2017), since someone who understands their behaviour as the product of a disease may feel disempowered, or come to believe attempts to change are doomed, or that further addictive behaviour is excusable. At the policy level, the implication of impaired agency could contribute to stigma or to policies that treat addicted people as incapable of choice, such as mandatory treatment policies.

A further implication of the disease label, that appears negative to us, is that all substance-related problems are thereby classified as problems of addiction. This is however untrue: some harms of substance use result from recreational use, for example accidents, injuries, or criminal behaviour linked to intoxication unrelated to addiction, or chronic health issues linked to long-term heavy use (Babor et al. 2010). Presenting all the problems of substance use as being about addiction removes attention from these harms. Companies that profit from substance use, such as alcohol companies, regard the disease model of addiction as helpful (Collins 2010; Fisher 2022): so long as the 'problems' of use reside in the bodies of a minority of diseased agents, other problems may remain unexamined and unaddressed. The pathologisation of addiction normalises other substance use at the border between addiction and recreational use.

Some critics argue that the pathologisation of addiction is an individualising move that reflects the values of neoliberal societies, such as productivity and individual responsibility, by pathologising failures of rational self-control. As Keane and colleagues put it, "[w]hile the model may refer to neurotransmitters rather than personality types or character flaws, it still operates to pathologize and regulate individuals, and to demonize certain forms of consumption" (Keane et al. 2022: 405). In other words, it reflects and reinforces social values that might themselves be questioned.[11]

There is dispute over whether some of the claimed positive effects of disease labelling are likely (Courtwright 2010; Keane et al. 2022; Pickard 2022). There are reasons to dispute whether pathologising addiction will destigmatise it: medicalisation can introduces forms of stigma even while disrupting others. As Kukla puts it medicalisation "[r]eplaces moral stigma with functional stigma" (2014: 518). Empirical studies indicate that while physiological explanations can reduce blame for psychological problems, they may increase other measures of stigma, such as pessimism that the person can change, or perception of the person's dangerousness (Kvaale et al. 2013; Pickard 2022). Nor has the disease model seemingly wrought much change on public attitudes or policy, as shown by the continued use of punitive models (Courtwright 2010). Courtwright argues that the disease model, in regarding addiction as neurological change, can motivate a zero-tolerance, punitive approach focused on prevention (2010: 142). Pickard points out that the implications of the disease model for treatment are so far limited; though there are some pharmacotherapies and treatments that aim to intervene in the brain, most treatments utilise social and psychological treatment and support (2022).

---

[11] It might thus be a case of the tyranny of the community via medicalisation that Kukla (Chap. 21, this volume) analyses.

Others argue that the disease model has limited impact on the suspension of blame or generation of compassion, because addicted people may still be seen as blameworthy for having become addicted. The disease model is premised on the understanding that the underlying physiological dysfunction that constitutes the disease develops over time as a result of repeated use, implying that the agent developed the addiction as a result of a cascade of individual free choices (Kennett et al. 2014).

Finally, some claim that labelling addiction a disease is not necessary for disrupting stigma or motivating compassionate responses (Pickard 2022; Levy 2013). Pickard argues that a psychosocial model is better for eliciting empathy, as personal stories also help to combat stigma (Pickard 2022; Rodriguez and Prestwood 2019). That is, we might motivate some of the positive effects of the disease model by emphasising that people are experiencing harm, and that explanations (and any dysfunctions) are not only physiological but psychological and social. These arguments further underscore the assumption present in the debate, that to claim addiction is a disease is to claim it is a physiological dysfunction.

15.5   How Should We Precisify 'Disease' in the Case of Addiction?

While the vague cluster approach motivates a turn to practical considerations, it is not certain exactly what the effects of pathologising addiction will be.[12] Further, and while acknowledging the limitations of empirical data, there are both positive and negative practical effects of disease designation. In this final section we argue that notwithstanding uncertainties and conflicting implications, the pragmatic considerations provide some guidance. In particular, they rule out some possible precisifications of disease in relation to addiction, namely precisifications of disease as physiological dysfunction. If we are to think of addiction as a disease, we should not take this to mean it is solely physiological dysfunction; as demonstrated in the previous section, this is a common conflation in the debate.

Before making this argument, however, we consider the pluralist option. An implication of the vague cluster approach was that we might adopt different precisifications in different contexts or with regard to different purposes. We might consider the various positive and negative effects of pathologisation in different contexts, and engineer our concept 'disease' differently per context to obtain the positive effects but avoid the negatives (to the greatest extent possible). For example, we might assume addiction is a disease for illicit drug law reform purposes, in view of the high costs of such laws for punished individuals and society.

---

[12] This might allow a pragmatic version of epistemicism in response to the vagueness of disease, where the concept is to be made more determinate based on its effects in deployment, but we lack the knowledge to make the required precisifications.

When considering a person's capacity as an agent, it may best protect people's medical autonomy to assume that an addicted person retains capacities for agency relevant to making decisions about treatment, unless there are reasons against this in particular cases. It might at the same time be best to allow judges to consider addiction to impair agency in criminal cases, if this best serves rehabilitative aims of the law. Then again, when thinking about policies to prevent addiction, viewing addiction as following from rational choices people make in given situations, such as adverse social environments, may be more useful if it promotes attention to social determinants of addiction. And so on: in each context a thorough examination of the evidence may give guidance as to whether or not pathologisation is likely to be overall beneficial.

The pluralist approach is intellectually satisfying: addiction could count as disease for one purpose but not for another, where each decision as to its status reflects a different precisification of our vague concept of disease. However, how this could work in practice is difficult. Some of the positive and negative consequences of pathological classification arise simultaneously: an intention to reduce blame by implying a person lacks control brings with it the implication of impaired agency; or an intention to destigmatise inadvertently results in addicted people being regarded as more dangerous (Kvaale et al. 2013). Further, pluralism will in practice be difficult because people are likely to import claims about addiction's disease status from one context to another (Van der Linden and Schermer 2022). When judges, legislators, or policymakers consider responses to addiction based on its disease status, they draw on advice from medical professionals (Fisher 2022), and this will likely continue. A public education campaign on conceptual vagueness is unlikely to be effective to prevent this.

Rather than taking the pluralist route, we therefore argue that the practical concerns canvassed in Sect. 4 have a further implication. The practical positives often claimed to follow from disease status are consistent with a range of conceptions of disease. There has been a tendency in the recent debate to focus on physiological dysfunction—an implicit 'brain model' precisification—but other precisifications are possible, based on dysfunctions such as lack of control, other kinds of psychological impairment, or dysfunctional adaptation to difficult social circumstances. Harm is another candidate for a precising definiton. It seems to be rarely discussed as indicating disease status, but might be implicit in the (seemingly substantial) agreement that compassion is an appropriate response to addiction.

However, the negative practical consequences—and the arguments that the positive effects are unlikely—are largely premised on a conception of disease more closely tied to physiological dysfunction.[13] This is problematic because it is the claimed physiological nature of the dysfunction that is linked to charges of oversimplification, functional stigma, neglect of psychological or social causes or interventions, and the implications that the condition is unchangeable and located in

---

[13] For further analysis of assumptions of meaning for disease and ways pragmatic approaches could resist it see Greco (Chap. 17, this volume).

the bodies of individual problematic agents. The pragmatic approach then has a further implication: that we should challenge the assumption behind this debate that disease means physiological dysfunction, or that physiological dysfunction is its central defining feature.

A similar move is already present in the debate, though it is differently described: some argue that instead of seeing addiction as a 'brain disease' it should be considered a 'biopsychosocial disease' (Pickard 2022). On this view we could regard addiction as a pathological state that involves not only (or even, not always) physiological dysfunction, but other features, including psychological and social features. In our terms, we could precisify our concept 'disease', in relation to addiction, in ways that incorporate not only physiological dysfunction, but other typical features of our vague concept of disease, such as harm, explainability in psychological (as well as biological) terms, and diminished control. Moving away from seeing disease primarily or solely as physiological dysfunction could help to counter negative practical implications noted above around oversimplification, functional stigma, unchangeability and individualisation, since many people still make biologically determinist assumptions about conditions resulting from biological states (especially neurological states). Inclusion of harm as a feature could open up other ways of inducing more compassion for people with addictions, such as recognising that they may have difficult personal histories or social circumstances, and/or are suffering in various ways because of their addiction. Potentially, though speculatively, seeing addiction as a disease not limited to physiological dysfunction, could help emphasise that although it can impair agency, the person does retain some agential capacities, and these can be fostered and strengthened. If this is right, then the vague cluster approach would not necessarily imply plural precising definitions in different contexts, at least for addiction. Instead it would imply a different route in conceptual engineering, to designate addiction a disease, but promote understanding that diseases may not be reducible to physiological dysfunction, and have other features.

This does not resolve all the issues. Pathologisation is still likely to induce some forms of stigma, given it implies that there is something wrong with the person or with their behaviour. Nonetheless, treating disease as a vague cluster concept, which may be precisified in light of practical considerations, provides some guidance in the debate about pathologising addiction. Viewing addiction as a disease where this is taken to mean that it is solely a physiological dysfunction, is unhelpful. Thus based on pragmatic considerations, this should be ruled out as a possible precisification of 'disease' in relation to addiction. Disease models of addiction that focus on naturalistically-describable dysfunction obscure more issues than they resolve.

15.6   Conclusion

We hope to have shown that the vague cluster approach is useful for understanding why a borderline state's disease status is unclear and contested. In relation to pragmatic accounts of disease, we have also shown that a pragmatic account can mediate between regarding disease status to be a discoverable or a decidable attribute, allowing for stipulation of disease status constrained by the vague borders of the concept.

In relation to addiction, the analysis thus shows that practical considerations do circumscribe some conceptions of disease in relation to addiction, specifically, that disease should not be precisified, in this case, to mean mere physiological dysfunction.

Further, it suggests that the debate about the disease status of addiction is not a substantive disagreement over facts. Rather the disagreements are partly semantic, as interlocutors in the debate sometimes have different underlying conceptions of disease; and partly normative, as interlocutors aim to motivate different strategies to address problems.

Acknowledgements Our thanks to the volume editors and participants at the "Health and disease as practical concepts" Conference at Erasmus University Medical Centre in Rotterdam in 2023 for their helpful comments and suggestions.

References

Alexander, Bruce K. 2022. Replacing the BDMA: A paradigm shift in the field of addiction. In *Evaluating the brain disease model of addiction*, 522–538. New York: Routledge.

Alston, William P. 1967. Vagueness. In *The Encyclopedia of philosophy*, ed. Paul Edwards, 218–221. New York: Macmillan.

Babor, Thomas F., Raul Caetano, Sally Casswell, Griffith Edwards, Norman Giesbrecht, Kathryn Graham, Joel Grube, Paul Grunewald, Linda Hill, Harold Holder, Ross Homel, Esa Osterberg, Jurgen Rehm, Robin Room, and Ingeborg Rossow. 2010. *Alcohol: No ordinary commodity: Research and public policy*. Oxford: Oxford University Press.

Barnett, Anthony I., Wayne Hall, Craig L. Fry, Ella Dilkes-Frayne, and Adrian Carter. 2018. Drug and alcohol treatment providers' views about the disease model of addiction and its impact on clinical practice: A systematic review. *Drug and Alcohol Review* 37 (6): 697–720.

Boorse, Christopher. 1976. What a theory of mental health should be. *Journal for the Theory of Social Behaviour* 6 (1): 61–84.

Collins, Peter. 2010. Defining addiction and identifying the public Interest in liberal democracies. In *What is addiction?* ed. Don Ross, Harold Kincaid, David Spurrett, and Peter Collins, 409–433. Oxford: Oxford University Press.

Courtwright, David T. 2010. The NIDA brain disease paradigm: History, resistance and spinoffs. *BioSocieties* 5: 137–147.

Epstein, David H. 2020. Let's agree to agree: A comment on Hogarth (2020), with a plea for not-so-competing theories of addiction. *Neuropsychopharmacology* 45 (5): 715–716.

Fingarette, Herbert. 1985. Alcoholism and self-deception. In *Self-deception and self-understanding: New essays in philosophy and psychology*, ed. Mike Martin, 52–67. Lawrence, Kansas: University Press of Kansas.

Fisher, Carl Erik. 2022. *The urge: Our history of addiction*. London: Penguin.

Foddy, Bennett, and Julian Savulescu. 2007. Addiction is not an affliction: Addictive desires are merely pleasure-oriented desires. *The American Journal of Bioethics* 7 (1): 29–32.

Frank, Lily E., and Saskia K. Nagel. 2017. Addiction and moralization: The role of the underlying model of addiction. *Neuroethics* 10: 129–139.

Gasking, Douglas. 1960. Clusters. *Australasian Journal of Philosophy* 38 (1): 1–36.

Graff, Delia. 2000. Shifting sands: An interest-relative theory of vagueness. *Philosophical Topics* 28 (1): 45–81.

Hammersley, Richard. 2022. Addiction is a human problem, but brain disease models divert attention and resources away from human-level solutions. In *Evaluating the Brain Disease Model of Addiction*, ed. Nick Heather, Matt Field, Antony Moss, and Sally Satel, 176–186. New York: Routledge.

Haverkamp, Beatrijs, Bernice Bovenkerk, and Marcel F. Verweij. 2018. A practice-oriented review of health concepts. *The Journal of Medicine and Philosophy: A Forum for Bioethics and Philosophy of Medicine* 43 (4): 381–401.

Keane, Helen, David Moore, and Suzanne Fraser. 2022. Multiple enactments of the brain disease model: Which model, when, for whom, and at what cost? In *Evaluating the brain disease model of addiction*, ed. Nick Heather, Matt Field, Antony Moss, and Sally Satel, 405–415. New York: Routledge.

Keil, Geert, and Ralf Stoecker. 2017. Disease as a vague and thick cluster concept. In *Vagueness in psychiatry*, ed. Geert Keil, Lara Keuck, and Rico Hauswald, 46–74. Oxford: Oxford University Press.

Keil, Geert, Lara Keuck, and Rico Hauswald. 2017. Vagueness in psychiatry: An overview. In *Vagueness in Psychiatry*, ed. Geert Keil, Lara Keuck, and Rico Hauswald, 3–23. Oxford: Oxford University Press.

Kennett, Jeanette, Steve Matthews, and Anke Snoek. 2013. Pleasure and addiction. *Frontiers in Psychiatry* 4: 117.

Kennett, Jeanette, Nicole A. Vincent, and Anke Snoek. 2014. Drug addiction and criminal responsibility. In *Handbook on neuroethics*, ed. Jens Clausen and Neil Levy, 1065–1083. Dordrecht: Springer.

Kingsbury, Justine, and Jonathan McKeown-Green. 2009. Definitions: Does disjunction mean dysfunction? *The Journal of Philosophy* 106 (10): 568–585.

Kukla, Quill. 2014. Medicalization, "normal function," and the definition of health. In *The Routledge companion to bioethics*, ed. John D. Arras, Elizabeth Fenton, and Quill Kukla, 539–554. New York: Routledge.

Kvaale, Erlend P., Nick Haslam, and William H. Gottdiener. 2013. The 'side effects' of medicalization: A meta-analytic review of how biogenetic explanations affect stigma. *Clinical Psychology Review* 33 (6): 782–794.

Leshner, Alan I. 1997. Addiction is a brain disease, and it matters. *Science* 278 (5335): 45–47.

Levy, Neil. 2013. Addiction is not a brain disease (and it matters). *Frontiers in Psychiatry* 4: 24.

Lewis, Marc. 2017. Addiction and the brain: Development, not disease. *Neuroethics* 10: 7–18.

Mackenzie, Catriona. 2018. Moral responsibility and the social dynamics of power and oppression. In *Social dimensions of moral responsibility*, ed. Katrina Hutchison, Catriona Mackenzie, and Marina Oshana, 59–80. Oxford: Oxford University Press.

Nordby, Halvor. 2006. The analytic–synthetic distinction and conceptual analyses of basic health concepts. *Medicine, Health Care and Philosophy* 9 (2): 169–180.

Papineau, David. 1994. Mental disorder, illness and biological disfunction. *Royal Institute of Philosophy Supplements* 37: 73–82.

Pickard, Hanna. 2016. Addiction. In *Routledge Companion to Free will*, ed. K. Timpe and M. Griffith. New York: Routledge.

———. 2022. Addiction and the meaning of disease. In *Evaluating the brain disease model of addiction*, ed. Nick Heather, Matt Field, Antony Moss, and Sally Satel, 321–338. New York: Routledge.

Richter, Linda, Lindsey Vuolo, and Mithra S. Salmassi. 2019. Stigma and addiction treatment. In *The stigma of addiction: An essential guide*, ed. Jonathan Avery and Joseph Avery, 93–130. Dordrecht: Springer.

Rodriguez, Lindsey M., and Lauren Prestwood. 2019. The stigma of addiction in romantic relationships. In *The stigma of addiction: An essential guide*, ed. Jonathan Avery and Joseph Avery, 55–69. Dordrecht: Springer.

Rogers, Wendy A., and Mary Jean Walker. 2017. The line-drawing problem in disease definition. *Journal of Medicine and Philosophy* 42 (4): 405–423.

———. 2018. Précising definitions as a way to combat overdiagnosis. *Journal of Evaluation in Clinical Practice* 24 (5): 1019–1025.

Sadegh-Zadeh, Kazem. 2000. Fuzzy health, illness, and disease. *The Journal of Medicine and Philosophy: A Forum for Bioethics and Philosophy of Medicine* 25 (5): 605–638.

———. 2015. *Handbook of analytic philosophy of medicine.* Dordrecht: Springer.

Schwartz, Peter H. 2004. An alternative to conceptual analysis in the function debate. *The Monist* 87 (1): 136–153.

———. 2007a. Decision and discovery in defining 'disease'. In *Establishing medical reality: Essays in the metaphysics and epistemology of biomedical science*, ed. Harold Kincaid and Jennifer McKitrick, 47–63. Dordrecht: Springer.

———. 2007b. Defining dysfunction: Natural selection, design, and drawing a line. *Philosophy of Science* 74 (3): 364–385.

Segal, Gabriel. 2022. Addiction is a brain disease: (But does it matter?). In *Evaluating the brain disease model of addiction*, ed. Nick Heather, Matt Field, Antony Moss, and Sally Satel, 87–98. New York: Routledge.

Sinnott-Armstrong, Walter, and Jesse S. Summers. 2018. Defining addiction: A pragmatic perspective. In *The Routledge handbook of philosophy and science of addiction*, ed. Hanna Pickard and Serge Ahmed, 123–131. New York: Routledge.

Slingerland, Edward. 2021. *Drunk: How we sipped, danced, and stumbled our way to civilization.* New York: Little, Brown Spark.

Sorensen, Roy. 2001. *Vagueness and contradiction.* Oxford: Clarendon Press.

Van der Linden, Rik, and Maartje H.N. Schermer. 2022. Health and disease as practical concepts: Exploring function in context-specific definitions. *Medicine, Health Care and Philosophy* 25 (1): 131–140.

Volkow, Nora D., George F. Koob, and A. Thomas McLellan. 2016. Neurobiologic advances from the brain disease model of addiction. *New England Journal of Medicine* 374 (4): 363–371.

Wakefield, Jerome C. 2017a. Addiction and the concept of disorder, Part 1: Why addiction is a medical disorder. *Neuroethics* 10: 39–53.

———. 2017b. Addiction and the concept of disorder, Part 2: Is every mental disorder a brain disorder? *Neuroethics* 10: 55–67.

Walker, Mary Jean. 2010. Addiction and self-deception: A method for self-control? *Journal of Applied Philosophy* 27 (3): 305–319.

Walker, Mary Jean, and Wendy A. Rogers. 2017. Defining disease in the context of overdiagnosis. *Medicine, Health Care and Philosophy* 20: 269–280.

———. 2018. A new approach to defining disease. *The Journal of Medicine and Philosophy: A Forum for Bioethics and Philosophy of Medicine* 43 (4): 402–420.

Chapter 16
Addiction and Its Ambiguities: Some Comments from History

Gemma Blok

16.1 Contested Nature of Disease Status

In their rich paper, Mary Jean Walker and Wendy Rogers state that addiction is a 'paradigmatic borderline case', its disease status having been contested for at least several centuries. Is addiction a sin, an individual weakness, or a brain disease? Is it caused by the substance, by the individual's vulnerability and psychology, or by social factors? As historians have demonstrated, ideas about addiction have changed significantly over time (Acker 2002; Carstairs 2006; Courtwright 2001). Walker and Rogers' conceptualization of the vague cluster approach of disease is very helpful in explaining addiction's contested status. First of all, addiction is a "vague" condition where the difference between health and disease is a matter of degree. Moreover, criteria that we might use to speak of a disease in spite of this vagueness, are not easily applied to addiction.

First of all, it is problematic to determine whether addiction counts as physiological or psychological dysfunction or not. For instance, the brain changes involved with addiction can also be seen as "an instance of a brain functioning normally, albeit in an environment different from that in which it was 'designed' to function". The harms of addiction are ambiguous as well, because we can theoretically conceive of a happily addicted person, who has unlimited and easy supply to a preferred intoxicating substance, and continues to function while not experiencing distress or harm. Also, harms related to substance use can be caused both by the drug itself, and by the fact that it is illegal, so by drug prohibition. Furthermore, views of experts differ on the topic of the medical or psychological origins of addictions.

G. Blok (✉)
History of Mental Health & Culture, Open University of the Netherlands,
Heerlen, The Netherlands
e-mail: Gemma.blok@ou.nl

© The Author(s) 2024   217
M. Schermer, N. Binney (eds.), *A Pragmatic Approach to Conceptualization
of Health and Disease*, Philosophy and Medicine 151,
https://doi.org/10.1007/978-3-031-62241-0_16

So, what arguments would be there to decide whether or not we should call this borderline disorder a disease, in spite of all these problems and ambiguities? I agree with Walker and Rogers' statement that the pathologising of addiction in the past centuries has had both positive and negative effects, and that a context-specific approach to addiction would be intellectually most satisfying—although it is hard to see how it would work in practice. Ideally, on a pragmatic approach, they write, "the solution would be to say that addiction can be legitimately considered a disease in some contexts, and not in others, and we could engineer our concept disease in order to obtain the positive effects but avoid the negatives." (Walker and Rogers, Chap. 15, this volume).

This quote resonates with views on substance use as they have been voiced for centuries by intensive users of alcohol and drugs (Blok 2017; Courtwright 1989; Oksanen 2012; Schmidt-Semisch 2019; Warhol 2002). Users of intoxicants have mixed attitudes to the disease concept of addiction, from the nineteenth century British writer Thomas de Quincey, writing about the pains and pleasures of opium use, to present-day users telling their stories to historians or writing memoirs and autobiographies. Some self-identify as addicts, as they feel it helps them and the people around them understand their actions. Others see the concept of addiction as a symbol of social exclusion and normative disapproval. It would indeed be wonderful to work with the addiction concept in a flexible way, as a narrative tool that can be helpful in some personal, social and institutional contexts, but not in others.

16.1.1   The Addiction—Or Substance Use Disorder— Spectrum

Recently, suggestions are increasingly being done as well to start speaking of an addiction spectrum, like we do with the autism spectrum (Thomas and Margulis 2018; Veach and Moro 2017). This is an interesting option to consider regarding the borderline case of addiction. The spectrum approach promises to explicitly acknowledge the wide variation in the type and severity of symptoms and social consequences users of intoxicating substances may experience. The DSM-5, the psychiatric handbook of psychiatry, in fact already distinguishes mild, moderate and severe substance use disorders (SUD). Substance use disorder in DSM-5 is a single disorder, but measured on a continuum from mild to severe. The "severity" of the cases is measured by a large variety of criteria: harm to self and others, physiological, psychological as well as social dysfunction, lack of control over urges to use, and so on.

This spectrum approach carries with it a danger of pathologising the whole continuum of intoxicant use: where would we draw the line on the spectrum? However, on the positive side, it would send a strong symbolic message: that addiction is not a black-or-white issue. The acceptance of addiction as a disease has created a divide tween 'addicts' and 'normal' people. Historians have analysed this as the 'moderationist paradigm' of the (late) twentieth century: the notion that drinking in moderation was not very harmful, and might even have some health benefits (Yokoe 2019). This notion gradually became more accepted. Public health campaigns encouraged citizens to drink in moderation, while alcoholics were increasingly treated in specialised centres and clinics.

This pathologisation of addiction, as Walker and Rogers state, normalises other substance use at the border between addiction and normal or recreational use. The binary divide that emerged over the past century between abnormal "addicts" and normal "moderate drinkers", has dangerous consequences: a growing acceptance of regular drinking and of drunken behaviour in the night time economy, in spite of the damage to users and the people around them.

Over the past few decades, the moderationist paradigm has come under attack by a grassroots anti-alcohol movement that is gaining fast in popularity on the internet, in social media and book publications (Lunnay 2022; McHugh 2019). Sober influencers avoid the word "addiction", partly because of the stigma that is attached to it. Research does in fact offer solid arguments for the statement that addiction is a very heavily stigmatised mental illness, even within the field of (mental) health care (Van Boekel 2014; Schomerus et al. 2011; Schomerus and Corrigan 2022). The word addiction, interestingly, is not used in the DSM-5 either, due to the negative implication associated with the term.

So, we might consider speaking of a SUD-spectrum instead of addiction as a disease. This framing could perhaps do useful work in deconstructing the binary divide between 'addicts' and 'normal people'.

16.2 Harmfulness of Drugs and Drug Policies

Another very important point to consider when we think of addiction as a disease, is the political context. In the history and science of addiction, discourses have focussed not only on the physiological and psychological dysfunctions of the addicts, but on their social dysfunctionalities as well. "Habitual drunkards" and "junkies" have often been described as antisocial, unproductive, hedonistic and manipulative citizens, in media, politics and scientific publications. Both alcohol and drugs have been prohibited in the twentieth century in the US and many European countries as well. With alcohol, this ban has lifted, but many intoxicating substances (opiates, cocaine, speed etc.) are still illegal. As Walker and Rogers argue, "many of the harms commonly experienced in addiction are attributable, not to the addiction itself, but to social arrangements surrounding a substance". They mention that criminalising drug use (and sale) tends to drive down drug purity, can lead to more dangerous forms of use, and increases the risks of overdose. I would add that criminalization of drug use and trade can result in substances being on the illegal market for extremely high prices. This contributes to the physical and social decay of dependent users, as has been described by historians working on various geographical areas and time periods.[1]

How can we incorporate this realization of the harms that can be caused by drug policies, in thinking about addiction as a disease? One way to go about it might be to emphasize the negative impact drug policies can have, in definitions of "harm reduction". This concept that is often used to refer to a treatment paradigm of addiction that is intended to minimize the negative physical and social impact of drug use, such as offering clean needles, condoms, methadone maintenance or heroin assisted treatment. The leading NGO Harm Reduction International, however, includes in its definition the impact of drug policies as well, speaking of harm reduction as "practices that aim to minimise the negative health, social and legal impacts associated with drug policies and drug laws".[2]

One thing seems sure: the vague cluster concept of disease Walker and Rogers propose is a highly useful and sophisticated tool for both philosophers, doctors, and historians, to enlarge our understanding of why addiction has always been such a heavily disputed condition, and to help us move forward in intellectual, medical and social directions that allow space for a wide variety of individual and social experiences with substance use.

References

Acker, Caroline Jean. 2002. *Creating the American junkie. Addiction research in the classic age of narcotic control*. Baltimore/London: The Johns Hopkins University Press.

Bänzinger, Peter-Paul, et al. 2022. *Die Schweiz auf Drogen*. Zürich: Chronos Verlag.

Blok, Gemma. 2008. Pampering needle freaks or caring for chronic addicts. Early debates on harm reduction in Amsterdam, 1972–82. *Social History of Alcohol and Drugs* 22: 243–261.

———. 2017. 'We, the avant-garde'. A history from below of Dutch Heroin use in the 1970s. *BMGN – Low Countries Historical Review* 132: 104–125. https://doi.org/10.18352/bmgn-lchr.10312.

Carstairs, Catherine. 2006. *Jailed for possession. Illegal drug use, regulation and power in Canada, 1920–1961*. Toronto: University of Toronto Press.

Courtwright, David Todd. 1989. *Addicts who survived. An oral history of narcotic use in America, 1923–1965*. Knoxville: University of Tennessee Press.

———. 2001. *Dark paradise. A history of opiate addiction in America*. Cambridge, MA: Harvard University Press.

Lunnay, B., et al. 2022. Sober curiosity: A qualitative study exploring women's preparedness to reduce alcohol by social class. *International Journal of Environmental Research and Public Health* 19: 14788. https://doi.org/10.3390/ijerph192214788.

---

[1] See for instance: P. Bänzinger et al., *Die Schweiz auf Drogen* (Chronos Verlag 2022); G. Blok, 'Pampering needle freaks or caring for chronic addicts. Early Debates on Harm Reduction in Amsterdam, 1972–82.', *Social History of Alcohol and Drugs* 2008; C. Carstairs, *Jailed for possession. Illegal drug use, regulation and power in Canada, 1920–1961* (2006); E.C. Schneider, *Smack. Heroin and the American city* (Philadelphia: University of Pennsylvania Press 2011).

[2] See: https://hri.global/what-is-harm-reduction/.

McHugh, Molly. 2019. *Sobriety is having a moment. Here come the influencers.* Vox.com.

Oksanen, Atte. 2012. To hell and back: Excessive drug use, addiction, and the process of recovery in mainstream rock autobiographies. *Substance Use & Misuse* 47: 143–154.

Schmidt-Semisch, Henning. 2019. "Sucht". Zur Pathologisierung und Medikalisierung von Alltagsverhalten. In *Handbuch Drogen in sozial-und kulturwissenschaftlicher Perspektive*, ed. Robert Feustel et al., 143–159. Wiesbaden: Springer.

Schneider, Eric C. 2011. *Smack. Heroin and the American city.* Philadelphia: University of Pennsylvania Press.

Schomerus, G., and P.W. Corrigan. 2022. *The stigma of substance use disorders.* Cambridge: Cambridge University Press.

Schomerus, Georg, et al. 2011. The stigma of alcohol dependence compared with other mental disorders: A review of population studies. *Alcohol and Alcoholism* 46: 105–112.

Thomas, P., and J. Margulis. 2018. *The addiction spectrum. A compassionate, holistic approach to recovery.* San Fransisco: HarperOne.

Van Boekel, Leonieke. 2014. *Stigmatization of people with substance use disorders: Attitudes and perceptions of clients, healthcare professionals and the general public.* Enschede: Ipskamp.

Veach, L., and L. Moro. 2017. *The Spectrum of addiction. Evidence-based assessment, prevention, and treatment across the lifespan.* Los Angeles: Sage.

Warhol, R. 2002. The rhetoric of addiction. From Victorian novels to AA. In *High anxieties. Cultural studies in addiction*, ed. M. Redfield and J.F. Brodie, 97–109. Berkeley/Los Angeles/London: University of California Press.

Yokoe, Ryo. 2019. Alcohol and politics in twentieth century Britain. *The Historical Journal* 62: 267–287.

Chapter 17
Pragmatism in the Fray: Constructing Futures for 'Medically Unexplained Symptoms'

Monica Greco

17.1 Introduction[1]

William James, a founding figure of pragmatism, described a 'sensation' as something "like a client who has given his case to a lawyer and then has passively to listen in the courtroom to whatever account of his affairs, pleasant or unpleasant, the lawyer finds it most expedient to give" (1910: 113). A sensation is a brute fact that is 'forced upon us'—but it does not speak for itself, and its meaning is subject to the whims of situated interpretation.

About a century after the publication of *Pragmatism*, anthropologist Joseph Dumit describes a social and epistemic configuration where the scene of a lawyer representing a set of sensations in a courtroom is not a mere figure of speech, but often a literal and concrete occurrence. The scene takes place in the United States and the truth in question concerns the sensations (or symptoms) associated with what Dumit calls the "new socio-medical disorders", a group of conditions whose aetiology is uncertain, and whose status as to whether they are "primarily mental, psychiatric or biological" is actively contested in "court battles, administrative categorization and legislative manoeuvering" (Dumit 2000: 210). At stake in these battles is the legitimation of the sick role in the context of the political economy of health in the United States, where access to healthcare and disability benefits is contingent on criteria set by corporate insurance agencies that rely on codified diagnoses. Underpinned by "a profound suspicion of malingering", policies

[1] The text of this chapter reproduces excerpts from Monica Greco, 'Pragmatics of explanation: creative accountability in the care of "medically unexplained symptoms"', *Sociological Review Monographs*, 65 (2017), 110–129.

M. Greco (✉)
University of Bath, Bath, UK
e-mail: mg2725@bath.ac.uk

© The Author(s) 2024
M. Schermer, N. Binney (eds.), *A Pragmatic Approach to Conceptualization of Health and Disease*, Philosophy and Medicine 151,
https://doi.org/10.1007/978-3-031-62241-0_17

typically offer very limited coverage for mental or psychiatric disorders (2000: 216). In such a context, arguing for the neurobiological or biomedical basis of symptoms is equivalent, in a very immediate and practical sense, to arguing for their recognition as true and real (illness), versus their dismissal as fake and unreal. This general background is also the context for what Dumit describes as an inevitable mobilisation of sufferers into activist groups, "forced (...) to advocate for the evidence of their illness in non-traditional medical settings: courtrooms, insurance offices, the mass media, and the Internet" (2000: 218).

Dumit's ethnographic study focuses on how brain imaging technologies are mobilised by patients and their lawyers as objective evidence for the claim that an illness is biomedical. Following how brain scans are used across a range of sites—medical labs, conferences, activist forums, courtrooms—he demonstrates that they are evaluated very differently as forms of evidence in each. Evidence that would be unsatisfactory to a scientific researcher can be 'good enough' in a courtroom setting, for example: not definitively, but long enough to secure a favourable judicial outcome. In other words, whether or not biomedical explanations based on brain imaging studies are held to be true by scientists, they fulfil practical functions for embattled patients. In explicit contrast to caution expressed by Dr. Helen Mayberg, a "respected mainstream PET researcher and neurologist", Dumit asks:

> Are these preliminary underfunded [brain imaging] studies, which are touted as proof, 'bad science'? Are interest groups pressing for specific research agendas biasing otherwise objective work? Or is it possible that there is a *need* for public relations research promoting these disorders as 'brain disorders'? (2000: 227)

Dumit's contention is that it doesn't matter if the brain science mobilised by patients and activist groups is 'bad science': what matters is that it facilitates access to healthcare while preventing the nature of symptoms from being treated as a matter of psychology or psychiatry, in the absence of adequate funding for research into their biomedical causes. Brain imaging, he writes, "will continue to play a key role in resisting the easy assignment of blame, stigma, and causation to the individual", even if this means that "these socio-medical disorders might only be 'explained' temporarily and locally" (2000: 227–8).

In the years since Dumit's original description the associated controversies have become momentous, reverberating strongly across anglophone communities and beyond, facilitated by the evolution of participatory online platforms. In a climate of communication that has become increasingly polemical and litigious—and that now includes controversies around the nature of 'Long Covid'—researchers investigating the value of psychological treatments for these contested illnesses report becoming targets of significant intimidation and abuse, leading many to abandon the field (see Hawkes 2011; Kelland 2019). Notably, abuse has also been directed at patients who testify to having been helped by non-biomedical treatments, pre-empting any simplistic reading of the struggle as one between the interests of 'underdog' patients and those of powerful professionals (see Chapman 2021).

What does James' philosophy have to offer to an understanding of this problematic? A historical and epistemic gulf separates the lawyer in James' figure of speech from the situation described by Dumit. Ostensibly, what their accounts hold in common is the reference to a pragmatist theory of truth, and specifically to the idea that "temporary resting states of varying lengths of time are a much better empirical description of 'truth' than atemporal, universal ones whose adjudication is not made clear" (Dumit 2000: 227). While James might have broadly agreed with this proposition, it is not at the level of abstract generalities that we should compare their respective approaches, but rather at the level of what concrete realities they deem it relevant to take into account, and with what consequences. James presented pragmatism as a "method only", citing as one of its merits the fact that "the most diverse metaphysics can use it as their foundation" (1910: 30, 241). Any abstract and general statement of a pragmatist position thus finds its concrete meaning through the (implicit or explicit) system of metaphysical postulates that puts it to work. Ultimately, the value of any such system of postulates is itself subject to the pragmatist test—to an assessment based on the "fruits, consequences, facts" to which it gives rise. In his mature philosophy, James gave the name "radical empiricism" to his preferred system, with "pure experience" as its main postulate and the "pragmatic method" as its main technique (James 2003: 82). In this context, for the pragmatic method "there is no difference of truth that doesn't make a difference of fact somewhere". As for the postulate of pure experience: "nothing shall be admitted as fact, it says, except what can be experienced at some definite time by some experient (...) and every kind of thing experienced must somewhere be real" (2003: 83).

Dumit's pragmatic approach to the 'new sociomedical disorders' is, as noted, tailored to the constraints imposed by the political economy of health in the United States at the turn of the twenty-first century. The observable social and economic consequences that follow, in that context, from the decision that an illness is biomedical (or not) are the realities ultimately deemed to matter for the purposes of assessing the value or 'truth' of an explanatory hypothesis. Consistently empirical in its attention to a multiplicity of social practices, this version of pragmatism resonates with what Richard Rorty called "epistemological behaviourism", an approach that sees its role as to "straighten out pointless quarrels between common sense and science, but not contribute any arguments of its own for the existence or inexistence of something" (Rorty 1979: 175). Like Rorty's, Dumit's pragmatism is agnostic—in this case concerning the nature of the illnesses (or the reality of sensations) about which he writes. It delegates to the literal lawyers and their conventions the representation of that reality. It neither ventures hypotheses of its own, nor does it take the risk of siding with socially or politically 'difficult' hypotheses, namely those that might not result in immediate practical gains for marginalised groups of patients. As such, it comes uncannily close to what James described as a reductive (mis)understanding of pragmatism, namely "a sort of bobtailed scheme of thought, excellently fitted for the man in the street, who naturally hates theory and wants cash returns immediately" (1910: 229).

Agnosticism with regards to the nature of symptoms is not untypical of social research on the 'new socio-medical disorders'. Much of this research has focused on the social consequences of living with an uncertain illness, while remaining silent on the question of how such illnesses might be truthfully characterised other than as 'uncertain' or 'contested'. In many cases, this reflects an implicit methodological commitment to a bifurcated division of labour that allocates social scientists to the study of 'culture', while forbidding them from entertaining hypotheses about 'nature'. By contrast, epistemological behaviourism, in the example offered here by Dumit, explicitly rejects foundational distinctions such as the one between mind and body, nature and culture, or indeed any explicit metaphysical commitment. Sceptical of the possibility of any truth consistent across any but the most temporary and local situations, it refuses to take first-hand responsibility for speculating about what any such truth might look like, but nevertheless lends its weight to whatever truth appears socially expedient from the immediate perspective of sufferers and their supporters. This approach can appear justified ethically and politically, as a piecemeal subversion of the political economy of health in the US, whose constraints produce effects that are often described as a greater source of suffering than the illnesses as such. While this justification might satisfy the conscience of the lawyers whose work secures access to the sick role for their clients, should others feel equally comfortable in endorsing the approach on this basis?

17.2 A Change of Scene

In what follows I will argue that a pragmatism at the service of political expediency is ultimately self-defeating, and indeed potentially harmful, specifically in relation to the types of illness that are the object of Dumit's description. I will develop this argument with reference to a set of practices that differ significantly from those we have just considered in the US context, but that are nevertheless related to that context by virtue of a background of ideas and polemics that resonates across both settings. The scene I move to now is that of a primary care clinic in the north of England, where a new intervention for patients with 'persistent physical (medically unexplained) symptoms' has been trialled by researchers working in the context of the UK National Health Service (NHS). In the reading I will propose, the Symptoms Clinic, as it is known, exemplifies a bolder version of pragmatism: one that follows the consequences of possible 'truths' beyond the domain of social and political relations, all the way down to infra-organic relations (or physiology). As we shall see, the Symptoms Clinic is an intervention that seeks first and foremost not to reproduce the epistemic dichotomies that inform the socio-political *status quo*, and that may be considered a significant factor in the emergence of this type of illness at both an individual and a societal level.

In order to appreciate the novelty of the Symptoms Clinic as an intervention, it is necessary to consider it in the context of a wider reflexive process that is currently taking place among a network of practitioners and researchers in the UK and other parts of Europe. I will describe this process as an emergent reconfiguration of how the problem of explanation is posed in relation to what are often called 'medically unexplained symptoms' (or 'MUS'). The Symptoms Clinic, and the model of intervention it pioneers, is a provisional culmination of this reconfiguration, whose genealogy I will shortly go on to describe. First, however, it is necessary to spend a few words on questions of nomenclature.

The expression 'medically unexplained symptoms', while increasingly contested and specifically avoided in the context of the Symptoms Clinic, is still widely used to indicate physical symptoms that are not attributable to any known conventionally defined disease. It is discursively related to several other concepts including *somatisation, somatoform disorders* (DSM-IV) *somatic symptom disorder* (DSM 5), and *functional somatic syndromes.* The latter correspond to the conditions that Dumit describes as "new socio-medical disorders", and they include the diagnoses of fibromyalgia, chronic fatigue syndrome, multiple chemical sensitivity, and irritable bowel syndrome, among others. This terminological multiplicity and the controversies surrounding the use of many of these terms reflect the fact that basic taxonomic questions have not been resolved.

The lack of consensus about nomenclature and classification, in combination with the supposedly 'unexplained' nature of the symptoms, constitutes an empirical phenomenon of psychological, social, and medical significance in its own right. If symptoms as such involve a certain degree of suffering, illnesses without a diagnosis, or with a contested diagnosis, involve an additional burden of suffering that stems from profound uncertainty, social stigma, and the potential denial of access to benefits and services. Most of the sociological and anthropological research on these conditions has focused on making this additional burden visible and discussable. The rise of the expression 'medically unexplained symptoms' in the anglophone medical literature arguably reflects an acknowledgment of the stigmatising connotations of 'somatisation'. The shift in nomenclature has been justified on pragmatic grounds: while 'somatisation' is a resented and stigmatising term because of its psychologising connotations, 'unexplained symptoms' appears, at least in principle, comparatively neutral, un-psychological, and benign. But there is something paradoxical and self-defeating in these terminological adjustments, in so far as they are intended *and* perceived to be catering to patient 'preferences' (that is, to their psychology) and to be informed by political or even cynical concerns (as a form of appeasement, to avoid conflict), rather than attending to the concrete reality or truth of the illness. The more care is taken not to offend patients, the more sensitive patients seem to become to the possibility of being duped, infantilised and offended. This is reflected in the notion, expressed colloquially to me by a senior clinician, that each new expression coming into use has a limited 'shelf life': its utility expires as soon as patients realize that it's 'just a new name for somatisation'.

The research developments to which I now turn emerge from and against this background. As we shall see, rather than discarding the concept of somatisation as scientifically obsolete or politically unacceptable, they reconfigure it in significant ways.

17.3   Turning the Tables on 'Somatisation'

In what is still a standard reference for the definition of the concept, Z. J. Lipowski describes somatisation as "a tendency to experience and communicate somatic distress and symptoms unaccounted for by pathological findings, to attribute them to physical illness, and to seek medical help for them (...) despite doctors' reassurances that physical illness cannot account for [the] symptoms". Lipowski further adds that "the appraisal, and hence the meaning, of the experienced symptoms needs to be in terms of an actual or threatened disease of or damage to the body for the term to apply" (1988: 1359). Core to the construct of somatisation is the notion that patients who somatise are committed to physical (biomedical) explanations of their condition. By inference, the construct also involves the assumption that patients cannot think psychologically and/or that they resist psychological explanations.

A series of studies published from the mid-2000s has tested these features of the construct empirically by analysing the transcripts of consultations for unexplained symptoms in primary care settings, with surprising results. One study showed that patients with unexplained symptoms in fact offered many different opportunities for their doctors to address them in a psychosocial register, but the cues patients presented about their emotional and social problems were blocked by most doctors, who reasserted the somatic agenda in various ways. "In responding to patients' cues for explanation by providing symptomatic treatments, investigations or referral", the authors write, "doctors effectively 'somatised' these patients" (Salmon et al. 2004: 175). Other studies pointed to multiple sources of doctors' anxiety when faced with symptoms they could not explain, ranging from fear of overlooking a serious physical disease, to feelings of inadequacy when faced with expressions of psychological distress that they did not feel trained to manage, to fear of litigation and media exposure. Interviews with GPs revealed intensely negative emotions towards patients, with some doctors admitting that these affected their clinical judgment, as GP10 does here:

> You can get yourself into the position where you will never spot an illness in this patient if it was staring you in the face and they were dead on the floor, because you will feel it's just their bloody somatising.

What emerged from the data of this study as a whole was a general "sense of powerlessness (...) in the face of apparently intractable symptoms rooted in the realm of the *social*" (Wileman et al. 2002: 181).

Taken all together, these propositions can be read as a reconfiguration of the problem of somatisation that does not simply dismiss the concept on epistemological grounds, or on grounds that patients find the term unacceptable. *This reconfiguration treats 'somatisation' as the name for a real process but turns the construct around, or reframes it fundamentally.* It does this by providing the elements for an empirically-based analysis of how patients are 'somatised' in the clinical interaction and by the medical system within which this takes place.

The proposition that somatisation is a systemic product of Western medicine is not new. Until recently, however, this remained an abstract proposition based on a discussion of the features of biomedical epistemology (see, e.g. Fabrega 1990). In anglophone biomedicine, what A. N. Whitehead called the "bifurcation of nature" translates into the conceptual distinction between objective 'disease' (physical, measurable pathology) and subjective 'illness' (the corresponding experience of symptoms). The reality of illness is assumed to be caused by—and therefore secondary to—the reality of disease. From the perspective of this epistemological orthodoxy, illnesses that are not supported by evidence of disease can in principle be dismissed or explained away as medically insignificant. The endurance of this orthodoxy, despite the proliferation of conditions that would seem to contradict it, is partly attributable to the practical value it has in relation to functions that are not strictly medical or therapeutic, such as gate-keeping (Greco 1998). Evidence of disease offers a baseline discriminating criterion for access to the sick role and a bulwark against moral ambiguity, however questionable this may have become in a social context defined by a prevalence of 'lifestyle' diseases, and by the redefinition of patients as consumers. In the absence of evidence of disease doctors still make gate-keeping decisions that admit patients into the sick role, but these become less transparent and more difficult to account for (Mik-Meyer and Obling 2012). In healthcare systems mediated by private insurance they may also become legally and politically contentious, as Dumit's work clearly demonstrates.

In this epistemic context, patients who present with unexplained symptoms can easily become caught in a dynamic well rendered in the famous dictum by the rheumatologist Nortin Hadler (1996): "If you have to prove you are ill, you can't get well". The dynamic goes something like this: since there is ostensibly 'nothing wrong' with them, patients with unexplained symptoms need to work hard in a clinical context to "fit in with normative, biomedical expectations" and become a "credible patient" (Werner and Malterud 2003). In this effort they adopt an idiom of explanation that focuses on the physical aetiology of symptoms at the expense of other, more nuanced idioms that are typically employed elsewhere, such as in conversations with family or friends (Bech-Risør 2009). Patients are encouraged to present in this way by doctors who, as we have seen, themselves tend to focus somatically and to ignore psychosocial cues. But the efforts made to behave as a credible patient, to the extent that they are perceived as such by others, will tend inevitably to backfire, because a true illness is supposed to be something that 'happens' rather than something we 'perform'. Therefore, while adopting a somatic idiom of explanation is facilitated and reinforced by the clinical setting and its structural constraints, doing so when the system thinks there is 'nothing wrong' actually renders the patient conspicuous from a psychobehavioural (and moral) point of view, prompting doubt or negative feelings in the doctor, and renewed effort to establish credibility on the part of the patient. The logic of this dynamic is that of a "pragmatic paradox" (Watzlawick et al. 1967). It is a socially embedded situational logic that tends towards a polemical polarisation of the positions of doctor and patient, and of physical versus psychological explanations, such that *physical and psychological explanations emerge as mutually exclusive alternatives, regardless of any more complex or nuanced understandings that the parties involved may privately hold.* This logic and process might be described as the 'iatrogenic vortex of somatisation'.

The social researchers whose work now allows us to describe the observable behavioural elements of this dynamic in rich empirical detail stop at this point. They do not venture further in speculating whether and how the interpersonal process that clearly produces detrimental effects on a psychosocial level might also feed into 'sensations', or the experience of symptoms. But explanatory accounts of such a relation exist elsewhere—in the growing field of placebo science, for example, and particularly in the rapidly proliferating research based on '4E cognition' and Bayesian models of brain function (see e.g. van den Bergh et al. 2017; Henningsen et al. 2018; Kirmayer and Gómez-Carrillo 2019). These accounts imply a feedback loop between the psychosocial suffering associated with delegitimation and the genesis of somatic symptoms. Scientifically plausible brain-based explanations of somatic symptoms therefore do exist, but they are explanations that emphatically do not support the linear models of causality that would attribute symptoms to a biomedical lesion- or disease-as-cause (Borsboom et al. 2018).

As Hadler already suggested on the basis of his own clinical experience with fibromyalgia, the effort to behave as a credible patient has effects beyond the social realm of interpersonal communication, reaching all the way down to the physiological capacity for self-regulation. As a result of having to prove they are ill, Hadler wrote, sufferers are "likely to lose the prerequisite skills for well being, the abilities to discern among the morbidities, and to cope" (1996: 2398). Hadler's observation captures what in Jamesian terms might be described as the mutual interference between two divergent imperatives—bearing in mind that these imperatives are also vectors of concrete experience. One is the social imperative for medical practice to be evidence-based, and thus for any action to be authorised by "truth as technically verified": this corresponds to what James, in *The Will to Believe*, called the duty to "shun error". The other is the imperative of trust, or the "will to believe" in the truth (or reality) of an illness, as the first step in verifying the possibility of healing, in so far as this possibility depends on facilitating a personal action or disposition in the patient (James 1956: 21, 18, 24–5).

Turning the tables on the concept of 'somatisation' refers to the reconfiguration of this concept (and phenomenon) as the iatrogenic product of a medicine organised around the distinction between disease and illness, objective evidence and subjective experience, and biomedical versus psychological explanations. As we have seen, the notion that somatisation is a product of Western medicine is not new, but it is significant that this reconfiguration is now taking place in the clinical research literature, with an explicit view to informing a change in medical interventions. This reconfiguration, in other words, also involves a *propositional* dimension—and it is to describing this that I now turn.

17.4 How to Take Symptoms Seriously?

17.4.1 'Forget (Biomedical) Explanation!'

The propositional dimension of the new discourse on 'unexplained' symptoms stems from two seemingly contrasting conclusions emerging from different strands of research. The first may be summarised by the injunction: 'forget explanation!', and we find it in the context of debates on classification and nomenclature that intensified as the fifth edition of the *Diagnostic and Statistical Manual of Mental Disorders* of the American Psychiatric Association (DSM 5) was being prepared. Beyond questioning specific aetiological assumptions reflected in terms such as 'somatisation', a group of researchers queried the privilege accorded to aetiology more generally, advocating a research and clinical focus on symptoms "in their own right" (Sharpe et al. 2006). They argued that existing diagnostic nomenclature is misleading whenever it implies hypothetical underlying pathology—be it physical (as in 'myalgic encephalomielitis') or psychological (as in 'somatoform disorder')— because this reinforces the assumption of a linear causal relation between pathology and symptoms. This proposition, articulated nearly two decades ago, stemmed from a new field of 'symptom research' where symptoms were approached as emergent phenomena reflecting the brain's integration of multiple aetiological factors—which might, but need not, include conventionally understood disease. Crucially, plausible explanatory frameworks based on '4E cognition' and on Bayesian models of brain function have since been developed to articulate the complex and emergent character of symptoms, whose specific aetiology therefore remains ultimately indeterminate. These new models shift clinical priorities away from establishing a primary causal truth of the illness as the basis on which to act, towards an observation of all interventions (including those designed to diagnose) as potential ingredients in the experience of symptoms.

17.4.2 'Patients Need (Good) Explanations!'

Alongside the proposition that invites clinical scientists to suspend the question of whether symptoms are biomedically explained, and to cease constructing taxonomies around it, we find the sum of consultation-based research which indicates that people who experience debilitating symptoms nevertheless *need* (good) explanations. A 'good' explanation, often implicit in a 'good' diagnosis, reassures and legitimates; without it, as we have seen, the interaction with the medical system can easily leave sufferers in a social and existential no-man's land. This research suggests that explanations and diagnoses should be regarded not merely as more or less accurate representations of the patient's condition, but as efficacious interventions with therapeutic and social value in their own right.

The question of what might constitute a 'good' explanation in this sense must take into account the social and cultural context that, to some extent, will determine its value. This is a context, as we have seen, characterised by the 'proto-professionalisation' of patients and the public at large, who have learned to adopt biomedical vocabularies and to expect biomedical explanations. It is also a context characterised by a 'democratisation of expertise', where multiple claims to epistemological authority as to the nature of symptoms co-exist, in potential tension with each other. In this context, evidence-based clinical algorithms, protocols, and guidelines typically represent a solution to the problem of "whose knowledge *counts* in the medical encounter", deflecting potential conflicts through the reference to a technologically-mediated, supposedly neutral ground (May et al. 2006: 1028). In the case of 'MUS' and associated diagnoses, however, protocols and guidelines have themselves become an object of heated controversy and conflict (see Smith and Wessely 2014).

What constitutes a 'good' explanation has been researched, as part of the reconfiguration outlined here, in ways that draw attention to a new set of contrasts: not the familiar contrast between the *physical* (or biomedically explained) and the *mental* (or biomedically unexplained); nor the contrast between *evidence* (or the authority of biomedicine) and direct *experience* (or the authority of the patient). Both of these contrasts are implicit in the classic model of somatisation, which renders them as a polemical contradiction between mutually exclusive alternatives. As we shall see, the new contrasts concern the pragmatic value of the explanations at play, and introduce important differentiations in terms of what it means to be 'pragmatic' in relation to the problem of explaining symptoms that are not susceptible to a linear causal narrative.

17.5   The Symptoms Clinic: Explanations as a Wager on an Unfinished Present

If it is true that the clinical consultation can have somatising effects, it is also true that not all patients are badly managed in the sense of being 'somatised'. On this basis, the same group of researchers who illustrated the somatising effects of the clinical consultation have also studied the communication between doctors and patients with 'medically unexplained symptoms' with a view to articulating what types of explanations are perceived as 'satisfying' and 'empowering', and can as such be considered clinically effective (Salmon et al. 1999; Dowrick et al. 2004). Once again, these studies turn the tables on how the problem of explanation had traditionally been posed.

The clinical process of reassuring patients, often described as an act of 'normalisation', is supposed to facilitate the "recognition that symptoms are part of the normal human experience" and do not necessarily represent disease, or even illness (Kessler and Hamilton 2004: 163). This process often fails, and the failure has traditionally been imputed to the psychological characteristics of patients. Indeed, the failure to be reassured is a key aspect of Lipowski's classic definition of somatisation. In contrast to the traditional approach, which focuses on how reassurance is received by patients, Dowrick and his team examined how reassurance is delivered by doctors in mainstream primary care practice. They found that the explanations that most succeeded in reassuring patients had three characteristics: first, they acknowledged and validated the patient's sense of suffering, without dismissing the reality or significance of the symptoms; second, they provided "tangible mechanisms" to explain the symptoms; and third, they offered patients the opportunity to link the physical symptoms to the psychosocial dimensions of their life. Their most important finding was that 'good' explanations were those that had the quality of being *co-constructed*: "[w]hat is emerging [from the data]", they write, "is a crucial difference between explanations drawn *a priori* from medical knowledge, and those developed by patients and practitioners within shared frameworks that (. . .) are more likely to provide a satisfactory representation of illness, and of the causes and consequences of symptoms" (Dowrick et al. 2004: 169).

The Symptoms Clinic Intervention (SCI) is a programme of research that takes all of this background as a point of departure for the development and testing of a new type of clinical intervention. The researchers describe the intervention in the following terms:

> The SCI consists of a series of three or four extended consultations. It comprises four key components of *Recognition, Explanation, Action,* and *Learning.* The first consultation lasts approximately 50 min and centres on *Recognition* – active listening and acknowledgement of the patient's account of illness and its impact on daily living. *Explanation*s for patients' symptoms are proposed in the first or in subsequent follow-up consultations (which last approximately 20 min each) and involve making constructive sense of a patient's reported symptoms in terms of physiological, and/or psychosocial mechanisms. GPs and patients then negotiate *Action:* symptom management strategies that are concordant with the explanations previously discussed. Throughout the consultations, GPs and patients *Learn* what does or does not make sense or work for the patient and their understanding and management of their symptoms. (Morton et al. 2017: 225; see also Fryer et al. 2023)

Note how the four components of the intervention—recognition, explanation, action, learning—yield the acronym '*REAL*'. Starting from the first extended consultation, the Symptoms Clinic marks a clear break with the conventions of routine mainstream practice. The consulting practitioner, for example, deliberately meets the patient without having had access to their previous medical notes, in a gesture that designates the space of *this* consultation as a new and different space. In this space, the patient's story can be told afresh, and virtually the whole hour is devoted to listening. Gradually, in the course of the first conversation and subsequent ones, explanations are not 'delivered', but rather—to use the authors' own term—they are *proposed*, in a cautious process that involves probing their resonance with the world of the patient, and whether—as we might say with James—the hypotheses offered constitute 'living options' for them (James 1956: 3).

In terms of their content, the explanations proposed have scientific plausibility, in the sense that they are implicitly based on the models from neuroscience and other disciplines already referenced above. By the very character of those models, however, they are explanations that do not pinpoint any exact cause(s) and that retain a lot of indeterminacy, vagueness, and flexibility. They lend themselves to playful improvisation, with abundant use of metaphor, and they are offered indeed as *propositions*, as lures for feeling, with the intent of mobilising the patient's own internal resources for self-healing. These internal resources ultimately consist in a patient's ability, based on a change in perspective, to practise new actions—new patterns of breathing, for example—with a view to forming new habits. In the formal parlance of healthcare research, these are 'symptom management strategies': but the formal language conveys nothing of what is arguably the most important factor mobilised by the intervention, which is (to use James' expression) the patient's *will to believe* in the possibility that the techniques can make a difference to their predicament. In this sense, the Symptoms Clinic instantiates the truth of its method as a truth that depends on faith, hope, and sense of promise for its verification: faith, hope, and promise whose activation, in turn, depend on the patient's experience that the doctor believes in the truth of their illness.

For this dynamic to become possible, an equivalent change of perspective is necessary on the part of the doctor. Clinicians who are newly trained to become Symptoms Clinic practitioners, after learning about the theoretical models that support this approach, are supplied with a manual that includes examples of effective imagery and turns of phrase they might want to employ with their patients. But they are also explicitly encouraged to explore their own style, to develop their own metaphors, and most of all to learn to trust a process whose success will depend as much on their capacities for listening and creative improvisation as on their biomedical knowledge and technical skills.

The effectiveness of the Symptoms Clinic Intervention has been tested in a pragmatic, randomised controlled clinical trial (Mooney et al. 2022; Burton et al. 2024). The question of what specific aspect of the intervention might explain its efficacy—'could it *just* be the extra time allowed for the consultation?'—is not part of the research protocol, but James' insights in *The Will to Believe* would caution against an attribution of efficacy to a specific ingredient. Instead, they would urge us to consider the orientation that underpins the intervention as a whole, which consists in a diametrical reversal of the mainstream biomedical imperative to avoid error. The Symptoms Clinic—designed for patients who have already undergone multiple tests to rule out underlying disease—steps in at the point where biomedicine finds 'nothing' and therefore can only suspect malingering, or cast doubt on a patient's sanity. In direct contrast to this, the Symptoms Clinic is conceived to trust the patient's experience of illness and, with it, to assume a richer and indeed indeterminate repertoire of onto-existential possibilities than biomedicine is usually able to acknowledge. In this sense, the Clinic instantiates pragmatism not as an expedient compromise, but rather as "wager on the unfinishedness of the present, (. . .) an (. . .) operation whose business is that of making thought *creative of an alternative future* by (. . .) putting experience to the test of its own *becoming*" (Savransky 2017: 35).

Since 'co-construction' has become a ubiquitous term in health policy discourse, it is worth taking a moment to appreciate the specific character of the co-construction of explanations in the Symptoms Clinic, and the creative difference this specificity implies. In a broader policy and cultural context characterised by values of patient-centredness and democratisation, explanations that are 'co-constructed' rather than unilaterally imposed appear inherently more desirable in so far as they instantiate values of patient participation. But when there is a conflict of interpretations involved, as is often the case in the domain of 'medically unexplained symptoms', the effort to include patient perspectives can result in seemingly intractable dilemmas and distortions. Medical professionals, when faced with potential conflict, often avoid it by 'colluding' with patients in endorsing biomedical theories proposed by them, even when these contradict existing evidence and/or the explanatory models favoured by the professionals themselves (Salmon et al. 1999). Equivalent dynamics have been highlighted at the level of clinical guideline development (Smith and Wessely 2014). These examples, which stem from research in the UK, bear a significant resemblance to the expediency advocated by Dumit in relation to endorsing brain imaging evidence that is "good enough" for the purpose of winning a legal battle, if not necessarily meaningful from a scientific or medical point of view. In all such cases, we might say that the pragmatic political value of honouring the patient perspective is allowed to trump the pragmatic medical value, when the former is privileged uncritically. This can further undermine confidence in the competences of each party involved, and in the process they are mutually engaged in.

It is therefore crucial to note, in this respect, that the practice of co-construction in the context of the Symptoms Clinic is something very different from the appeal to a form of political (and procedural) accountability in the name of patient-centredness or democratisation. The Symptoms Clinic instantiates co-construction not primarily as a political and democratic value, but rather as a process that is *medically* appropriate on account of how the *nature* of the phenomenon it addresses is conceived. The intervention as a whole is borne out of a recognition that the concrete nature of the illness—and not just the patient *qua* citizen, consumer, and bearer of preferences—is such as not to be indifferent to what explanations (or theories) are addressed to it. It is therefore an approach that differs profoundly from the epistemological structure of biomedicine, where diagnostic acts are separated from therapeutic acts, on the assumption that disease is "a biological reality, independent of any therapeutic relationship or intervention, that is simply waiting to be discovered and correctly labeled" (Kirmayer 1994: 184). Conceived in this way, the explanations of the Symptoms Clinic resemble forms of psychotherapeutic truth that Kirmayer describes as prospective, proactive and prescriptive: truths whose value relies not on their ability to accurately describe causal mechanisms and predict the future, but on their ability to lure events—and the embodied experience of the patient, in this case—in the direction of new possibilities. Importantly, however, the *de facto psycho*-therapeutic character of the explanations, in this case, emphatically does not imply an attribution of psychopathology (as distinct from organic pathology) or a psychiatric diagnosis (as distinct from a biomedical one). Again, the reason for this stems less from 'political' concerns—a concern not to offend, for example—than from how the nature of the problem at hand is conceived, and the exigencies or obligations this generates in terms of constructing a valid explanation, one that satisfies from a scientific as much as from a clinical and political perspective. If the explanations 'ring true', and are thereby effective, this is not because they simply confirm pre-existing (and possibly dysfunctional) beliefs, but because there is no longer an irreducible contradiction between the medical truth and that of the patient's experience.

17.6   Conclusion: Choose Your Pragmatism Carefully

The whole originality of pragmatism, the whole point in it, is its use of the concrete way of seeing. (...) It may be, however, that concreteness as radical as ours is not so obvious.
(James 1910: 241)

The example of the Symptoms Clinic and the Jamesian pragmatism that it quietly enacts offer a meta-perspective on pragmatism in the vein of 'epistemological behaviourism'—or the version of pragmatism exemplified in Dumit's approach to the "new medical disorders". If what matters, as Dumit suggests, is the pragmatic value of the biomedical explanations favoured by patient interest groups rather than their scientific truth, it is still possible to put such explanations to the pragmatic test of the difference they make to the phenomenon they address. It is still possible, in other words, to consider how different theoretical/explanatory characterisations of the nature of symptoms, as a factor of experience, may infect and affect the becoming of symptoms themselves. To state the matter in the broadest terms, the value of 'good enough' biomedical explanations—in so far as these 'work' when they are mobilised by patients and their advocates to ensure access to the sick role—is *to purchase legitimacy on an immediate and piecemeal basis at the expense of reinforcing a bifurcated mode of thought* that, as we have seen, is an important factor in (re)producing the experience and predicament of 'medically unexplained symptoms' in the first place. Pragmatism wedded to 'epistemological behaviourism' is metaphysically neutral only in appearance: in practice, by its very logic, it endorses whatever metaphysical postulates are dominant in the social context where it hopes to intervene. In this case, it leaves unchallenged the mainstream tendency to regard the alternative between psychogenic or biomedical explanations as one between what James called "genuine options", and indeed contributes to intensifying the momentous or consequential character of the alternative.

From the perspective here articulated, the genuine options that are relevant to the contemporary problematic of 'medically unexplained symptoms' lie somewhere else entirely. In general terms, they lie in the choice between a timid pragmatism that bows to the constraints of the socio-political *status quo*, and a bolder pragmatism proper to a radical empiricism. More specifically, they lie in the choice between limiting oneself to considering pragmatic consequences at the level of observable social relations, on the one hand, or alternatively doing "*full justice to conjunctive relations*" (James 2003: 23)—which means following consequences for relations that go "all the way down", to include deep relations between social structure, embodied experience, and physiology. Recall the postulate of James' pragmatic method: "there is no difference of truth that doesn't make a difference of fact *somewhere*" (italics added). The difference between the two pragmatics of truth here contrasted lies in *where and how deeply they look* for the facts made different by the truths.

The momentous character of this alternative lies, I propose, in how each of these versions of pragmatism configures the problem of agency in relation to the problem of illness. In the context of modern dualism, as Dumit rightly observed, biomedical explanations do not only function as a warrant of the *material reality* of an illness but, by the very same token, they also function as a warrant of its *moral insignificance, or neutrality*. To recall Dumit's words, brain imaging techniques are valued as evidence because they "play a key role in resisting the easy assignment of blame, stigma, and causation to the individual" (2000: 227). But the pragmatic advantage of a guarantee against blame comes at a very high cost: the cost of positing a metaphysical chasm between a material domain of blind physical processes, and a spiritual domain at once unconstrained by, and disconnected from, all physical necessity. Investing in the truth of this chasm means investing in the image of a world—and of a living body—that is indifferent to our dispositions and our actions. Followed to its ultimate conclusions, this postulate denies the very possibility of medicine as purposeful and value-driven human activity: it is an expression of what the philosopher Alfred North Whitehead diagnosed as a "radical inconsistency at the basis of modern thought" (1985: 94). Modern biomedicine can afford to gloss over the inconsistency in so far as its methods work—but the domain of 'medically unexplained symptoms' and the iatrogenic vortex of somatisation are limit-cases, epistemically marginal yet empirically ubiquitous, that betray its *hubris*.

Aside from questions of logical inconsistency, investing in the truth of the chasm is not without consequence or neutral: it too implies a disposition, an outlook on experience that is also a vector of experience. In doing *full justice to conjunctive relations*, radical empiricism invites us to consider the continuity between our dispositions and the possibilities that are, in turn, enabled or foreclosed by them. From this perspective, as James insisted, the value of positing 'free will' as a form of causality has little to do with functions of apportioning responsibility, blame or praise. The true significance of the possibility of free will is *prospective*: it is to *enable the feeling of the possible* as such, or "the right to expect that (...) the future may not identically repeat and imitate the past" (James 1910: 59). Without it, there would be no reason to act at all or, in relation to a situation of illness, to suppose or trust we can do anything to make it better. Here we can appreciate quite how costly the investment in a social convention that guarantees against blame can be, if this guarantee is purchased at the expense of foreclosing the experiential and therapeutic value of hope and of positive expectancy. If the Symptoms Clinic is effective, it is at least partly because it carefully crafts a space where the continuity between personal agency and illness can be re-established, while being divorced from the question of blame.

To conclude, the reconfiguration of the problematic of 'unexplained symptoms' that we see currently occurring is facilitated by the existence of healthcare systems that do not as yet require the reality of an illness to be established as objective, codifiable and itemisable as a condition of access to services. This means that cultivating individuals' capacity to imagine themselves (and their symptoms) differently, in an attempt to activate their potential for self-healing, can occur in the safety of a context of care and under the aegis of care. In a different system, as the US case illustrates, the same capacity to imagine oneself differently might constitute a reason to be excluded from the system, and in this sense the system acts as a strong deterrent against developing it. This suggests, perhaps counter-intuitively, that responsibilisation for health—the fostering of the ability to 'take ownership' of health factors that are beyond the remit or control of medical practice—may be most effective when it is premised on the existence of a system that provides universal coverage, that is capable of tolerating a relatively greater degree of indeterminacy or uncertainty at the point of entry, and that is premised not on the suspicion of malingering but on a positive cultivation of trust.

References

Bech-Risør, Mette. 2009. Illness explanations among patients with medically unexplained symptoms: Different idioms for different contexts. *Health* 13: 505–521.

Borsboom, Denny, Angélique O.J. Cramer, and Annemarie Kalis. 2018. Brain disorders? Not really. Why network structures block reductionism in psychophathology research. *Behavioral and Brain Sciences* 42: e2. https://doi.org/10.1017/S0140525X17002266.

Burton, Christopher, Cara Mooney, Laura Sutton, David White, Jeremy Dawson, Aileen R. Neilson, Gillian Rowlands, Steve Thomas, Michelle Horspool, Kate Fryer, Monica Greco, Tom Sanders, Ruth E. Thomas, Cindy Cooper, Emily Turton, Waquas Waheed, Jonathan Woodward, Ellen Mallender, and Vincent Deary. 2024. Effectiveness of a symptom-clinic intervention delivered by general practitioners with an extended role for people with multiple and persistent physical symptoms in England: the Multiple Symptoms Study 3 pragmatic, multicentre, parallel-group, individually randomised controlled trial. *The Lancet* 403 10444: 2619–2629

Chapman, James David. 2021. In summary. *blogpost*. https://batteredoldbook.blogspot.com/2021/04/in-summary.html. Accessed 10 Aug 2023.

den Bergh, Van, Michael Witthöft Omer, Sybille Petersen, and Richard J. Brown. 2017. Symptoms and the body: Taking the inferential leap. *Neuroscience & Biobehavioural Reviews* 74: 185–203.

Dowrick, Christopher, Adele Ring, Gerry M. Humphris, and Peter Salmon. 2004. Normalisation of unexplained symptoms by general practitioners: A functional typology. *British Journal of Medical Practice* 54: 165–170.

Dumit, Joseph. 2000. When explanations rest: 'Good enough' brain science and the new socio-medical disorders. In *Living and working with the new medical technologies*, ed. Margaret Lock, Allan Young, and Alberto Cambrosio. Cambridge: Cambridge University Press.

Fabrega, Horacio. 1990. The concept of somatization as a cultural and historical product of Western medicine. *Psychosomatic Medicine* 52: 653–672.

Fryer Kate, Thomas Sanders, Monica Greco, Cara Mooney, Vincent Deary, and Christopher Burton. 2023. Recognition, explanation, action, learning: teaching and delivery of a consultation model for persistent physical symptoms. *Patient Education and Counselling* 115: 107870

Greco, Monica. 1998. Between social and organic norms: Reading Canguilhem and 'somatisation'. *Economy & Society* 27: 234–248.

Hadler, Nortin. 1996. If you have to prove you are ill you can't get well: The object lesson in fibromyalgia. *Spine* 21: 2397–2400.

Hawkes, Nigel. 2011. Dangers of research into chronic fatigue syndrome. *British Medical Journal* 342: d3780.

Henningsen, Peter, Harald Gründel, Willem J. Kop, Bernd Löwe, Alexandra Martin, Winred Rief, Judith G.M. Rosmalen, Andreas Schröder, Christina van der Feltz-Cornelis, Omer van den Bergh, and EURONET-SOMA Group. 2018. Persistent physical symptoms as perceptual dysregulation: A neuropsychobehavioural model and its clinical implications. *Psychosomatic Medicine* 80: 422–431. https://doi.org/10.1097/PSY.0000000000000588.

James, William. 1910. *Pragmatism & The meaning of truth.* Gearhart: Watchmaker Publishing.

———. 1956. *The will to believe and other essays in popular philosophy.* New York: Dover Publications.

———. 2003. *Essays in radical empiricism.* Mineola: Dover Publications.

Kelland, Kate. 2019. *Online activists are silencing us, scientists say.* Reuters Special Report. https://www.reuters.com/article/us-science-social-media-specialreport-idUSKBN1QU1EI. Accessed 10 Aug 2023.

Kessler, David, and William Hamilton. 2004. Normalisation: Horrible word, useful idea. *British Journal of General Practice* 54: 163–164.

Kirmayer, Laurence. 1994. Improvisation and authority in illness meaning. *Culture, Medicine and Psychiatry* 18: 183–214.

Kirmayer, Laurence, and Ana Gómez-Carrillo. 2019. Agency, embodiment and enactment in psychosomatic theory and practice. *BMJ Medical Humanities* 45: 169–182.

Lipowski, Zbigniew J. 1988. Somatisation: The concept and its clinical application. *American Journal of Psychiatry* 145: 1358–1368.

May, Carl, Tim Rapley, Tiago Moreira, Tracy Finch, and Ben Heaven. 2006. Technogovernance: Evidence, subjectivity and the clinical encounter in primary care medicine. *Social Science and Medicine.* 62: 1022–1030.

Mik-Meyer, Nanna, and Anne Roelsgaard Obling. 2012. The negotiation of the sick role: General practitioners' classification of patients with medically unexplained symptoms. *Sociology of Health and Illness* 34: 1025–1038.

Mooney, Cara, David Alexander White, Jeremy Dawson, Vincent Deary, Kate Fryer, Monica Greco, Michelle Horspool, Aileen Neilson, Gillian Rowlands, Tom Sanders, Ruth E. Thomas, Steve Thomas, Waqas Waheed, and Christopher D. Burton. 2022. Study protocol for the Multiple Symptoms Study 3: A pragmatic, randomised controlled trial of a clinic for patients with persistent (medically unexplained) physical symptoms. *BMJ Open* 12. https://doi.org/10.1136/bmjopen-2022-066511.

Morton, LaKrista, Alison Elliott, Jennifer Cleland, Vincent Deary, and Christopher Burton. 2017. A taxonomy of explanation in a general practitioner clinic for patients with persistent 'medically unexplained' physical symptoms. *Patient Education and Counseling* 100: 224–230.

Rorty, Richard. 1979. *Philosophy and the mirror of nature.* Princeton: Princeton University Press.

Salmon, Peter, Sarah Peters, and Ian Stanley. 1999. Patients' perceptions of medical explanations for somatisation disorders: Qualitative analysis. *British Medical Journal* 318: 372–376.

Salmon, Peter, Christopher Dowrick, Adele Ring, and Gerry M. Humphris. 2004. Voiced but unheard agendas: Qualitative analysis of the psychosocial cues that patients with unexplained symptoms present to general practitioners. *British Journal of General Practice* 54: 171–176.

Savransky, Martin. 2017. The wager of an unfinished present: Notes on speculative pragmatism. In *Speculative research: The lure of possible futures,* ed. Martin Savransky, Alex Wilkie, and Marsha Rosengarten, 25–38. London: Routledge.

Sharpe, Michael, Richard Mayou, and Jane Walker. 2006. Bodily symptoms: New approaches to classification. *Journal of Psychosomatic Research* 60: 353–356.

Smith, Charlotte, and Simon Wessely. 2014. Unity of opposites? Chronic Fatigue Syndrome and the challenge of divergent perspectives in guideline development. *Journal of Neurology, Neurosurgery and Psychiatry* 85: 214–219.

Watzlawick, Paul, Janet Bevan Bavelas, and Don D. Jackson. 1967. *Pragmatics of human communication: A Study of interactional patterns, pathologies, and paradoxes*. New York: Norton.
Werner, Anne, and Kirsti Malterud. 2003. It is hard work behaving as a credible patient. *Social Science and Medicine* 57: 1409.
Whitehead, Alfred North. 1985. *Science and the modern world*. London: Free Association Books.
Wileman, Lindsey, May, Carl, Chew-Graham, Caroline A. 2002. Medically unexplained symptoms and the problem of power in the medical consultation: A qualitative study. *Family Practice* 19: 178–182.

Chapter 18
The Bodily Deficit in Contemporary Healthcare

Jenny Slatman

In her chapter 'Pragmatism in the Fray', Monica Greco convincingly argues that 'somatization' is the iatrogenic product of Western medicine (Greco, Chap. 17, this volume). The term somatization is used for the phenomenon that people still suffer from physical complaints even after doctors have been unable to find pathology. For example, you have suffered from pain in your joints for a long time, but all diagnostic tests are negative (the blood values are good, nothing shows up on scans). When somatization occurs, there is an illness without a disease. Because no underlying pathology is found in the biological body, it is assumed that the reason for, or cause of the complaints must be psychological. After all, within Western medicine we assume a distinction between body and mind. If there is not something wrong in the body, there must be something wrong in the mind. People who somatize are then said to be unable to psychologize—they resist a psychological explanation. Greco shows that this somatization is systemic within Western medicine. Doctors are trained to search for pathology in the biological body with the purpose of providing a pathological explanation for every condition. Physicians are not well prepared for situations where pathological explanations are not so obvious. Uncertainty is no part of the doctor's habitus.

I agree with Greco that this is a systemic phenomenon, something deeply rooted in modern western medicine, in its theory, practice and training. What I would add here is that this systemic problem is based on a very narrow understanding of human embodiment. It is often stated that body-mind dualism can be traced back to the thinking of French philosopher René Descartes (1596–1650). This, however, is a widespread misrepresentation. It is true that this philosopher demonstrated that a body-mind distinction can occur when someone (usually a philosopher) performs a methodical doubt experiment. During that experiment, everything must be doubted

J. Slatman (✉)
Department of Culture Studies, Tilburg University, Tilburg, The Netherlands
e-mail: j.slatman@tilburguniversity.edu

© The Author(s) 2024
M. Schermer, N. Binney (eds.), *A Pragmatic Approach to Conceptualization of Health and Disease*, Philosophy and Medicine 151,
https://doi.org/10.1007/978-3-031-62241-0_18

until you stumble upon something that cannot be doubted. According to Descartes (2008), that unquestionable foundation is the doubting or thinking I, the *res cogitans*, the thinking thing. Whereas the existence of one's own body can be doubted, the existence of one's own thinking or doubting (the mind) cannot be doubted. This is what Descartes' dualism entails. The seventeenth century thinker, however, does not apply this way of thinking to health problems. In the description of his doubt experiment, in the first Meditation, we can read that Descartes still embraced the four body fluids theory that had its origins in theory and practice of Hippocrates (460 BC–377 BC) and Galen (129 AD–216 AD). According to this theory, health problems are caused by too much or too little of one of the humors: yellow bile, black bile, blood, phlegm. In his texts Descartes compares his own excessive doubting with the behavior of insane people, people who according to Descartes suffer from fumes in their brains caused by an excess of black bile. Even after the Flemish anatomist Andreas Vesalius (1514–1564) had already established in the sixteenth century that there is no such thing as black bile at all, many people (including Descartes) continued to cling to the four humors theory for centuries. For example, at the beginning of the nineteenth century, the French physician François-Joseph Broussais (1772–1838) still performed bloodlettings with the idea of restoring the balance of the four humors. So, the four humors theory has been dominant in European (and Arab) thought about disease and health for over two thousand years. The interesting thing about this humoral thinking is that the boundaries of health problems did not simply coincide with the boundaries of the biological body. A disturbance of the humors had everything to do with temperature and humidity, with climate, with food, with lifestyle. According to humoral reasoning, disease does not simply exist in an individual body.

All this changed around 1800, as described by Michel Foucault in his *Birth of the Clinic* (1973). In his historical analysis, Foucault distinguishes between "classifying medicine" (eighteenth century) and "anatomico-clinical medicine" (as of 1800). According to classifying medicine (or nosology), diseases can be classified into different classes, species and types. The view of the nosographer, Foucault says, is that of a gardener classifying plants. Just as Carl Linnaeus drew up a taxonomy of nature in the eighteenth century, so did the physicians of the eighteenth century assume that there was some kind of natural objective taxonomy of diseases. A disease was then understood as something that exists in a certain way in this order of rank and is manifested by certain symptoms in someone who has that disease. The eighteenth-century doctor would read the symptoms of a sick person—for example, cough, rapid pulse, fever, pain in the side—and thereby look for the disease in his classification table, in this case, for example, inflammation of the pleura. Thus, in interpreting an illness, the classifying physician begins with a particular 'class' or 'type' of disease as it was described in the taxonomy of the time and tries to match it with a sick person's symptoms. By contrast, the gaze of the nineteenth-century clinician is no longer that of a classifying horticulturist, but that of a chemist who disassembles, analyzes individual elements. The instrument of that analysis becomes anatomy, which can take apart single tissues. The anatomist Xavier Bichat (1771–1802) called for opening up corpses to identify diseases in tissues. It is by means of this epistemological shift that the idea of class is replaced by 'seat' (*siège du mal*). Disease is seated in the body. This seat—headquarter—is the spatial and temporal starting point of disease.

It is through this way of clinical thinking, rather than Descartes' doubt experiment, that the humoral idea of embodiment turns into solid anatomy. The body becomes the map for diseases. Instead of doctors trying to make sense of a humoral imbalance against the backdrop of climate, lifestyle and diet, or doctors trying to classify diseases by a pre-given taxonomy, as of the beginning of the nineteenth century doctors start searching for the seat of disease in the body. This is what Foucault calls the "medical gaze." By the way, this gaze does not consist only of looking; on the contrary, it can also involve other types of sense perception. The stethoscope, for instance, invented at the beginning of the nineteenth century by Rene Laennec, was crucial for the performance of the medical gaze in its early period. Throughout time, this gaze has continued to be refined owing to all sorts of scientific and technological developments. While Bichat's gaze remained at the level of tissues, from Rudolf Virchow (1821–1902) onward it penetrates further into the cell, to enter the nucleus and DNA in the twentieth century. Supported by all kinds of technological diagnostic tests, including the whole pallet of imaging technology, the medical gaze invades the very smallest structures of the biological body. There must be a biomarker—the contemporary term for the *siège du mal*—for every disease (cf. Boenink and van der Molen, Chap. 11, this volume). And, following the logic of elimination, it is quickly inferred that if nothing can be found in the body, it must be psychological. When it is said that someone is somatizing, it is suggested that the problems someone experiences in her body cannot be traced back to the map of the biological body and are therefore of a different order.

Illnesses without disease, medically unexplained physical symptoms (MUPS) or persistent physical symptoms (PSS), reveal that the anatomical medical gaze has reduced people's bodily existence to the existence of a body as an object to be analyzed. In contemporary medicine, then, we are dealing with a very poor understanding of embodiment, a body as a thing, an object, conceived, moreover, as something individual, disconnected from everything around it. This is what I would call the bodily deficit of contemporary healthcare. If we foreground the idea that it is because of our embodiment that we are always in direct contact with our environment, we might be able to overcome this deficit. For the German neurologist and psychiatrist Kurt Goldstein (1934), the relationship between body (organism) and environment was central to his evaluation and treatment of WWI veterans with brain injuries. It was not the size or localization of their injury somewhere in the brain tissue that determined everything about their health problems. The degree of ill-health, he said, was defined by the ability these men had to deal with their world. And these possibilities are shaped by the *relationship* between environment and body. The French philosopher Maurice Merleau-Ponty (1962) adopts this relational way of thinking about embodiment from Goldstein, adding that the experience people have of their own bodies cannot be reduced to an experience of object-ness. One's own body is also a body-as-subject. And this body-as-subject does not stick to the boundaries of the biological body at all. The lived subject body incorporates all that is appropriated by means of habit—cloths, tools, aids, prosthesis vehicles, but also skills become part of the subject body (Slatman 2014). The boundaries of the subject body extend as far as one's capabilities reach. Where in mainstream modern philosophy the subject was defined by an 'I think,' Merleau-Ponty defines the body subject as 'I can.'

Within the phenomenology of health and illness, the idea of 'I can' is often used to make sense of health problems. As with Goldstein, the strength or weakness of an 'I can' is not determined by the organism alone; its relationship to material and social environment is equally decisive. Monica Greco describes a hopeful trial in which doctors and patients search together for a possible interpretation of the health problem. Within this Symptom Clinic, much space is given to metaphors and imagery to break the common medical gaze paradigm. If we indeed want to move away from this narrow view, I think it would be wise to also look for words and images that connect to embodiment as a relationship between body and everything around that body. Then a psychologizing discourse may eventually become redundant, and we will no longer need stigmatizing descriptions such as somatization.

References

Descartes, René. 2008. *Meditations on first philosophy. With selections from the objections and replies*. Trans. M. Moriarty. Oxford: Oxford University Press.
Foucault, Michel. 1973. *Birth of the clinic: An archaeology of medical perception*. Trans. A.M. Sheridan. London: Routledge.
Goldstein, Kurt. 1934. *The organism: A holistic approach to biology derived from pathological data in man*. (With a foreword by Oliver Sacks, 1995). New York: Zone Books.
Merleau-Ponty, Maurice. 1962. *Phenomenology of perception*. Trans. C. Smith. London/ New York: Routledge.
Slatman, Jenny. 2014. *Our strange body: Philosophical reflections on identity and medical interventions*. Amsterdam: Amsterdam University Press.

Chapter 19
Conceptual Engineering Health: A Historical-Philosophical Analysis of the Concept of Positive Health

Rik van der Linden and Maartje Schermer (ID)

19.1 Introduction

In the philosophy of medicine, debates about the meaning of concepts of health and disease have been going on for decades. Conceptual analysis has long been the methods of choice, but in recent years, we witness a shift in approach towards more pluralistic and pragmatic approaches (Cooper 2020; De Vreese 2017; Haverkamp et al. 2018; Nordby 2019; Schwartz 2017; Van der Linden and Schermer 2022; Walker and Rogers 2018). Instead of defining health and disease in a traditional analytic fashion, the medical-philosophical debate now seems to be refocusing on explicating, specifying and contextualizing concepts. If currently used definitions are no longer useful or are considered to be problematic, they can be adjusted, changed or even replaced by ones that are better fitting. This new view on pragmatically formulating definitions for health and disease concepts seems to be in line with what has recently become known as *conceptual engineering* (Belleri 2021; Cappelen 2018; Cappelen and Plunkett 2020).

While conceptual engineering is first and foremost a philosophical theory and method, it has also been argued that conceptual engineering is that what researchers (of any kind) have been doing all along: adjust concepts to accommodate new observations or generalizations, to enable better understanding or improve practical utility, and/or to influence or change social practices.[1] A good example of such a conceptual engineering project is the development of the concept *Positive Health*,

---

[1] For example, one could think of conceptual debates on 'planet' in theoretical physics, 'justice' in philosophy, and 'race' in social sciences.

R. van der Linden (✉) · M. Schermer
Section Medical Ethics, Philosophy and History of Medicine, Erasmus MC University Medical Center, Rotterdam, The Netherlands
e-mail: r.r.vanderlinden@erasmusmc.nl; m.schermer@erasmusmc.nl

M. Schermer, N. Binney (eds.), *A Pragmatic Approach to Conceptualization of Health and Disease*, Philosophy and Medicine 151,
https://doi.org/10.1007/978-3-031-62241-0_19

that has gained significant popularity within the Dutch medical community and beyond. In a publication in the British Medical Journal, following an international interdisciplinary conference, health was re-defined in contrast to the WHO definition as "the ability to adapt and self-manage in the face of social, physical and emotional challenges" (Huber et al. 2011, 2016). Positive Health has been further developed since then, and—despite so far limited international recognition—has become known as an important concept in both health policy and medical practice in The Netherlands.

In this paper we will first analyze the brief history of Positive Health and explore what were the reasons for re-engineering the concept of health and which actors were involved. What were considered to be problems with the WHO definition and for what purposes was this new concept formulated? By investigating the historical development and associated practices of a concept ('genealogy'), we aim to clarify their current use and elucidate their desired and acquired functions (Binney 2021; Dutilh Novaes 2020a). Next, after providing this historical background, we will discuss the developments in the field of conceptual engineering and discuss its implications for our case study. In particular, we will apply the methods of *Carnapian explication* and *Ameliorative analysis* to assess the adequacy of the newly engineered concept of Positive Health. Next to evaluating this new health concept, we thus also hope to contribute to the philosophy of medicine and the promising new field of conceptual engineering, by providing a real-world example of conceptual engineering in the healthcare domain.

19.2   A Brief History of the Concept of Positive Health

19.2.1   Prelude

The concept of Positive Health was conceived and developed in the Netherlands, within the domain of healthcare and health-policy, from 2010 onwards. Of course, the birth of this new concept did not take place in a vacuum. As we will see, the 1948 WHO definition was taken as an important point of reference and other actors in the field of medicine and health policy had already questioned and criticized this definition and proposed other views—some in much the same vein as the proponents of Positive Health. A notable example is a Lancet editorial (2009), showcasing Canguilhem's notion of health as "the ability to adapt to one's environment", making health something—according to the editorial—that is not defined by the doctor but by the person, according to his or her functional needs. By replacing the notion of perfection as included in the WHO definition with adaptation, the editor notes, "we get closer to a more compassionate, comforting, and creative programme for medicine" (2009: 781). Understanding health as adaptability creates new possibilities and a new role for the doctor: to contribute to the self-defined health needs of the patient and help them adjust and adapt to their conditions. This sounds remarkably similar to what Positive Health has come to stand for.

19.2.2   A Conference and a Position Paper

In December 2009, an international invitational conference entitled 'Is health a state or an ability?' was organized by researcher and former general physician Machteld Huber on behalf of the Dutch funding agency ZonMw[2] and the Health Council of the Netherlands.[3] The conference report explains that the initiative for a conference arose because "in different domains it became apparent that there is a need for a revision, or at least discussion, of the widely known WHO definition of health" (Huber 2010: 3).

The favored formulation of health by the experts at the 2009 conference was a dynamic one, "based on the resilience or capacity to cope and maintain and restore one's integrity, equilibrium, and sense of wellbeing" (Huber et al. 2011: 236). Reference is made to the notion of 'allostasis': the maintenance of physiological homeostasis through changing circumstances.[4] The outcomes of the conferences were laid out in an article entitled 'How should we define health?', published in the British Medical Journal (Huber et al. 2011). This article conceptualizes health "as the ability to adapt and to self-manage in the face of social, physical and emotional challenges". The authors explicitly state that this is not a new definition, but rather what they call "a general concept", "a characterization of a generally agreed direction in which to look" (Huber et al. 2011: 236). They also use the terms "conceptual framework" and "formulation" to stress this point. Next to this general concept, operational definitions are also deemed necessary for measurement purposes.

The article takes the WHO definition, which is said to be no longer *fit for purpose,*[5] as their counter-point. Given the rise of chronic disease, the authors write, "the WHO definition becomes *counterproductive* (italics ours) as it declares people with chronic diseases and disabilities definitively ill. It minimizes the role of human capacity to cope autonomously with life's ever changing physical, emotional, and social challenges and to function with fulfilment and a feeling of wellbeing with a chronic disease or disability" (Huber et al. 2011: 236). Other reasons given in the article for the need to abandon the WHO definition are the medicalizing effects of the qualification 'complete' wellbeing, and the problems of operationalization and measurability.

---

[2] ZonMw is the main Dutch public funding agency in the domain of medical scientific research and healthcare research.

[3] The Health Council of the Netherlands is an important independent scientific advisory board, advising the Dutch Government.

[4] No references are made to the Lancet editiorial or Canghuilhems notion of adaptability, nor to older notions of health as equilibrium, e.g. in humorism.

[5] The authors use remarkably functional language in their discussion of the concepts. The limitations of the WHO definition are said to affect health policy in undesirable ways, but the brief article doesn't spell out how this works, exactly. Nevertheless, the alleged function of the health concept is to steer policy.

19.2.3   Developing the 'New, Dynamic Concept' into Positive Health

Between 2011 and 2013 Huber led a research project commissioned and funded by ZonMw to look into the support base for the new notion of health, and into ways to operationalize it. This research resulted in a report coining the notion of positive health (2013) and a national and international scientific publication (2013, 2016). This 2016 paper is the first introduction of the term 'Positive Health'—here still written in undercast, later written with capitals to distinguish it from other positive notions of health—and of the so-called 'spiderweb' (Fig. 19.1) in the international scientific literature.[6]

To further develop the new dynamic concept of 2011 from a 'general concept' into a 'definitive concept', qualitative and quantitative research was conducted amongst seven groups of stakeholders. Healthcare providers, patients, citizens, policy makers, health insurers, public health professionals, and researchers were included. They were asked their opinion about the new health concept, what they considered to be indicators of health and how these indicators fit the new concept.

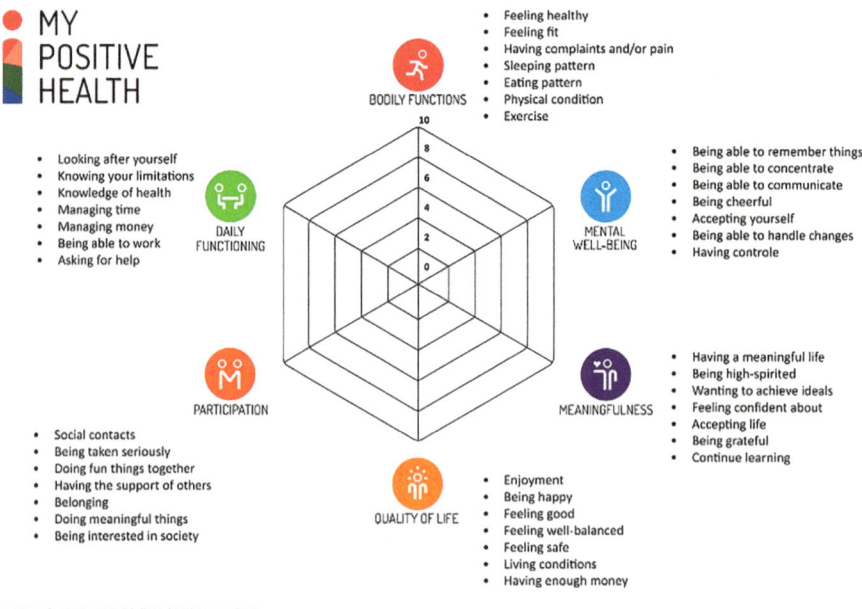

Fig. 19.1   The six-dimensional 'spiderweb' of Positive Health (IPH)

---

[6]According to Huber (2019) the term 'Positive Health' was chosen partly because some of the dimensions of Positive Health touch on 'positive psychology', and partly because Huber found that the WHO in their founding minutes discussed the notion of positive health (positive definition here meant as: defining health by what it is, rather than what it isn't.)

This resulted in a broad range of 556 health indicators, which were grouped together in 6 dimensions—bodily functions, mental functions and perception, spiritual/existential dimension, quality of life, social and societal participation and daily functioning—covering 32 'aspects'.

Huber and her colleagues developed a visual representation of these dimensions that can be scored on a subjective scale, indicating an estimation of a person's state of positive health. This spiderweb represents Positive Health as a subjective state of functioning and wellbeing on the six dimensions. One remarkable finding of this research was that physicians held a much narrower notion of health than patients. Whereas physicians mainly emphasized bodily function and quality of life, patients considered all six dimensions to be equally important. Huber et al. (2016) argue that because of the "prevailing policy trend of 'patient-centered care'" this was to be taken seriously. They therefore propose to use the term Positive Health for this broad perception, as preferred by patients. This concept is claimed to be useful for healthcare providers since it can support shared decision-making, and for policy makers since it can bridge the gap between healthcare and the social domain.

Another research project, aimed at turning the 6-dimensional model of positive health into a measurement instrument, did not deliver the hoped-for results. The authors concluded that "because of major concerns regarding the conceptual model of 'positive health'", it was not possible to develop a valid questionnaire to measure 'positive health' (Prinsen and Terwee 2019: 76). A later attempt to develop a measurement instrument was somewhat more successful. This study concluded that "the current MyPositiveHealth dialogue tool seems reliable as a dialogue, but it is not suitable as a measurement scale" (van Vliet et al. 2021: 1) It proposed a 17-item model with improved, acceptable psychometric properties.[7] In a subsequent study, this 17-item model was further tested for validity and reliability, demonstrating mixed results: only 3 of the 6 dimensions could be considered as valid scales of measurement (Doornenbal et al. 2022). While being moderately optimistic about the results, the researchers call for further development and for more validation through empirical studies.

Between the 2011 and 2016 papers, a number of changes can be observed. There is a major change from a 'general' to a 'definitive' concept, which effectively means that a more general direction-providing description 'health as the ability to adapt and self-manage' has been turned into a six-dimensional construct describing a subjective state of individuals. Self-management and resilience are still mentioned, but now as indicators of health and not as health itself, and as part of the dimension of 'mental functions and perception'. The subjective evaluation of health and the emphasis on what patients say they find important in health has been given more emphasis. The nature of health as dynamic (equilibrium) and as an ability (rather

---

[7] This proposed measurement scale still exists of 6 dimensions, but these have been renamed: physical fitness, mental functions, future perspective, contentment, social relations and daily life-management. Instead of the 32 aspects of the PH model and the 42 items of the dialogue tool, the proposed measurement instrument contains 17 items.

than as a state) appears to have drifted a little into the background in 2016 as compared to 2011.

19.2.4   Uptake and Implementation

The implementation of the Positive Health concept into health policy and practice in the Netherlands has gone very fast. As early as 2012, Huber gets an important prize from ZonMw (a 'Pearl') for her work on a new health concept. This creates quite some attention in the Dutch healthcare world. ZonMw announces they will adopt the new concept of Positive Health and will "stimulate discussion about it on national and international level for policy, practice and research." (ZonMw 2014). Apart from funding Huber's research to further develop the concept, ZonMw also explicitly stimulates and funds other research and activities that use the Positive Health concept. In their 2016–2020 policy plan, ZonMw states it "embraces the concept of Positive Health. This concept – health as one's ability to adapt and self-manage – plays a generally directive role in all ZonMw activities."[8] It is interesting to note that in these policy documents, the general concept of 2011 and the definitive concept of 2016 are used almost interchangeably.

Around the same time, in 2015, Huber founds the Institute for Positive health (iPH),[9] with the mission to create a paradigm switch, shifting the focus from disease and illness, to stimulating resilience and meaning in life (Huber 2019). To this purpose the 'spiderweb' conversation tool is made available for healthcare providers and is said to be increasingly used in medical practice, especially by GPs. By assessing how a patient is doing in terms of the 6 dimensions, it is hoped that healthcare will be directed in a more 'patient-centered' way, which should lead to empowerment and more self-management. Other supporting tools are developed as well, such as information leaflets and guidance documents, for different target audiences such as doctors, patients, and HR professionals.[10]

In 2016, ZonMw together with iPH issue an inventory report on Positive Health in the Netherlands, concluding that it is "both a movement and a method", and urging further implementation starting from practice-based research, development and education. The report is intended for the Minister of Health, to facilitate further policy and to strengthen the societal initiatives around Positive Health (Van Steekelenburg et al. 2016). The inventory shows that many municipalities, health organizations and other institutions are already working with Positive Health, mainly

---

[8] "Research and results: Positive Health", ZonMw, assessed August 7, 2023, www.zonmw.nl/en/research-and-results/positive-health/.

[9] See the website www.iph.nl for more information.

[10] On the webpage of the Institute of Positive Health one can find various educative articles, blogs, videoclips and more, that are made available to help people and organizations to learn about Positive Health and to promote its implementation. For example, see: https://www.iph.nl/toolbox-professional/.

through 'learning by doing'. There appears to be a large variety in activities and interpretations of the 'ideas' or the 'philosophy' of Positive Health. Many parties report that cooperation between domains (especially the medical and the social) could be enhanced or facilitated by Positive Health (Yaron et al. 2021).

In the same year, Huber and the iPH are asked to help the province of Limburg in the Netherlands (with 1,1 million inhabitants) become the first 'Positive Healthy province'. They start an action center ('Limburg Positief Gezond') and implementation project, and the entire process is described and researched by the University of Maastricht, with ZonMw funding (Lemmen et al. 2021; Yaron et al. 2021). By 2018, 51% of Dutch municipalities mention Positive Health in their health policies[11]; it is embraced by many healthcare insurers, healthcare organizations and institutions, and the Federation of Medical Specialists; the concept is explicitly mentioned as a learning objective in the national learning plan for medical students (NFU 2020), it is part of the official professional profile of nurses and physiotherapists, among others, and it is the central concept in the Dutch National Health Policy Note 2020–2024 of the Ministry of Health.

In November 2021, thirteen years after the original invitational conference, a celebration-conference took place under the banner "10 years 'health as the ability to adapt and self-manage'". At this conference, keynotes were given by the Secretary of health, the president of the Health Council, the (vice)presidents of the main funding bodies (ZonMw and NWO), and the CEO of one of the largest healthcare insurers in The Netherlands. Furthermore, the brand new 'Handbook Positive Health in primary care' was presented (Huber et al. 2022).

In recent years, the iPH has entered a strategic partnership with the organization 'Alles is Gezondheid' (which translates as: Everything is Health[12]). They aim to further develop and implement the concept of Positive Health, envisioning The Netherlands to become the 'largest Blue Zone of the world'.

19.2.5 Critique

From the early days on, there has also been critique on the concept of Positive Health. We discuss some frequently mentioned issues here. Jambroes et al. (2015) argue that the new concept of 'health as the ability to adapt and self-manage' may increase inequalities in health within the general population, since it stimulates seeing health as an individual asset and tends to neglect the socio-economic, cultural and environmental determinants of health. Van der Stel (2016) claims it confuses

---

[11]"Positive Health", iresearch, assessed August 7, 2023, https://www.iresearch.nl/positieve-gezondheid.

[12]This is an organization that focuses on health promotion through stimulating and creating regional networks of societal and healthcare organizations.

health with behavior. Moreover, it may lead to absurd consequences, since even a person who is seriously ill may adapt to their situation and hence be deemed healthy.

A similar conclusion is drawn by Stronks, a professor of Public Health, who says that since socio-economically disadvantaged groups often use smoking and drinking alcohol as ways of self-managing the stress of everyday life, this behavior should be seen as 'healthy' according to the general definition of Huber. Moreover, she emphasizes that notions of health that include aspects such as self-management and meaning in life are predominantly recognized by higher socio-economic groups, whereas lower socio-economic groups understand health mainly in terms of absence of physical impairments (Buijs 2017; Stronks et al. 2008). The general concept of health as ability to adapt and self-manage therefore risks increasing socio-economic inequalities and neglects collective dimensions of health. In her article for the Dutch Medical Journal Buijs also voices some surprise that ZonMW embraces and promotes the concept without any evidence for its effectiveness.

Poiesz et al. (2016) point out that the general concept is not very new: Law and Widdows (2008) and Bircher (2005) have earlier formulated very similar concepts. They also list numerous unclarities and confusing aspects of—what they call the "definition"—of health as ability. Moreover, they critique the muddled relationship between the general concept (ability to adapt) and the 'definitive' six-dimension concept of Positive Health in which, they claim, aspects, determinants and dimensions of health are mixed up. "The spiderweb diagram explodes into a firework of concepts (...) which makes it seem the health concept includes the whole of life" (Poiesz et al. 2016: 254). This promotes medicalization, and also makes the concept so broad it "effectively can legitimize almost any intervention (...) there is something in there for everyone – that is, in our view, the primary reason for its popularity" (Poiesz et al. 2016: 254). A more positive framing of this observation follows from empirical research by Lemmen et al. (2021), who conclude that primary healthcare professionals working with Positive Health experienced increased job satisfaction. The concept "helped them to legitimize and give substance to their vision (...) Positive Health's malleability allows for the frame's customization and the creation of the match. Simultaneously, malleability introduces ambiguity on what the concept entails" (Lemmen et al. 2021: 159).

Others have also criticized the concept of Positive Health (Van der Staa et al. 2017; Kingma 2017; Schermer and Van der Horst 2022; Van Boven and Versteegde 2019). Returning issues are the conceptual unclarity and incoherence, and the (presumed or feared) unintentional negative consequences of the concept: medicalization of life, victim blaming ('if you don't adapt to your disease, you are unhealthy'), denial of the negative experiences and limitations of illness (by framing them as challenges to be overcome); increased health inequalities, neglect of social and environmental determinants of health, individualization of health. Some critiques focus on the general concept, some on the six-dimensional approach, some on both taken together. Some critique is on the conceptual level, other on the ways in which Positive Health is interpreted and used in practice, still other on its potential effects.

Finally, it is also questioned whether it is really necessary, in order to achieve some of the laudable goals the Positive Health movement promotes, to introduce a new concept. "A change in course does not require a new ship" (Van der Staa et al. 2017: 34). This, however, remains to be seen. According to the conceptual engineering approach, a new—or at least a refurbished—ship may be exactly what is needed to change course.

19.3 Conceptual Engineering

19.3.1 A New Meta-semantical Theory and Philosophical Methodology

In recent years, the field of analytic philosophy has welcomed *conceptual engineering* as a new meta-semantical theory and philosophical methodology that focuses on the analysis, assessment and revision of concepts (Cappelen and Plunkett 2020). Conceptual (re-)engineering projects typically involve evaluative questions on which concept is best suited for serving a particular function or purpose. The term 'conceptual engineering' originates from the work of Richard Creath (1990), who primarily discussed the form of Carnapian *explication* (Carnap 1950). Explication can be described as the process of replacing (or revising) an inexact concept (the explicandum) into a more exact concept (the explicatum) that is better equipped for a specific theoretical purpose.[13] While conceptual engineering entails much more than adjusting concepts for theoretical purposes, the Carnapian framework is often used as a method in conceptual engineering (Nado 2021). Another type of conceptual engineering can be found in the work by Sally Haslanger (2000, 2012), who speaks of concept *amelioration*. In contrast to the originally more narrowly focused explication, amelioration refers to the process of improving concepts for a broad range of social and political purposes.

Although conceptual engineering is primarily an area of philosophical theory and method, the scope of conceptual engineering can also be extended further, viewing it as something that all researchers do: to adapt their concepts to fit their scientific purposes. It can even be viewed as a ubiquitous human process: concepts are construed in specific times and cultures and adapted throughout time due to scientific and societal developments. So, importantly, while we use the term 'conceptual engineering' to refer to the philosophical theory and method here, we use it for the purpose of evaluating a concept that has been 'engineered' not by philosophers but by professionals involved in the healthcare field.

---

[13] Two examples that are mentioned in Carnap (1950) include the explication of 'fish' as 'piscis', for biology, and the explication of the qualitative concepts 'warm' and 'cold' into the quantitative concept 'temperature', for chemistry. Explications are not necessarily directed towards 'scientific' purposes, however (Brun 2016).

19.3.2   The Target of Conceptual Engineering

In the conceptual engineering literature, there is currently no consensus on how we should define a 'concept' and thus what the target of conceptual engineering is, exactly (Isaac 2021; Koch 2021). However, two main views can be distinguished: semanticism and psychologism (Koch 2021). For semanticism, conceptual engineering is not more than advocating and implementing changes in what our words mean. For psychologism, conceptual engineering is about changing the psychological structures and thereby changing our cognitive processes and linguistic behavior (i.e., the way we classify things in the world, the inferences we make, and the linguistic expressions we use).

In his paper titled 'Engineering what? On concepts in conceptual engineering', Koch argues that conceptual engineering is typically described as a means of achieving at least two goals at the same time: the semantic goal of changing the meaning of terms and expressions in a language, and the practical goal of making real-world changes in our practices and classifications. Neither the semantic nor the psychological view can sufficiently achieve both goals on their own. Therefore, Koch proposes a hybrid view, the "dual content view", in which both positions are deemed necessary for the overall aims of conceptual engineering: "In order to achieve the semantic goal, we ought to engineer referential contents; in order to achieve the practical goal, we ought to engineer cognitive contents" (Koch 2021: 1956). From a pragmatist perspective, in which theory and practice, conceptualization and observation, are viewed as highly interdependent, this makes sense. Koch's view resonates with recent contributions to the philosophy of medicine that demonstrate the performativity of disease classification (e.g., Greco 2012; Hacking 2007; Hyman 2010; Van der Linden et al. 2022). By making changes to the way we define disease and health (semantically), we do not only describe them in a different way but also actively change the way we observe them and deal with them (psychologically).

Isaac et al. (2022), who provide a first sketch for a 'roadmap to practice', take a pluralistic view towards the goals and targets of conceptual engineering. They argue that engineering projects rarely propose representational changes for their own sake—usually, such changes are means towards achieving a further practical purpose, or they are the result of making changes for such practical purposes. In this regard, the nature of concepts, meta-semantically speaking, becomes less relevant. Rather, the motivation behind conceptual engineering proposals is what matters most. What is the problem with the current conceptualization? To whom is this a problem, and why? Does the (re-)engineered concept live up to its task of solving the problem at stake? So, the 'target' of conceptual engineering can be better explained in functional than in ontological terms. It is more about *why* we want to change a concept and less about *what* we are changing. This view is in line with our pragmatic approach to health and disease concepts (see Chap. 3, this volume).

19.3.3   Conceptual Engineering in and for Medicine

The idea of conceptual engineering fits quite well with recent calls in the philosophy of medicine for a pragmatic approach towards health and disease concepts that foreground the function and usefulness of concepts withing specific contexts (De Vreese 2017; Haverkamp et al. 2018; Van der Linden and Schermer 2022; Walker and Rogers 2018). Nevertheless, the methods of conceptual engineering have not yet been applied much within the philosophy of medicine. A notable exception is the work of Elisabetta Lalumera (forthcoming), who has shown how shifting cancer classifications and the debate on the disease status of obesity can be understood from a conceptual engineering framework. She argues that medical research and healthcare organization are active fields of conceptual engineering— conceptual engineering is not merely a philosophical project, but also actually takes place in medical practice, driven by pragmatic considerations. Lalumera shows, for example, that the classification of cancer is pragmatic as the different (sub)types are based not so much on epistemic grounds but rather based on clinical utility—which, she argues, is value-based rather than evidence-based. Regarding obesity, she remarks that the goal of classifying it as a disease has much to do with practical aspects, such as stimulating more research, fostering medical care, and reducing stigma. She concludes that medicine and philosophy of medicine can benefit from insights and methods from the philosophy of conceptual engineering:

> "First, we need to investigate the details of how medical concepts or conceptions are formed and changed in various cases (a task which is close to the social epistemology of science and medicine). Here, a reflection on the non-epistemic basis of medical knowledge is in place: what is the role of values in such decisions? Which and whose values are involved? Starting from such investigations, then, there is room for a normative project, that of individuating the principles according to which such changes should be accomplished in general – including norms and criteria about inclusion and public participation" (Lalumera forthcoming: 20)

We agree with Lalumera that the philosophy of medicine could benefit from incorporating insights from the field of conceptual engineering, especially for analyzing and evaluating conceptual shifts. Conceptual engineering provides us with useful tools for giving further interpretation to this approach. In the next sections, we will discuss two methods to assess the adequacy of (re-)engineered concepts in a way that accounts for context-specificity, and apply them to the evaluation of the concept of Positive Health.

19.4   Methods for Concept Evaluation

19.4.1   A Functional Approach

Conceptual engineering projects may have different aims and goals. Therefore, it has been argued, we do not need to adhere to one specific method or framework for concept evaluation (Nado 2021; Dutilh Novaes 2020b). Whereas some engineering projects may need to be evaluated in more theoretical terms, others may rely more or even completely on empirical studies—or on 'experimental philosophy' (Andow 2020). As we have argued elsewhere (Van der Linden and Schermer 2022), it is key to evaluate health and disease concepts in relation to the function they should serve and the context they ought to be used in. Therefore, we take a 'functionalist approach': assessing the concept in terms of how well it performs in serving the desired or expected function (Nado 2021; Fisher 2015; Brun 2016). In doing so, we will take seriously the warning by Jorem (2022) that a functionalist approach should also account for the moral and epistemological dimensions of concepts.

Here, we describe two main methods of concept evaluation in more detail: *Carnapian explication*, and Haslanger's *ameliorative analysis*. Whereas Carnapian explication is primarily used for the analysis of theoretical—mostly scientifically relevant—concepts, the ameliorative method described by Haslanger is always tied to political goals (Dutilh Novaes 2020b). Yet, both methods could be adjusted to some extent, which enables them to be used for the evaluation of either a broader or a narrower scope. They can complement each other for the analysis of the engineering project we are concerned with—that of Positive Health.

19.4.2 Carnapian Explication

In his book 'Logical Foundations of Probability' (1950), Carnap distinguished between "explicandum" (the broad and 'pre-theoretical' concept used in everyday language) and "explicatum" (the concept explicated in more specific 'theoretical' terms for our target use). Although explications are often formulated in the form of (precising) definitions they can also include other forms. In any form, the explicatum is assumed to replace the explicandum in the targeted context.[14] Furthermore, it is important to note that the explicandum can thus be explicated in multiple ways, possibly resulting in a plurality of explicata. Although the original method of Carnapian explication has been criticized for several reasons (Brun 2016; Nado 2021), the framework is still useful in conceptual engineering when a couple of assumptions and rules of use are made clear. For this, we may turn to the pragmatic interpretation by Georg Brun (2016). Brun makes it clear that we can use Carnapian explication as a method without having to accept Carnap's meta-semantical theory of meaning and reference. The method can be separated from its theory.

In the original work by Carnap, a concept explication is evaluated on basis of four criteria: (1) similarity to the explicandum, (2) exactness, (3) fruitfulness, and

---

[14]Brun reads Carnap here as follows, that the explicatum and explicandum are two different concepts. However, there is plenty discussion about this interpretation. Others for example speak about 'transforming' concepts instead of replacing. For our use of Carnapian explication as a method for evaluation it is not necessary to take a specific position.

(4) simplicity. The first criterion requires that the explicatum must retain some similarity with the explicandum. In Carnap (1950) it is not described in detail how much similarity is exactly required, but the overall idea is that the explication cannot be a process of complete change of meaning. "Close similarity is not required and considerable differences are permitted", Carnap notes (1950: 7). The second criterion requires the explicatum to be more exact than the explicandum, which includes that it should be "unambiguous" and "consistent".[15] Third, the explicatum must also be fruitful. This means that it can be used for new or better generalizations and for the formulation of universal statements—this can also include normative statements. Finally, the fourth criterion can be used to distinguish between explicata in case they score equally well on the other three criteria. The explicatum of choice is the one that has the simplest form and rules for use. It should be clear how and when to use the concept, also in relation to other concepts.

Brun (2016) provides some important reflections on the original method and makes some suggestions that should be taken into account when using Carnapian explication. First and foremost, the four criteria should not be understood as representing fixed scales. It is not possible to measure them in strictly objective ways and it is a priori not decided how to weigh one criterion against the other. In order to do so, Brun stresses that we should use the four criteria in light of the role the explicatum is expected to play in the target theory. Such judgements are context-dependent and usually one type of explicatum cannot be expected to 'outperform' all competitors. Incorporating a (pragmatic) functionalist approach may thus successfully overcome some of the problems with the original method of Carnapian explication (Brun 2016; Nado 2019; Fisher 2015).[16]

19.4.3 Ameliorative Analysis

Ameliorative analysis is another main pillar in the conceptual engineering literature. Although the term 'amelioration' is also discussed in the work of W.V.O. Quine (Two Dogmas of Empiricism 1951), it has become particularly important as a way of analysis and type of conceptual engineering in recent work by Sally Haslanger on the concepts race and gender (2005; 2006; 2012). Haslanger describes her method as a project that seeks to identify what legitimate purposes we might have in making categories of people, and to develop concepts that help us achieve these ends.

---

[15] Some scholars also view 'preciseness' as a subcategory of this criterion. For example, see Isaac (2021).

[16] Jennifer Nado and Justin Fischer hold fundamentally different views regarding the functional approach. Whereas Fischer tends to view concepts by definition *as* (having) functions, Nado disagrees and argues concepts often have (acquired) functions but cannot be reduced to functions: *"concepts do not always need to be in service of some normative end"* (Nado 2019: 87).

"Ameliorative projects, in contrast [to the conceptual- and descriptive approach], begin by asking: What is the point of having the concept in question—for example, why do we have a concept of knowledge or a concept of belief? What concept (if any) would do the work best? In the limit case, a theoretical concept is introduced by stipulating the meaning of a new term, and its content is determined entirely by the role it plays in the theory. If we allow that our everyday vocabularies serve both cognitive and practical purposes that might be well-served by our theorizing, then those pursuing an ameliorative approach might reasonably represent themselves as providing an account of our concept—or perhaps the concept we are reaching for—by enhancing our conceptual resources to serve our (critically examined) purposes." (2005: 12).

So, in ameliorative analysis, one always starts with a critical examination of the current use of a concept and identification of the practical (i.e., primarily social and political) goals it has in our practices and discourses. When this is clarified, one may subsequently investigate what kind of conceptualization would promote such goals or to assess a new revision in light of them.[17] Haslanger distinguishes between three important concepts: the manifest concept, the operative concept, and the target concept (Haslanger 2012; Dutilh Novaes 2020b). The manifest concept is explicit, public and intuitive (our concepts in 'everyday language'). In contrast, while more implicit and 'hidden', the operative concept is the one that is usually practiced. Finally, the target concept concerns the one we *should* use.

According to Haslanger, a successfully ameliorated concept should meet both a "semantic condition" and a "political condition". Meeting the semantic condition implies that the proposed shift in meaning is viewed as appropriate if the 'central functions' of the term remain the same. Meeting the political condition means that the ameliorated concept is successful in the sense of being implemented and used for its desired political goals. However, Haslanger acknowledges that formulating this political condition in general terms is very difficult, as this is significantly influenced by the acceptability of the goals being served, the intended and unintended effects of the change, the politics of the speech context, and whether the underlying values are justified. Dutilh Novaes (2020b), comparing ameliorative analysis with Carnapian explication, notes that the semantic condition can be viewed as the counterpart of the 'similarity' in Carnap, just as the political condition can be viewed as the counterpart of the 'fruitfulness' criterion.

19.5   Assessing the Adequacy of Positive Health

19.5.1   Method of Choice

Now that we have discussed Brun's pragmatic interpretation of Carnapian explication and Haslanger's ameliorative analysis, it is time to put the two methods to work

---

[17]It is argued that both ameliorative analysis as well as Carnapian explication should be viewed as non-linear processes instead of 'step-by-step' linear procedures (Dutilh Novaes 2020b; Brun 2017).

for our own purposes: the analysis and evaluation of the concept Positive Health. As we have demonstrated in our historical analysis, the concept of Positive Health is expected to fulfill a variety of goals. For the goal of improving measurements in medical research and/or improving clinical evaluation and decision-making, the re-engineering project may be considered as an explication. However, the broader goals of Positive Health, such as changing the direction of healthcare, shifting from a biomedical towards a holistic approach, or empowering patients, could be better described as an ameliorative project. Therefore, it seems only reasonable to assess the adequacy of Positive Health in light of both methods. Going through this procedure requires us to go back to our historical analysis but now look through the lens of conceptual engineering.

19.5.2 Through the Lens of Pragmatic Carnapian Explication

Clarifying the Explicandum and Identifying the Task

In order to evaluate the Positive Health concept within the pragmatic Carnapian framework, a first step is to clarify the explicandum and to explore the motivational reasons behind its development. What is the desired function and in what type of context should the concept be deployed? As our brief historical analysis shows, the ideas about the concept and its desired function changed significantly through time. In the prelude to the birth of Positive Health, the suggestions for rethinking the concept of health were primarily for social and political purposes. First and foremost, it was proposed as a response to the WHO definition[18] of health, which was considered to be too demanding and leading to problems of medicalization. However, not long after the conference in 2010, in which the idea of re-engineering the concept of health was born, the aims of the re-engineering project already started to diverge.

Importantly, in the first academic publications by Huber and colleagues, it was made clear that there was a perceived need for a general concept, but also for operationalization for measurement purposes in medical research and policy. This operationalization can be seen as a legitimate attempt for Carnapian explication. In later publications, where the term Positive Health was now explicitly used, a six-dimensional model was developed for measurement purposes. In addition to this, the model was visualized as a spiderweb, so it could also be used as a 'dialogical tool' for clinical practice. This tool was intended to be used by clinicians to have 'the other conversation' (Dutch: 'het andere gesprek') with their patients about their health. By using the spiderweb, as well as the spiderweb-based MyPositiveHealth digital application, clinicians should be able to better understand their patients'

---

[18]The WHO definition can be viewed as a competing explicatum for 'health'.

suffering and needs beyond the usual narrow biomedical scope, and to plan and evaluate medical interventions in a more holistic way. Although the application of the 6-dimensional model as a dialogical tool does not have a scientific purpose, we could still speak of epistemic function: to increase knowledge about the patient's health status and to inform clinical decision-making in a more holistic and patient-centered manner. We may view the measurement tool and the clinical tool as two different applications of the same explicatum.

Assessing the Adequacy of the Proposed Explicatum

Let us now assess the explicatum, the six-dimensional model of Positive Health, in terms of the four Carnapian criteria: (1) similarity to the explicandum, (2) exactness, (3) fruitfulness, and (4) simplicity. With regard to similarity, the explicatum seems to be reasonably similar to the explicandum ('health' in our everyday usage). As the six dimensions were the result of a mixed-method study among various stakeholders in interviews and focus group sessions, it seems that sufficient effort was made to secure a match between the Positive Health model (the explicatum) and the intuitions and beliefs of people on health (the explicandum).

Exactness may be viewed as perhaps the most important criterion if we look at the application for measurement purposes. As described in the historical analysis, studies have been conducted to further develop and test the spiderweb as a measurement instrument. In several methodological studies, the six-dimensional model of Positive Health has been further developed and tested for its validity and soundness of (Prinsen and Terwee 2019; Van Vliet et al. 2021; Doornenbal et al. 2022). The model tested in Prinsen and Terwee was considered to be problematic regarding both construct validity as well as content validity. Doornenbal et al. were more optimistic about the model, after several changes were made. However, still, mixed results were demonstrated: not all dimensions could be sufficiently validated. Notably, 'resilience', one of the core aspects of Positive Health, only showed a moderate relation with the validation scale that was used. Furthermore, 'mental functioning' and 'daily functioning' even showed a weak relation with validation scales.[19]

Doornenbal and colleagues give a couple of possible explanations for the mixed findings. One of their suggestions is that the weak/moderate correlations may be correct findings but perhaps not so relevant because the "explanatory power of these domains may reach beyond that of our validation scales" (2022: 8). In other words, the relation is weak because the dimensions measure something different than the validation scales do, and for good reasons, as this was the goal of Positive Health, they suggest. This is an interesting suggestion that needs further consideration. Yet, we still need to establish somehow the 'exactness' and 'fruitfulness' of the

---

[19] Doornenbal and colleagues remark that "This is an important finding given that patients and citizens in a previous large panel study rated mental functioning and daily functioning as about equally important aspects of health as the other dimensions" (2022: 6).

explicatum before concluding that the explicatum is adequate. The researchers themselves therefore suggest that the usefulness of the Positive Health model must be demonstrated through its successful application in clinical intervention studies. In terms of fruitfulness, the important question that needs to be answered is if the use of the Positive Health measurement instrument actually leads to new and better insights or observations. The same goes for the application of the spiderweb and the MyPositiveHealth app as epistemic tools within the physician-patient relationship.

This also relates to the last criterion: simplicity. Although simplicity is described by Carnap as only being relevant in the situation we have to choose between two or more good explicata, for the sake of completeness we may briefly say something about this criterion as well. Even if the explicatum as proposed may be developed further and may proof to be sufficiently exact and fruitful for its desired applications, than we still need to compare it with competing (currently used) explicata. For instance, for the purpose of scientific measurements, it could be compared with the internationally used HRQL-5D-EL questionnaire (World Health Organization) for measuring 'health related quality of life'. For clinical purposes, we might want to compare it with the biopsychosocial model or the ICF classification system.

Although the proposed explicatum may thus share sufficient similarity with the explicandum, it does not seem to meet the criteria of exactness (yet?). Regarding fruitfulness, more research is needed. This counts for both its application as a measurement instrument for scientific purposes as well as for being used as an evaluation tool for clinical practice. Our analysis and evaluation are based on information that is currently available. As explication is a non-linear and open-ended process (Brun 2016; Dutilh Novaes 2020b), future research and developments may lead to new assessment and evaluation. Prinsen and Terwee (2019) suggest that at least it should be made clear what is purported to be measured, to whom the measurements will be relevant (target population), and clarify whether the instrument is intended for evaluative or diagnostic/prognostic purposes. Indeed, getting clear on such questions can help to improve the process of explication.

19.5.3 Through the Lens of Ameliorative Analysis

Critical Analysis

Although explication for scientific and clinical purposes is one of the goals of this conceptual engineering of health, the original goal of Huber and colleagues may be primarily viewed as a social and political project: to change the way we think about health, the ways we treat patients (with chronic diseases), and how we shape our healthcare system. As described in our historical analysis, these first steps of activism could be viewed as echoing a broader biopolitical discussion that was going on internationally, in which re-defining health was viewed as essential for securing a future of healthcare that was more compassionate and adapted to the needs of the patient (The Lancet 2009; Jadad and O'Grady 2008).

In their seminal article, Huber et al. claimed that the WHO definition of health was no longer "fit for purpose" and that it had become "counterproductive" in the conduct of modern-day medicine. In particular, its negative effects for people living with chronic diseases or disability were emphasized. The WHO definition would disregard their capacities as autonomous human beings to cope with physical, social and emotional challenges in life. By shifting the focus to resilience and adaptive capabilities, Positive Health was meant to empower patients and invoke a paradigm shift in thinking about the relation between health and disease. Therefore, not considered as a concept definition in a narrow sense, but as a 'general concept' and as "a characterization of a generally agreed direction in which to look" (Huber et al. 2011: 2). Thus, it appears that the main goal of Positive Health is primarily social and political. With these goals in mind, the overall project may therefore be best understood in ameliorative spirit.

Assessing the Adequacy of the Ameliorative Concept

As an ameliorative project, the movement of Positive Health departs from the idea that health should be exclusive to people free from diseases or other physical limitations. To elucidate the problem and target of the ameliorative project further, it is useful to apply Haslanger's distinction between *manifest-*, *operative-*, and *target concept*. As for the manifest concept, we are talking about 'health' in the implicit and broad sense. It is the manifest concept of health that we refer to in our everyday life. According to Haslanger, however, there is often an operative concept, which is more explicit yet silent, that actually does the work in specific contexts.[20] The operative concept, in our case, is the one that is actually used in medical practice and healthcare policy. Interestingly, the 6-dimenional spiderweb, was developed with empirical input from healthcare providers, patients, citizens, policy makers, health insurers, public health professionals, and researchers. While physicians prioritized bodily function and quality of life, patients evaluated all six dimensions as equally important. As physicians have authority on our health concept in medical context, we could say that the operative concept of health appears to be that of bodily function and quality of life, while the manifest concept as employed by non-physicians is much broader. While there can be good reasons for the operative concept differing from the manifest concept (i.e., practices often need a more specific concept), such a difference can also become a problem.

In the case of health, one may question if it is a good thing when the concept used in medicine is too distanced from our everyday concept. There can be miscommunications in which the type of health that is promoted by a medical intervention is not in line with the health views of the patient. On the other hand, there can be good reasons for clinicians to have a different understanding of health, prioritizing the

---

[20] In Haslanger (2005), 'tardiness' is given as an example for demonstrating a mismatch between the manifest concept and the operational concept of tardiness in a school.

operative concept. It is not without reason that we consult a medical professional when we feel ill instead of a random person on the streets. We often want our suffering to be understood through the medical lens. However, this may be, in the development of Positive Health the difference between the manifest and operative concept was clearly viewed as a mismatch that needed to be resolved. The concept of Positive Health, that takes the perspective of the patient as central, can therefore be viewed as the target concept (i.e., the concept that is proposed to be used). The further socio-political goal of this target concept is to empower patients and to broaden our approach to health problems.

While this goal seems laudable and may indeed improve healthcare, we should, first, critically reflect on the question which and whose values are involved (cf Lalumera forthcoming), and second, critically assess whether the proposed target concept is up to its task. Besides assessing the intended effects of Positive Health, it is also important to assess its unintended effects. If we understand Positive Health as 'the ability to adapt and self-manage in the face of social, physical and emotional challenges', there is clearly a strong focus on promoting autonomy in people. Although this indeed supports the idea of empowering patients and to focus on resilience and adaptation, it has also been argued that the strong emphasis on the individual's 'self-management' may have adverse effects (Franssen and van Geelen 2017), and that it may resonate more with people from higher socio-economic groups than with lower ones (Stronks et al. 2008). Although the external (social) environment is viewed as an important part of one's experience of health (e.g., 'participation' is one of the six dimensions), the focus is on the individual rather than the environment. Positive Health as a movement has therefore also been criticized for putting too much pressure on the individual, while determinants of health are also in part beyond our personal influence (Schermer and Van der Horst 2022).

If we assess the adequacy of this ameliorative project in light of Haslanger's two conditions, the semantic and the political condition, we may say the following. Regarding the semantic condition, our evaluation here is not very different from the 'similarity' assessment in the previous section. Although the change in meaning is significant, it seems that the 'central functions' of the original concept are maintained sufficiently. For evaluating whether the political condition is met, however, we may not have sufficient data to come to a firm conclusion yet. While the wide uptake and implementation seems to indicate a successful conceptual shift in the Dutch healthcare system (Johansen et al. 2018, 2023), it still remains to be seen if Positive Health will in fact improve the way in which the healthcare system deals with health problems and approaches patients. The saying 'the proof of the pudding is in the eating' strongly counts here, and leaning too much on critiques made in the academic realm (see Sect. 19.2.5) will keep us from answering the most important question: does it work in practice? Does the new conceptualization of health actually change anything, for the better?

One good example of the type of research needed here is a study by Bock et al. (2021). They found that medical residents considered using Positive Health to be beneficial for consultations with patients having multidimensional problems, particularly in cases of chronic conditions and generalist care: "In these situations, the tool yielded valuable patient information beyond physical health, helped foster patient engagement, and enabled tailoring the treatment plan to individual patients' needs. On the other hand, the PH-tool was not a good fit for simple problems, clearly demarcated help requests, periodic follow-up consultations, or verbose patients. In addition, it was not suitable for super-specialised care, because it yielded an abundance of general information." (Bock et al. 2021: 9). We believe that these kinds of studies are important to investigate in what instances Positive Health could be useful, why it is considered to be useful, and to whom it is of use.

19.6   Conclusion

In this paper, we have analyzed the development of the concept of Positive Health and assessed its adequacy through the lens of conceptual engineering. A combined historical-philosophical approach was considered necessary, as the pragmatist approach we use requires a concept to be evaluated in terms of how well it serves the desired functions in a particular context. The historical analysis provides insights into the reasons behind the development of Positive Health, sheds light on the actors involved (and their interests) and provides important information about the context in which the conceptual shift is occurring. In this regard, providing historical background has an epistemic function in our analysis.

We have shown that the concept of Positive Health was born in practice rather than from theoretical debate, developed by medical practitioners instead of philosophers, as a response to problems they encountered with the currently held 1948 WHO definition of health. We also showed that the aims and goals of Positive Health have changed and diverged significantly over time. Because of the variety of aims and goals of Positive Health that we observed, we assessed the concept of Positive Health by two different types of analysis and method: pragmatic *Carnapian explication* and Haslanger's *ameliorative analysis*. On the basis of Carnapian explication, we conclude that applications of Positive Health as a scientific measurement instrument and as a clinical tool, at least for now, seem to fall short in terms of exactness. Partly because of that, its fruitfulness is as yet unclear. Understood as an ameliorative project, that aims for a holistic approach to health and healthcare, and that strives for empowerment of patients, Positive Health appears to fare somewhat better. However, whether Positive Health can really live up to its ameliorative goals, considering both its intentional as well as unintentional effects, is yet to be seen and needs further empirical research, we have argued.

We hope to have shown the benefits of using insights and methods from the field of conceptual engineering for dealing with conceptual issues in medicine and healthcare. It enables us to analyze and evaluate conceptual shifts in ways that can account for complexity, contextuality and plurality. This supports and strengthens the pragmatist turn in the philosophy of medicine. At the same time, we believe that the field of conceptual engineering may also benefit from studies as these that apply their methods to real-world cases. For example, our historical analysis shows something important about how a concept develops and is shaped and implemented in actual practice. That is, the successful uptake and implementation of Positive Health was for a significant part due to external, socio-political factors—such as the strong support of funding organizations—and not necessarily because of its conceptual strength. Investigating case-studies may give important input to meta-semantical issues that are debated in the field of conceptual engineering, such as the 'externalism-internalism debate' and the 'implementation problem' (Pollock 2021). In this regard, we hope to have made a contribution to the philosophy of medicine as well as to the promising field of conceptual engineering.

Acknowledgements   The authors are thankful for the brainstorm sessions with Dr. Mirela Fuš, Dr. Joey Pollock and Dr. Sigurd Jorem in the early stage of writing this chapter, and for the constructive feedback of Dr. Gili Yaron. The research for this chapter was supported by the Dutch Research Council (NWO) as part of the project 'Health and disease as practical concepts', project number 406.18.FT.002.

References

Andow, James. 2020. Fully experimental conceptual engineering. *Inquiry*. https://doi.org/10.1080/0020174X.2020.1850339.

Belleri, Delia. 2021. On pluralism and conceptual engineering: Introduction and overview. *Inquiry*. https://doi.org/10.1080/0020174X.2021.1983457.

Binney, Nicholas. 2021. Using medical history to study disease concepts in the present: Lessons from georges canguilhem. *Teorema XL*, 1: 67–89. Retrieved April 4, 2022, from: https://www.jstor.org/stable/27094754.

Bircher, Johannes. 2005. Towards a dynamic definition of health and disease. *Medicine, Health Care and Philosophy* 8: 335–341. https://doi.org/10.1007/s11019-005-0538-y.

Bock, Lotte A., Cindy Y.G. Noben, Gili Yaron, Erwin L.J. George, Adrian A.M. Masclee, and Brigitte A.B. Essers. 2021. Positive Health dialogue tool and value-based healthcare: a qualitative exploratory study during residents' outpatient consultations. *BMJ Open* 11: e052688. https://doi.org/10.1136/bmjopen-2021-052688.

Brun, Georg. 2016. Explication as a method of conceptual re-engineering. *Erkenntnis* 81: 1211–1241. https://doi.org/10.1007/s10670-015-9791-5.

———. 2017. Conceptual re-engineering: From explication to reflective equilibrium. *Synthese* 197: 925–954. https://doi.org/10.1007/s11229-017-1596-4.

Buijs, Marieke. 2017. Een nieuwe kijk op gezondheid. Staat de patiënt centraal of vooral verwarring? *NTvG Nieuws*. Retrieved from: https://www.ntvg.nl/artikelen/een-nieuwe-kijk-op-gezondheid.

Cappelen, Herman. 2018. *Fixing language. An essay on conceptual engineering*. Oxford: Oxford University Press.

Cappelen, Herman, and David Plunkett. 2020. A guided tour of conceptual engineering and conceptual ethics. In *Conceptual engineering and conceptual ethics*, ed. A. Burgess, H. Cappelen, and D. Plunkett, 1–26. Oxford: Oxford University Press. https://doi.org/10.1093/oso/9780198801856.003.0001.

Carnap, Rudolf. 1950. *Logical foundations of probability*. Chicago: University of Chicago Press.

Cooper, Rachel I. 2020. The concept of disorder revisited: Robustly Value-Laden despite change. *Aristotelian Society Supplementary* 94 (1): 141–161. https://doi.org/10.1093/arisup/akaa010.

Creath, Richard. 1990. *Dear Carnap, Dear Van: The Quine-Carnap correspondence and related work: Edited and with an introduction by Richard Creath*. Berkeley: University of California Press.

De Vreese, Leen. 2017. How to proceed in the disease concept debate? A pragmatic approach. *The Journal of Medicine and Philosophy: A Forum for Bioethics and Philosophy of Medicine* 42 (4): 424–446. https://doi.org/10.1093/jmp/jhx011.

Doornenbal, Brian M., Rimke C. Vos, Marja Van Vliet, Jessica C. Kiefte-De Jong, and M. Elske van den Akker-van Marle. 2022. Measuring positive health: Concurrent and factorial validity based on a representative Dutch sample. *Health & social care in the community* 30 (5): e2109–e2117. https://doi.org/10.1111/hsc.13649.

Dutilh Novaes, Catarina. 2020a. Carnap meets Foucault: conceptual engineering and genealogical investigations. *Inquiry.* https://doi.org/10.1080/0020174X.2020.1860122.

———. 2020b. Carnapian explication and ameliorative analysis: A systematic comparison. *Synthese* 197: 1011–1034. https://doi.org/10.1007/s11229-018-1732-9.

Fisher, Justin C. 2015. Pragmatic experimental philosophy. *Philosophical Psychology* 28: 412–433. https://doi.org/10.1080/09515089.2013.870546.

Franssen, Gaston, and Stefan van Geelen. 2017. Self-management as management of the self: Future directions for healthcare and the promotion of mental health. *Philosophy, Psychiatry, and Psychology* 24 (2): 179–184. https://muse-jhu-edu.eur.idm.oclc.org/pub/1/article/660581/pdf.

Greco, Monica. 2012. The classification and nomenclature of 'medically unexplained symptoms': Conflict, performativity and critique. *Social Science and Medicine* 75: 2362–2369. https://doi.org/10.1016/j.socscimed.2012.09.010.

Hacking, Ian. 2007. Kinds of people: Moving targets. *Proceedings of the British Academy* 151: 285–318. https://doi.org/10.5871/bacad/9780197264249.003.0010.

Haslanger, Sally. 2000. Gender and race: What are they? What do we want them to be? *Noûs* 34: 31–55. https://doi-org.eur.idm.oclc.org/10.1093/acprof:oso/9780199892631.001.0001.

———. 2005. What are we talking about? The semantics and politics of social kinds. *Hypatia* 20 (4): 10–26. https://doi.org/10.1111/j.1527-2001.2005.tb00533.x.

———. 2006. What good are our intuitions? *The Aristotelian Society Supplementary* 80: 89–118. https://doi.org/10.1111/j.1467-8349.2006.00139.x.

———. 2012. *Resisting reality.* Oxford: Oxford University Press.

Haverkamp, Beatrijs, Bernice Bovenkerk, and Marcel F. Verweij. 2018. A practice-oriented review of health concepts. *Journal of Medicine and Philosophy* 43 (4): 381–401. https://doi.org/10.1093/jmp/jhy011.

Huber, Machteld. 2010. Invitational conference 'Is health a state or an ability? Towards a dynamic concept of health' report of the meeting December 10–11, 2009. Retrieved from https://www.healthcouncil.nl/documents/advisory-reports/2010/07/13/invitational-conference-is-health-a-state-or-an-ability-towards-a-dynamic-concept-of-health.

———. 2013. Naar een nieuw begrip van gezondheid: Pijlers voor Positieve Gezondheid. *Tijdschrift voor Gezondheidswetenschappen* 91: 133–134. https://doi.org/10.1007/s12508-013-0046-z.

———. 2019. Positieve Gezondheid – de status anno 2019. *Bijblijven* 8. Retrieved from: https://mijn.bsl.nl/positieve-gezondheid-de-status-anno-2019/17243844.

Huber, Machteld, André Knottnerus, Lawrence Green, Henriëtte van der Horst, Alejandro R. Jadad, Daan Kromhout, Brian Leonard, et al. 2011. How should we define health? *British medical Journal* 343 (7817): 235–237. https://doi.org/10.1136/bmj.d4163.

Huber, Machteld, Marja van Vliet, Michèle Giezenberg, and André Knottnerus. 2013. *Towards a conceptual framework relating to 'Health as the ability to adapt and to self-manage'.* Louis Bolk Institute. Retrieved from: https://www.louisbolk.nl/publicaties/towards-conceptual-framework-relating-health-ability-adapt-and-self-manage.

Huber, Machteld, Marja van Vliet, Michèle Giezenberg, Bjorn Winkens, Yvonne Heerkens, Pieter C. Dagnelie, and André Knottnerus. 2016. Towards a 'patient- centred' operationalisation of the new dynamic concept of health: A mixed methods study. *British Medical Journal Open* 6 (1): 1–12. https://doi.org/10.1136/bmjopen-2015-010091.

Huber, Machteld, Hans P. Jung, and Karolien van den Brekel-Dijkstra. 2022. Introduction. In *Handbook positive health in primary care.* Houten: Bohn Stafleu van Loghum. https://doi.org/10.1007/978-90-368-2729-4_1.

Hyman, Steven E. 2010. The diagnosis of mental disorders: The problem of reification. *Annual Review of Clinical Psychology* 6: 155–179. https://doi-org.eur.idm.oclc.org/10.1146/annurev.clinpsy.3.022806.091532.

Isaac, Manuel G. 2021. Which concept of concept for conceptual engineering? *Erkenntnis*. https://doi.org/10.1007/s10670-021-00447-0.

Isaac, Manuel G., Steffen Koch, and Ryan Nefdt. 2022. Conceptual engineering: A road map to practice. *Philosophy Compass* 17 (10): e12879. https://doi.org/10.1111/phc3.12879.

Jadad, Alejandro R., and Laura O'Grady. 2008. How should health be defined? *British Medical Journal* 337: a2900. https://doi.org/10.1136/bmj.a2900.

Jambroes, Marielle, Trudi Mederland, Marian Kaljouw, Katja van Vliet, Marie-Louise Essink-Bot, and Dirk Ruwaard. 2015. Implications of health as 'the ability to adapt and self-manage' for public health policy: A qualitative study. *The European Journal of Public Health* 26 (3): 412–416. https://doi.org/10.1093/eurpub/ckv206.

Johansen, Françoise, Derk Loorbach, and Annemiek Stoopendaal. 2018. Exploring a transition in Dutch healthcare. *Journal of Health Organization and Management* 32 (7): 875–890. https://doi.org/10.1108/JHOM-07-2018-0185.

———. 2023. Positieve Gezondheid: Verandering van taal in de gezondheidszorg. *Beleid en Maatschappij* 50 (1). https://doi.org/10.5553/BenM/138900692022009.

Jorem, Sigurd. 2022. The good, the bad and the insignificant—Assessing concept functions for conceptual engineering. *Synthese* 200: 106. https://doi.org/10.1007/s11229-022-03548-7.

Kingma, Elisabeth. 2017. Kritische Vragen bij Positieve Gezondheid. *Tijdschrift voor Gezondheidszorg en Ethiek* 3: 81–83. Retrieved from: http://eprints.soton.ac.uk/id/eprint/418913.

Koch, Steffen. 2021. Engineering what? On concepts in conceptual engineering. *Synthese* 199: 1955–1975. https://doi.org/10.1007/s11229-020-02868-w.

Lalumera, Elisabetta. forthcoming. Conceptual engineering of medical concepts. In *New perspectives on conceptual engineering*, ed. Manuel Gustavo Isaac, and Kevin Scharp.

Law, Ian, and Heather Widdows. 2008. Conceptualizing health: Insights from the capability approach. *Health Care Analysis* 16: 303–314. https://doi.org/10.1007/s10728-007-0070-8.

Lemmen, Caro H.C., Gili Yaron, and Rachel Gifford. 2021. Positive Health and the happy professional: A qualitative case study. *BMC Family Practice* 22: 159. https://doi.org/10.1186/s12875-021-01509-6.

Nado, Jennifer. 2019. Conceptual engineering via experimental philosophy. *Inquiry* 64 (1–2): 76–96. https://doi.org/10.1080/0020174X.2019.1667870.

Nado, Jennifer. 2021. Conceptual engineering, truth, and efficacy. *Synthese* 198: 1507–1527. https://doi.org/10.1080/0020174X.2019.1667870.

NFU (The Netherlands Federation of University Medical Centers). 2020. *Raamplan Artsopleiding 2020*. Retrieved from: https://www.nfu.nl/sites/default/files/2020-08/20.1577_Raamplan_Artsenopleiding_-_maart_2020.pdf.

Nordby, Halvor. 2019. Who are the rightful owners of the concepts disease, illness and sickness? A pluralistic analysis of basic health concepts. *Open Journal of Philosophy* 9: 470–492. https://doi.org/10.4236/ojpp.2019.94029.

Poiesz, Theo, Jo Caris, and Freek Lapré. 2016. Gezondheid: een definitie? *Tijdschrift voor Gezondheidswetenschappen* 94 (7): 252–255. Retrieved from: https://www.iph.nl/assets/uploads/2021/01/TSG-8-2016-reactie-M.-Huber-op-def-gezondheid.pdf.

Pollock, Joey. 2021. Content internalism and conceptual engineering. *Synthese* 198: 11587–11605. https://doi.org/10.1007/s11229-020-02815-9.

Prinsen, Cecilia A.C., and Caroline B. Terwee. 2019. Measuring positive health: For now, a bridge too far. *Public Health* 170: 70–77. https://doi.org/10.1016/j.puhe.2019.02.024.

Schermer, Maartje H.N., and Henriëtte van der Horst. 2022. Positieve gezondheid nader bekeken. *Nederlands Tijdschrift voor Geneeskunde* 165 (26–27): D5897. Retrieved from: https://www.ntvg.nl/artikelen/het-concept-positieve-gezondheid-nader-bekeken.

Schwartz, Peter H. 2017. Progress in defining disease: Improved approaches and increased impact. *Journal of Medicine and Philosophy* 42: 485–502. https://doi.org/10.1093/jmp/jhx012.

Stronks, Karien, Nancy Hoeymans, Beatrijs Haverkamp, Frank R.J. Den Hertog, Marja J.H. Van Bon-Martens, Henrike Galenkamp, Marcel Verweij, and Hans A.M. Van Oers. 2008. Do conceptualisations of health differ across social strata? A concept mapping study among lay people. *BMJ Open* 8: e020210. https://doi.org/10.1136/bmjopen-2017-020210.

The Lancet. 2009. What is health? The ability to adapt (editorial). *The Lancet* 373 (9666): 781. Retrieved from. https://doi.org/10.1016/S0140-6736(09)60456-6.

Van Boven, Kees, and Tessa Versteegde. 2019. Positieve Gezondheid een onsamenhangend concept. *Bijblijven* 35 (8): 55–58. https://doi.org/10.1007/s12414-019-0078-7.

Van der Linden, Rik R., and Maartje H.N. Schermer. 2022. Health and disease as practical concepts: Exploring function in context-specific definition. *Medicine, Health Care and Philosophy* 25: 131–140. https://doi.org/10.1007/s11019-021-10058-9.

Van der Linden, Rik, Timo Bolt, and Mario Veen. 2022. 'If it can't be coded, it doesn't exist'. A historical-philosophical analysis of the new ICD-11 classification of chronic pain. *Studies in History and Philosophy of Science* 94: 121–132. https://doi.org/10.1016/j.shpsa.2022.06.003.

Van der Stel, Jaap. 2016. Definitie 'gezondheid' aan herziening toe. *Medisch Contact* 23: 18–19. Retrieved from: https://www.medischcontact.nl/nieuws/laatste-nieuws/artikel/definitie-gezondheid-aan-herziening-toe.

Van Staa, Anne L., Mieke Cardol, and Angelique Van Dam. 2017. Positieve gezondheid kritisch beschouwd: Niet nieuw, onduidelijk, misleidend en niet zonder risico. *Positieve Psychologie* 4: 33–39. Retrieved from: https://www.hogeschoolrotterdam.nl/onderzoek/projecten-en-publicaties/pub/positieve-gezondheid-kritisch-beschouwd/a984b2f6-d19c-45b9-b9d1-0797342651eb/.

Van Steekelenburg, Ellen, Ingrid Kersten, and Machteld Huber. 2016. 'Positieve gezondheid in Nederland' Wie, wat, waarom en hoe? Een inventarisatie. iPH en ZonMW. Retrieved from: https://www.zonmw.nl/sites/zonmw/files/2023-03/Inventarisatie_Positieve_gezondheid_in_Nederland.pdf.

Van Vliet, Marja, Brian M. Doornenbal, Simone Boerema, and Elske M. van den Akker-van Marle. 2021. Development and psychometric evaluation of a Positive Health measurement scale: A factor analysis study based on a Dutch population. *BMJ Open* 11: e040816. https://doi.org/10.1136/bmjopen-2020-040816.

Walker, Mary Jean, and Wendy A. Rogers. 2018. A new approach to defining disease. *The Journal of Medicine and Philosophy: A Forum for Bioethics and Philosophy of Medicine* 43 (4): 402–420. https://doi.org/10.1093/jmp/jhy014.

Yaron, Gili, Marieke Spreeuwenberg, and Dirk Ruwaard. 2021. Praktijkhandreiking: Werken met Positieve Gezondheid. Lessen uit Limburg. Retrieved from: https://www.iph.nl/assets/uploads/2021/09/handreiking_werken_met_positieve_gezondheid.pdf.

ZonMw. 2014. *ZonMw in The Netherlands: 2016–2020 policy plan*. Retrieved from: https://www.zonmw.nl/en/about-zonmw/zonmw-in-the-netherlands/.

Chapter 20
On the Social and Material Lives of Health Concepts in the Wild

Gili Yaron

The widely accepted definition of health upheld by the World Health Organization (WHO) is no longer fitting and should be replaced—that, in a nutshell, is the position of a group of international healthcare experts headed by Machteld Huber, a Dutch general practitioner, researcher and activist (Huber et al. 2011). Accordingly, Huber and her colleagues proposed a new health concept called Positive Health (Huber et al. 2016). This broader and more dynamic alternative, in their view, will solve the problems they identify in the WHO definition. In the last decade, Positive Health has taken the Netherlands by storm; the concept has been adopted by a growing number of individuals and organizations operating in Dutch healthcare research, policy and practice (Lemmens et al. 2019). Moreover: it is increasingly taken up in the public sector at large.

But does Positive Health deliver on its promises? In their chapter, Van der Linden and Schermer's tackle this question—a timely and important undertaking given the new concept's rising impact (Van der Linden and Schermer, Chap. 19, this volume). To answer it, Van der Linden and Schermer approach Positive Health as a conceptual engineering project. Conceptual engineering, they explain, is a budding theory and method in the field of analytical philosophy, that pragmatically focuses on analyzing, assessing, and revising concepts in practice. Applying conceptual engineering to Positive Health, Van der Linden and Schermer first reconstruct the new concept's brief history. Next, they clarify its intended functions: (1) to provide a scientifically sound alternative to the existing, WHO concept of health, and (2) to empower patients and broaden the prevailing approach to health issues. Positive Health, Van der Linden and Schermer conclude, ultimately falls short in achieving

G. Yaron (✉)
Research Group Living Well with Dementia, Windesheim University of Applied Sciences, Zwolle, The Netherlands
e-mail: g.yaron@windesheim.nl

© The Author(s) 2024
M. Schermer, N. Binney (eds.), *A Pragmatic Approach to Conceptualization of Health and Disease*, Philosophy and Medicine 151,
https://doi.org/10.1007/978-3-031-62241-0_20

the former, theoretical-epistemological goal. And given that the concept is still young, it is yet unclear whether it succeeds in the latter, social-political aspiration.

Van der Linden and Schermer's meticulous evaluation of Positive Health as a conceptual engineering project provides a fresh new angle to the somewhat tired debate on this new health concept. But given their disappointing verdict, how can the concept's quick dissemination be explained? Apparently, evaluating Positive Health in terms of attempts to achieve knowledge or justice cannot account for its success. As Van der Linden and Schermer argue, the rise of Positive Health can be viewed as part of a conceptual shift in healthcare, which I believe to be associated with the field's increasing dissatisfaction with the biomedical model of health. The concept's resonance with the zeitgeist may therefore provide something of an answer. But ideas rarely spread out of their own accord, which means the question as to the popularity of Positive Health still stands. In this reflection, I will draw on my ethnographic fieldwork into Positive Health as well as on key works in Science and Technology Studies to gain insight in the new concept's eager uptake. As will become clear, following Positive Health 'in the wild' affords a deeper understanding of its social and material life—an understanding which not only explains the concept's widespread adoption, but also opens up new vistas for conceptual engineering as a scholarly approach.

A first explanation of the success of Positive Health can be found by examining initiator Machteld Huber's strategic abilities. Indeed, Van der Linden and Schermer's historiography demonstrates how Huber repeatedly manages to rally powerful allies in research, policy, and government to her cause of conceptually re-engineering health. Meanwhile, she also succeeds in finding and binding professionals, managers, and policy makers in practice. In this, Huber calls to mind famous biologist Louis Pasteur's ability to gather a vast network of collaborators, including doctors, public health officials, and farmers, in his campaign to pasteurize France (Latour 1988). Huber's personal charisma and deep conviction, which I have witnessed on various occasions, prove vital for such sophisticated networking. But her appeal also includes her ability to credibly inhabit different social worlds. In interactions with her allies, she frequently draws on her personal experiences as a former patient, doctor, researcher, policy advisor and activist (Institute for Positive health 2023). These distinct identities grant Huber a unique position as a 'multiple insider' who can speak to the particular realities and interests of various stakeholders and thereby gain their trust and support. All the while, she skillfully translates these realities and interests to her overall project of conceptually re-engineering health—and links both to the commonly felt desire to transform healthcare. In this, Huber in fact creates adhesion among associates with divergent agenda's (Akrich et al. 2002).

Huber's unique skillset, however, is not the whole of such strategizing. Focusing on what Positive Health does for Huber's allies, reveals a second explanation for the concept's popularity: the fact it allows stakeholders to further their own interests. One example my colleagues and I studied in-depth was a general practice which leveraged the new concept to accelerate desired organizational changes (Lemmen et al. 2021). By embracing the new concept of health, the practice was able to participate in a health insurer's pilot with a new financial model, which translated into more time per patient and a reduction of healthcare professionals' workload. In addition, the practice was able to intensify its collaborations with other public sector organizations (e.g. the municipality, social housing association). Finally, the general practice was able to present itself to employees, patients, and partners as a distinctive, innovative organization. My fieldwork therefore suggests Positive Heath effectively functions as an adaptable narrative frame that allows individuals and organizations to tap into new resources, bonds, and identities.[1] Interestingly, stakeholders' ability to strategically mobilize the concept in this way is predicated on the widespread (though somewhat misguided) belief in its scientific legitimacy in healthcare and the public sector. The theoretical-epistemological and social-political aspects of Positive Health therefore seem interconnected.

A third explanation of the wide distribution of Positive Health, finally, has to do with the material changes it affords. Adopting this concept, as discussed above, allows organizations to alter their financial models and employees' timetables. Other examples of such material changes from my fieldwork include rearranging administrative procedures, reframing professionals' job descriptions, and redesigning nursing guidelines. Mobilizing the concept—and being mobilized by it—takes place by and through the realities of healthcare. Thus, Positive Health resides as much in things (money flows, bureaucracies, protocols) as it does in ideas (propositions, attitudes, beliefs). Of course, scholars of conceptual engineering, too, link conceptual change to real-world transformation (Van der Linden and Schermer, Chap. 19 this volume). Still, these scholars seem to understand concepts as essentially psychological and linguistic phenomena—as psychic objects, in other words. But as practice theory teaches us, mind and matter cannot be divorced (Reckwitz 2002). In fact, it is the very possibility of realizing actual change that accounts for the appeal of concepts.[2] To understand the rapid uptake of Positive Health, we must therefore look at the various materializations it enables.

Positive Health may not actually be as scientifically sound as its advocates believe, and we do not yet know whether it truly empowers patients and extends the prevailing, biomedical approach to health. Nevertheless, this concept clearly leads vibrant social and material lives within the field of healthcare and beyond. In this, Positive Health is by no means unique. The last decades have seen the emergence of numerous new concepts that aspire to improve healthcare. Examples

---

[1] To take this point closer to our scholarly practices: we all develop and leverage particular concepts as part of a social game geared towards accumulating credibility in our various epistemic cultures (Latour and Woolgar 1986; Hessels et al. 2019). All in the name of (theoretical-epistemological and social-political) goals we genuinely believe in—but our strategic interests undeniably also play a part here.

[2] Note that conceptual change is not always associated with real change, and that real change is not always good. As various critics argue, Positive Health may contribute to entrenching health inequalities through its focus on self-management (Jambroes et al. 2016; Van Staa et al. 2017). Moreover: many concepts escape their designers' intentions to go on leading unpredictable sociomaterial lives.

include person-centered care, evidence-based medicine, vitality, patient participation, and resilience. In each case, concepts come into play in ever-changing assemblages of research, policy, practice, government, education—and often industry, too. And in each case, stakeholders continuously re-frame and re-negotiate the meaning, scope and use of concepts according to their interests. The tools of conceptual engineering may help us analyze, evaluate, and steer such conceptual change-in-action. But in order to do so effectively, we need to acknowledge the strategic dimensions of conceptual change. Meanwhile, we must also accept that the theoretical-epistemological and socio-political cannot be thought apart. Finally, we need to recognize and further explore the entanglement of conceptual and material change. Conceptual engineering scholars should therefore continue following concepts in the wild, while attending to how they help us enact and re-enact our worlds.

References

Akrich, Madeleine, Michel Callon, and Bruno Latour. 2002. The key to success in innovation part I: The art of interessement. *International Journal of Innovation Management* 6 (2): 187–206.

Hessels, Laurens K., Thomas Franssen, Wout Scholten, and Sarah de Rijcke. 2019. Variation in valuation: How research groups accumulate credibility in four epistemic cultures. *Minerva* 57: 127–149.

Huber, Machteld, Johannes A. Knottnerus, Lawrence Green, Henriëtte van der Horst, Alejandro R. Jadad, Daan Kromhout, Brian Leonard, et al. 2011. How should we define health? *British Medical Journal* 343 (7817): 235–237.

Huber, M., M. van Vliet, M. Giezenberg, B. Winkens, Y. Heerkens, P.C. Dagnelie, and J.A. Knottnerus. 2016. Towards a 'patient- centred' operationalisation of the new dynamic concept of health: A mixed methods study. *British Medical Journal Open* 6 (1): 1–12. https://doi.org/10.1136/bmjopen-2015-010091.

Institute for Positive Health. 2023. *Over ons* [about us]. Institute for Positive Health. http://www.iph.nl/organisatie/over-ons. Accessed 21 June 2023.

Jambroes, Marielle, Trudi Nederland, Marian Kaljouw, Katja van Vliet, Marie-Louise Essink-Bot, and Dirk Ruwaard. 2016. Implications of health as 'the ability to adapt and self-manage' for public health policy: A qualitative study. *European Journal of Public Health* 26 (3): 412–416.

Latour, Bruno. 1988. *The pasteurization of France*. Cambridge, MA: Harvard University Press.

Latour, Bruno, and Steve Woolgar. 1986. *Laboratory life: The construction of scientific facts*. London: Sage.

Lemmen, C., G. Yaron, R. Gifford, and M.D. Spreeuwenberg. 2021. Positive health and the happy professional: A qualitative case study. *BMC Family Practice* 22 (159): 1–12. https://doi.org/10.1186/s12875-021-01509-6.

Lemmens, Lidwien, Simone de Bruin, Maarten Beijer, Roy Hendrikx, and Caroline Baan. 2019. *Het gebruik van brede gezondheidsconcepten: inspirerend en uitdagend voor de praktijk* [The usage of broad health concepts: Inspiring and challenging for practice]. RIVM. https://www.rivm.nl/sites/default/files/2020-01/Factsheet%20brede%20gezondheidsconcepten_RIVM_dec%202019a.pdf. Accessed 21 June 2023.

Reckwitz, Andreas. 2002. Toward a theory of social practices: A development in culturalist theorizing. *European Journal of Social Theory* 5 (2): 243–263.

Van Staa, A.L., M. Cardol, and A. Van Dam. 2017. Positieve gezondheid kritisch beschouwd: Niet nieuw, onduidelijk, misleidend en niet zonder risico. *Positieve Psychologie* 4: 33–33.

Chapter 21
Healthism, Elite Capture, and the Pitfalls of an Expansive Concept of Health

Quill R. Kukla

21.1 Introduction

After a long history of attempts to analyze the essence of the concept of health, it is becoming popular for philosophers to turn to a more fluid, strategic, pragmatist conception of health: one that recognizes health and disease as ineliminably pluralist and contextual terms without unified essential cores, which can be mobilized for different purposes by different sorts of stakeholders (Barnes 2023; Binney et al., Prologue, Chap. 2, this volume; Valles 2018; Kukla 2015, 2022a, b; Van der Linden and Schermer 2022). Indeed, I have been one of the defenders of this move, and I continue to think it is the right way to approach understanding both health and disease. For example, in "What Counts as A Disease and Why Does It Matter" (Kukla 2022b), I argued that the concept of disease serves radically different strategic purposes for different stakeholders and is essentially contested, and that coming up with a unified philosophical definition of disease is hopeless and will at best delegitimize from the armchair some of the practically useful deployments of disease discourse. I claimed there that it is appropriate to count a condition as a disease when it furthers legitimate strategic goals to at least partially medicalize the disease, and to understand it as pathological from inside the epistemology and metaphysics of medicine. According to the account I develop there, some conditions, such as pancreatic cancer, will be diseases from every reasonable stakeholder's point of view, while other conditions, such as autism, infertility, and deafness, will be irreducibly context-dependent and perspective-dependent in their disease status. In a similar vein, I argued in Kukla (2015) that health should be defined strategically and counterfactually, in terms of what it is useful to bring under the auspices of medicine.

Q. R. Kukla (✉)
Department of Philosophy, Georgetown University, Washington, DC, USA

Leibniz Universität Hannover, Hannover, Germany

© The Author(s) 2024
M. Schermer, N. Binney (eds.), *A Pragmatic Approach to Conceptualization of Health and Disease*, Philosophy and Medicine 151, https://doi.org/10.1007/978-3-031-62241-0_21

While not abandoning my earlier work, this paper urges caution when it comes to the development of a pragmatist, pluralist conception of health. It explores the harms that can happen when we allow health to be too open-textured and all-encompassing, and when we allow different groups to define health to suit their own strategic needs, at least in our current ideological climate. When we allow health discourse and a health framework to spread unchecked, it can be and often is weaponized in ways that harm well-being and justice. The concept of health should not, as philosopher Sean Valles (2018) urges it should, simply be a "big tent" without controlled boundaries. In particular, I claim here that we need to understand health as a narrower concept than well-being more generally, otherwise all domains of life become subject to the specific sorts of social discipline, resource prioritization, evaluation, and surveillance that we collectively accept for health issues.

The most well-known "big tent" definition of health is the often-cited and often-mocked World Health Organization definition, which aims to equate health with an expansive conception of well-being that goes way beyond the narrowly medical: Health, for the WHO, is "a complete state of physical, mental, and social well-being, and not merely the absence of disease or infirmity."[1] As has been pointed out often, this broad definition seems to be so all-encompassing that it makes pretty much any dimension of well-being into a part of health. Despite this breadth, the definition remains popular in a variety of contexts. Meanwhile, Huber et al. (2011) influentially suggested replacing the WHO definition with an understanding of "health as the ability to adapt and to self manage, in the face of social, physical and emotional challenges." Here again, health exceeds the medical to include the social and the emotional, with no explicit boundaries around what counts as successful adaptation or self-management. Valles (2018) has defended a similarly unbounded and expansive approach to understanding health. He argues that we should not circumscribe the notion of health, but rather should allow different communities to define health for their local needs, enabling them to use the concept as a tool of social empowerment. He tweaks the WHO definition, proposing that "health is a life course trajectory of complete well-being in social context" (see 57 and throughout), but thereby retains the conceptual link between health and 'complete well-being.'

An explicit motivation for Valles, and for others who want to understand health in terms of a broad and flexible conception of well-being, is the conceptual decoupling of health from medicine. The dangers of over-medicalization, diagnostic creep, disease-mongering, and the like are well-documented. If our definition of health is tied too tightly to medical notions such as disease and treatment, the worry goes, then we are unable to strategically mobilize health discourse without inviting the immediate spectre of medicalization. Theorists like Valles want to be able to, for instance, talk about stress as a health crisis among poor people without invoking the idea that poor people's woes can be addressed through anxiety medication or the like.[2]

---

[1] www.who.int/about/governance/constitution

[2] I do see this risk, although in "Medicalization, 'Normal Function,' and the Definition of Health" (2015) I argued for a conceptual link between health and appropriate medicalization. There, I

Another explicit motivation behind many pragmatist, pluralist definitions of health is to make health a more flexible notion so that it can better serve the ends of social justice. Framing a systematic failure of community well-being as a health issue is a way of both giving it weight and detaching it from certain forms of moralizing and victim-blaming. But while it is true that framing something as a health issue removes certain kinds of moral meanings and gives it social and political heft, there are social dangers that come along with allowing the notion of health to swallow up our notion of well-being and insert itself into all sorts of corners of daily life far removed from the domain of medicine. Indeed, in a deep, often implicit way, our social concept of health is already a concept that has all sorts of complex and tenacious moral and aesthetic meanings. Allowing these moral and aesthetic meanings free reign over all aspects of our life can result in oppression, inappropriate surveillance and discipline, stigmatization, and an amplification of power differences, as I hope to show. In short, while population and public health scientists and some philosophers have argued for an expanded and more flexible definition of health for social justice ends, in fact the concept of health, when allowed to expand unchecked, has a tendency to become a weapon used to moralize and discipline all aspects of daily living.

Valles makes the seemingly common-sense claim that "promoting a population's health is part and parcel of the social empowerment of that population" (2018: 32). While this is often true, I want to stress the ways in which "health" promotion can turn into a tool of social control that imposes health standards and imperatives on people which can in fact disempower them in dramatic ways. To give two quick examples: The earliest versions of South African apartheid originated in British attempts to promote 'public health' by segregating the races, to prevent 'social mixing' as an epidemic mitigation measure (recounted and documented in Carr 1990). And pregnant people often find themselves subject to involuntary treatments such as psychiatric and substance abuse treatments, involuntary caesarean sections, social surveillance of their every decision, regressive restrictions of their autonomy, and so forth.[3] We will see more detailed examples below.

My central claim will be that while I agree that there is no strict or unified definition of health, and that the concept must ultimately be contextually understood through its pragmatic uses, allowing the concept to expand indefinitely enables it to be weaponized by stakeholders with social power and by institutions with specific interests in ways that undermine social justice and become tyrannical and elitist. This weaponization is enabled by our society-wide deeply entrenched healthism, which is the view that health is indefeasibly an end we all must value no matter what, and which we are individually responsible for pursuing and socially responsible for enforcing. In short, when every dimension of our well-being starts to count as part

---

argued that not everything that is appropriately medicalized should be treated. But I do acknowledge that even marshalling a medical framework to understand a condition can lay it open to pharmaceutical capture and other kinds of institutional control.

[3] The documentation and discussion here is extensive, but for example see Morris and Robinson (2017) and Paltrow and Flavin (2013).

of health, then healthism starts to control every dimension of well-being. I look at three case studies, each of which is a domain in which the concept of health has expanded and become weaponized in toxic ways: healthy eating, healthy sexuality, and healthy gender identity. I end by arguing that we can understand the weaponization of health as a kind of elite capture, to use the phrase that Olúfẹ́mi Táíwò (2022) recently popularized, wherein elite group members redirect energy and efforts that nominally are designed to promote group interests towards the re-entrenchment and furthering of their own privileged position.

21.2 Healthism and the Moral and Aesthetic Meanings of Health

Healthism is roughly the (pervasive and penetrating) ideology according to which people are socially obligated to prioritize and value health and "healthy living" over other goals and without exception (Metzl et al. 2010). Once something gets labelled as a "health issue," there emerges a social imperative to do something about it, and a social license to institute regimes of surveillance and discipline to force people to be well-behaved with respect to the issue. Under our healthist regime, there is a tendency to measure all choices and situations in terms of how healthy or unhealthy they supposedly are, and to immediately conflate 'good' choices and environments with 'healthy' choices and environments. Under healthism, people are expected to modulate and discipline multiple dimensions of their life through the lens of health: we should eat healthy food, maintain a healthy weight, built healthy relationships, which in turn sustain healthy boundaries, make sure to keep a healthy attitude towards life, buy products to ensure healthy skin, safeguard healthy finances, and so forth. If we make a decision that does not maximize health in a narrow physical sense, like eating a piece of chocolate cake or skipping a workout, we are socially encouraged to justify that choice, not in terms of reasonable competing values, but in terms of our 'mental health' or our need for 'self-care.' Meanwhile, protecting people's health is treated as such an absolute value that it justifies unending social practices of top-down and interpersonal control and surveillance. For example, food assistance programs such as WIC and SNAP in the United States control the diets of poor people by allowing only specific food purchases. Until recently, hormonal birth control was made available to North Americans only if they agreed to regular irrelevant cervical cancer screening (Henderson et al. 2010).

Talia Welch is eloquent on the tyranny of healthism:

> The tightest noose around bodily pleasures today is the set of norms regarding health. I cannot engage in any behavior without processing it as healthy/good or unhealthy/bad. The spread of the knowledge about health now extends far beyond the endless prescriptions regarding diet and exercise. Everything from watching TV, getting an education, and working late at night, to having friends is codified. Each activity is studied for how it is correlated with one's health. No matter how apparently removed a behavior, thought, or feeling is from the physical operations of one's body, they are accompanied by a sometimes

quiet, sometimes loud running commentary on their relative health risks and benefits (Welch 2011: 43)

The healthist imperative is not equally distributed. For example, people who are (or even might become) pregnant are not only subject to exceptional discipline and surveillance in the name of health, but they are expected to engage in practices of extreme self-monitoring and self-discipline in the name of health maximization, to the point where 'good mothering' is associated with assessing literally every move, activity, product choice, and emotion for its health risks and benefits (Kukla 2005). The omnipresent pregnancy advice book, *What to Expect When You're Expecting*, insists, "Every bite counts: before you close your mouth on a forkful of food, consider, 'Is this the best bite I can give my baby?' If it will benefit your baby, chew away. If it'll only benefit your sweet tooth or appease your appetite, put your fork down" (2002: 81). Poor people, fat people, and disabled people endure layers of added scrutiny and calls for self-discipline in the name of health.

As should already be clear, healthism and its pressures are not restricted to the domain of the narrowly and traditionally medical. Consider two very recent examples of health language and healthist behavioural prescriptions that have spread into unexpected places. A recent exposé on timeshare companies showed a manager explaining to his sales reps that "Time shares save lives" because taking vacations is "healthy" (Last Week Tonight with John Oliver, March 20, 2023). And in a recent article in Frontiers in Public Health, Keyes et al. (2022) show that there is a small negative correlation between self-reported loneliness and going to watch live sporting events; they suggest that this shows that going to live sporting events causes us to be happier and less lonely, and thus that getting people to go to sporting events is healthy, and should be systematically encouraged in the name of public health. Not only does this reframe loneliness and how we use our free time and disposable income as 'health issues,' but it ignores the obvious likelihood that people with the time, extra money, and friends to go to sporting events with are likely already happier and less lonely people. Thus the healthist social imperative is not just to avoid disease. Indeed, part of its distinctive power comes from its ability to expand and absorb virtually all dimensions of daily life into its orbit.

'Health' has become infused at the level of the social imaginary with a wide range of aesthetic meanings and characterological meanings. Our image of a 'healthy' lifestyle is not merely of one that avoids morbidity, but of an attractive and appealing lifestyle that manifests a virtuously disciplined and socially appropriate character, infused with just the right amount of properly tempered joy. Unsurprisingly, as I will discuss in more detail and more concretely below, our pre-theoretical notions of what counts as an attractive and virtuous lifestyle are infused with racist, classist, sexist, heterocentric, ciscentric, and other elitist meanings and social prejudices. My Google image search for 'healthy' yielded a parade of thin, conventionally attractive and conventionally feminine women, most of them white, sporting ponytails and light-coloured clothing, grinning wildly or laughing as they eat salad or do gentle exercises with small weights or stand in yoga poses in beautiful natural settings. This merging together of traditional conceptions of health with aesthetic and moral meanings is often detached from any realistic tie to well-being, and often serves capitalism or oppressive gendered and raced body norms: makeup is advertised as giving skin a "healthy glow"; models with BMIs in the "overweight" category— which is arguably the category associated with the best health outcomes in the traditional, narrowly medical sense of health (see for instance Afzal et al. 2016)— are criticized in comments sections and the like for "promoting unhealthy living" through their mere visibility; and so forth.[4]

My point so far is twofold. First, there is already a tendency for our notion of health to expand to swallow up all dimensions of everyday life. Second, this goes along with the enforcement of norms at the personal, interpersonal, and social level that may well not further well-being at all, but may instead reinscribe and amplify conformist, oppressive, and elitist norms. Critics of healthism have pointed out that while health is valuable, it is not the only value, nor need it be the default or privileged value. It is perfectly rational for people to prioritize values other than health sometimes (Barnhill et al. 2014; Metzl et al. 2010). But when we insist on the privileging and the enforcement of health as a value, and we also allow 'health' to expand without check to encompass almost every dimension of life and perceived well-being, health becomes an almost limitlessly powerful tool for sculpting people's behaviour and enforcing dominant values and conceptions of the 'good life.'

People who care about furthering social justice often favour a broad pragmatist conception of health, so that they can marshal this tool to insist that various forms of suffering and inequality should be social and policy priorities. For example, reframing the stress that comes from poverty as itself a health issue helps move such stress up our social agenda (for instance see Thoits 2010). Likewise, there has been all sorts of research seeking to cast social stigma from racism and other bigotry as not just a moral issue or a psychological issue but a health issue (for example see Williams et al. 2019). But while these sorts of reframings may well serve social justice, it is all too easy, once we cast something as a health issue under a healthist regime, to hold people personally responsible for addressing it. The right to health can so easily transmute into a responsibility to be healthy. And the more our conception of health expands, the more domains of life become subject to self-discipline imperatives. Jonathan Kaplan (2019) explores how 'self-care' has become a broad individualized duty, for example. The self-care imperative, he demonstrates, means much more than the imperative to protect one's physical health narrowly defined; it includes everything from "connecting with a loved one" to "meditation" to "purchasing gifts for yourself" (Kaplan 2019: 110). Likewise, we often label various activities as self-care in order to carve out social space for them, when they are just everyday activities that everyone deserves to enjoy without their health

---

[4]To see just one example, consider the blog post https://www.healthyisthenewskinny.com/blog-all/2018/2/11/why-plus-size-models-dont-promote-obesity, featuring an American size 16 model who clearly is not obese. The comments section is filled with claims such as "Modeling overweight models are [sic] unhealthy" and "Let's not lie and call obesity healthy." When Victoria's Secret introduced apparently average weight "plus size models" in 2019, twitter (now X) was filled with threads complaining about how these women promoted unhealthy bodies and habits.

necessarily depending on them. Having a hobby outside of work, making time for friends, having an occasional glass of wine: all of these get legitimized by being labelled 'self-care' activities, as if we would be wrong to make time for these perfectly reasonable pleasures otherwise.

Our relationships with others also often get framed as 'health' issues for which we are personally responsible; women in particular are instructed to take steps, including pharmaceutical steps, to ensure proper, 'healthy', virtuous relationships. In the United States, direct-to-consumer drug advertisements aimed at women often portray a progression from relationship breakdown, to a pharmaceutical fix, to social healing. Woven into this narrative is the idea that a woman has a moral responsibility to fix her health, out of duty to her partner and children; conversely, if she isn't sustaining traditional relationships, it must be because she has let herself be 'unhealthy'. For example, an advertisement for Zoloft begins with a cartoon mom shaped like a drug pill, watching her children play, saying "When my daughter said, 'Mommy, you're no fun anymore,' it hit me, it was time to get help." The advertisement ends with her saying, "You get one chance to raise your kids. Why do it with depression?" Here, not being sufficiently "fun" is treated as a health issue, and women are given the message that they owe their children their own good health (sending a rather chilling message to disabled mothers along the way). Of course, there are many reasons why having fun with your child is good, and worth pursuing. But what interests me at the moment is how the enforcement of conventional gender and social roles, and the expansion of health, come together. I do not mean to trivialize depression or to deny the value of its medicalization, but here we see a very broad swath of everyday interactions being interpreted through a diagnostic lens. Part of the point of broadening health is to decouple it from medicalization, but as this last example shows, this broadening can also open the door to expanding medicalization and pharmaceutical capture. When we follow the WHO in including 'social well-being' as part of health, then many new dimensions of our social interactions become vulnerable to healthist imperatives, and sometimes in turn to pharmaceutical cooptation. It is not surprising that the power of health discourse and our social valuation of health can be fairly easily co-opted by pharmaceutical companies, the self-care and weight loss industries, and other stakeholders with motives other than social justice.

21.3 The Tyranny of the Community

It is tempting to push back on my concerns about healthism by saying that while health can be weaponized when it is wielded as a tool top-down or by for-profit industries, the real goal of a pragmatic, expansive view of health is to allow local communities to set their own standards for health and well-being, so that they can use health as a tool for community empowerment. For example, Valles argues that communities, when they are allowed to define health to meet their own needs, will be better able to access the empowering potential of health (see for instance Valles 2018: 70).

While I have so far focused on top-down impositions of conceptions of health and lifestyle standards, I don't actually see any particular reason to trust communities to set their own standards of health and well-being in ways that will be empowering rather than oppressive. When we think about 'local communities,' we tend to invoke images of scrappy underdog communities fighting back in the name of social justice. But plenty of local communities are themselves regressive and oppressive, and, left to their own devices, will impose conceptions of health that deeply damage the well-being of many of their members. Plenty of local communities are homophobic, sexually puritanical, misogynist, and so forth. Given free rein, many communities will define 'healthy' sexuality and gender expression regressively, and force 'treatment' on those who do not comply. Or they may enforce ableist conceptions of health, demanding that autistic people be trained to mask their autistic traits, for example. Local communities are just as good as society writ large at pathologizing devalued and aesthetically non-normative behaviours and traits, and at using health as a tool to justify disciplining, subjugating, and surveilling those who display these behaviours and traits.

Moreover, communities are rarely homogeneous, and are instead typically intersectional and internally complex. There is little reason to think they will share a uniform conception of health or what protecting it demands. Consider for example community views on policing. Diverse communities in the United States often have extremely sharp within-community disagreement over whether police presence furthers or threatens public health.[5] Given a flexible conception of health, there isn't a clear metric along which we can settle this disagreement. The presence of police officers may reduce some people's stress and increase others, reduce some kinds of violence and increase others, and so forth.

At the level of population health rather than individual health, there is a long history of communities and nations generating morally corrupt and oppressive conceptions of what 'the health of the community' should look like, especially when health talk is allowed to spread into domains removed from the narrowly medical. In the twentieth and twenty-first centuries, we have witnessed social visions of "healthy" communities as immigrant-free, as racially unmixed, as exemplifying "traditional family values", as purged of fat and disabled people through eugenics, body discipline, or exclusion from public life, and so forth, and we have seen these visions implemented, often to horrific effect. Hence a challenge for any pragmatist conception of health is that in order to prevent serious harm and oppression, we have to have some standards for assessing whose pragmatic goals and uses of the concept are legitimate, as well as mechanisms for protecting those standards. It is not at all

---

[5] For example, in Washington, DC, which is a highly racially diverse city, a March 2023 opinion piece in the local newspaper, entitled "D.C. needs hundreds more police officers. Here's how to do it right." (Washington Post, March 24, 2023), engendered over 400 comments, many of them debating this question.

clear who the 'we' is who should get to establish these standards. But history shows that leaving this up to the internal formal and informal democratic processes of communities is not morally trustworthy.

In the next three sections, I will explore three case studies where an expansive and strategic notion of health, in the context of healthism, has led to an exacerbation of inequality, oppression, and a reduction rather than an increase in social power and well-being. In each case we will see how understandings of health become infused with aesthetic and moral meanings that reflect and further dominant values and agendas.

21.4 Healthy Eating

It is not breaking news that healthist social imperatives are particularly intense when it comes to food and eating, nor that 'healthy eating' is deeply moralized and socially surveilled and controlled. People who eat 'unhealthy' diets are scorned, especially but not only if they are fat, and fatness is taken to be a direct expression of unhealthy and unvirtuous eating (despite the correlations between eating and weight and between weight and health being actually quite complex).[6] Healthist social imperatives have led to various policies aimed at controlling food choices, many of which are only dubiously connected to traditional health or well-being, such as calorie counts on restaurant menus. We saw earlier that food assistance programs control what low-income people can purchase to eat. There is a literature critiquing such practices, pointing out both that there are good reasons to eat things other than for their health value (Dean 2021; Barnhill et al. 2014) and that these policies are paternalistic and only speculatively related to better medical outcomes (see for instance Kiszko et al. 2014).

For my purposes, the most important point is that our conception of 'healthy eating' is one infused by all sort of aesthetic and ideological values and imagery that has little to do with morbidity or mortality. Our broadened conception of health to include 'well-being' more generally has, in the domain of food and eating, lead to the social enforcement of a picture of well-being that is culturally specific and arguably has little to do with people's ability to flourish and function. Meanwhile, people who fail to engage in 'healthy eating' are seen as aesthetically unappealing, at a minimum, and as either vicious and weak or as pawns of their circumstances without agency.

Anna Kirkland points out how deeply aestheticized our conception of healthy eating is:

---

[6]There is a huge scientific literature on this complexity. Lenz et al. (2009) and Nuttall (2015) provide helpful critical reviews of the relationship between weight and health. Allison and Goel (2018) privides one recent example among many of a study of the indirect and complicated relationship between food intake and weight.

There is a highly specific and evolved set of social rules governing the hierarchy of foods. A baguette is not junk food, but sliced white bread is; the sugar in honey and fruits is healthy while white granular sugar is junk. Modes and moments of consumption are also hierarchically arranged: growing or obtaining local fresh produce is the best way to eat, while eating outside the home and on the move is the worst. Why not exhort inner city mothers to consume canned or frozen vegetables, presumably much easier to provide than access to farmer's markets? One is hard pressed to find anti-obesity recommendations that do not include "fresh" modifying "fruits and vegetables," after all. Could it be that shopping at farmer's markets is simply the most virtuous mode of food consumption for the upper classes at this moment in history? (2011: 474)

Those of us in a position to impose cultural values through policy and public discourse want people not only to get nutrients they need, but also to live a certain kind of 'food lifestyle', eating beautiful foods attached to cultural traditions we value, obtained in a pleasant and atmospheric way, and consumed in a leisurely fashion with loved ones, in a dedicated eating space. That is, we want eating to be an aesthetic practice of a specific sort, and we see people who engage in such practice as virtuous and those who don't as gross. It is of course fine, even rewarding, to enjoy the aesthetics of buying, preparing, and eating food! But privileged folks generally treat our own aesthetic food and eating preferences as objective, and treat those who don't have the time, money, or inclination to share them as defective and in need of social control. I suggest that it is our imagistic linking of all this aesthetics with health that makes this socially acceptable, and that it is the flexibility and expandability of our notion of health that makes this slippage happen.

Those who promote an 'environmental approach' to explaining people's eating habits and promoting 'proper' eating try to shift the emphasis away from individual blame and responsibility, arguing instead that people's food choices are controlled by their environment. This approach is typically favoured by people who see themselves as socially progressive and attuned to structural inequality. Poor people and people of colour, the line goes, eat wrong because they are stuck in 'obesogenic environments' that determine or all-but-determine their choices. It is of course true that living in a neighbourhood without farmers markets or baguettes makes it very difficult to buy food at the farmer's market or to eat baguettes. But proponents of the environmental approach rarely question whether we ought to be trying to elicit such food choices from people in the first place. Kirkland writes,

I argue that this environmental approach to obesity has been sold as a progressive, structurally focused alternative to stigmatization, but it actually embeds and reproduces a persistent tension in feminist approaches to social problems: well-meant efforts to improve poor women's living conditions at a collective level often end up as intrusive, moralizing, and punitive direction of their lives. In this case, the environmental argument seems structural, but it ultimately redounds to a micropolitics of food choice dominated by elite norms of consumption and movement... The environmental mode of intervention aims to shift the ground beneath the most basic habits and choices during daily life, with the aim of making poor environments more like elite ones (2011: 464, 466)

Pulling no punches, Kirkland concludes:

But what if it is the case that many elites find the terms of the environmental account to be simply a more palatable way to express their disgust at fat people, the tacky, low class foods they eat, and the indolent ways they spend their time? Proper practices of food, eating, and exercise have been raised to the status of absolutely correct rules for good health rather than simple features of human cultural variety. (2011: 474)

I want to take Kirkland's analysis on board wholesale. But for the purposes of this paper, I also want to insert her critique into my larger argument: First, the fluidity of our conception of health helps enable this kind of elitist redefinition and moralization of appropriate eating. No one is called upon to justify why vegetables should be fresh and bread should be in baguette form from a health point of view, because health can just mean 'well-being,' and well-being is vague enough that aesthetic imagery can define our imagination of it. This is especially so when our conception of who has well-being tracks our intuitions about who is generally socially appealing and successful. Second, this entrenchment of an ideological and aesthetic picture of what counts as 'healthy eating' generally penalizes those who are already under-privileged: poor people, fat people, and people of colour, who may not share or are unreflectively assumed not to share the aesthetic food lifestyle valued within wealthy white culture.

21.5 Healthy Sexuality

"Healthy sexuality" is a common phrase, but its meaning is wildly variable. It can suggest anything from the ability to enjoy one's own sexuality without shame or fear, all the way to conformity with social norms that entrench heterosexual, kink-free, device-unassisted sex within a monogamous relationship. Sex and sexual pleasure are, for most (though importantly not all) people, essential dimensions of well-being. But it is not at all obvious how, whether, and when to include the sexual dimensions of well-being as part of health. Moreover, our social conceptions of 'healthy sexuality' are deeply gendered, in ways that don't have any clear tie to the medical needs of different genders. Healthy sexuality for cisgender men is typically understood as the ability to get and keep an erection on demand. Healthy sexuality for cisgender women is fraught and murky in its conception, but most often seems to have something to do with having appropriate sexual responses within a loving, intimate relationship, and being able to maintain a 'normal' heterosexual sex life. Meanwhile, healthy sexuality for trans folk is rarely discussed.

Cisgender men's sexuality has been heavily medicalized since the US Federal Drug Administration's approval of Viagra in 1998. The rise of Viagra and various knock-off drugs as a treatment for 'erectile dysfunction' is a classic example of the generation of a disease to suit the needs of the pharmaceutical industry (Carpiano 2001; Lexchin 2006). While Viagra, whose erection-sustaining properties were discovered by accident, was originally pitched as a drug to help older men with other morbidities and little to no capacity for erection, over time erectile dysfunction was developed into a self-standing 'disease', and any failure to get or keep an erection on demand was framed as a 'symptom,' rather than perhaps as a normal fluctuation of mood, situation, attraction, and so forth. This meant that virtually any cisgender man, over the course of his normal sex life, could find himself with a 'health problem' requiring medical intervention and a pharmaceutical solution; in effect, Viagra and erectile dysfunction turned all men into potential patients with 'sexual health' issues. What we see here is the expansion of a health framework into the fine-grained details of men's sexual lives, pathologizing normal variations in sexual response. The medicalization of men's sexuality has contributed to a depressingly narrow conception of that sexuality as defined almost entirely by erections, with little room left for more nuanced conversations about cis men's sexual well-being, desire, and fulfilment.

Meanwhile, there have been intense arguments for decades now over what 'healthy sexuality' for cisgender (and generally presumed heterosexual) women even means. Various 'disorders' of female sexuality have been named, including "sexual interest disorder", "orgasmic disorder", and "subjective sexual arousal disorder." Unlike for men, there is no agreement where in the body or at what stage of the sexual narrative we should be looking for 'healthy sexuality' in women (Moynihan 2003). Women's purported sexual disorders range from physical pain, to muscular response, to lack of 'normal' propositional attitudes such as desire for intercourse, to emotions such as fear, to general existential malaise (see Anastasiadis et al. 2002 for an overview). Attempts to medicalize women's sexuality and develop pharmaceuticals for 'female sexual disorder' have been largely unsuccessful, and much more controversial than their male counterparts. (For an in-depth discussion, see Kukla 2016). All the same, there seems to be widespread agreement that cis women's generally lacklustre response to heterosexual sex under patriarchy is 'unhealthy' in one way or another.

In "The sexualization of the medical", Judy Segal (2012) takes relatively familiar concerns about the perils of medicalizing sexuality and turns them on their head. She demonstrates that we are taken to be responsible for having good sex, as a signal of and for the purpose of maintaining good health. Thus the healthist regime not only penetrates our sex lives, but, as is its way, demands sexual health from us, while conceiving of health in broad and ideological terms. Segal argues that for people with chronic illnesses, particularly women, having sex–indeed, having 'good', pleasurable sex–is treated as an imperative, required to prove that you are striving to overcome illness. She writes, "[W]ith the equation of sex and health, popular representations of cancer, in particular, have added 'have sex' to the imperatives of the illness experience, along with 'be positive,' 'be a fighter,' 'look good,' and 'don't lose your sense of humor'" (Segal 2012: 369). In short, the duty to have and enjoy (traditional heterosexual) sex in order to prove yourself 'life affirming' and worthy of cure is itself framed as 'health work.' This phenomenon, Segal argues, is part of the larger "cooptation of sex by the discourse and the institutions of health and health promotion" (2012: 371).

I am neither denying that there is such a thing as sexual health, nor that the conceptual and technological toolbox of health can be helpful in the context of sexuality. But there is nothing like social agreement over what healthy sexuality looks like, and our images of it seem infused by ideology at every turn. And when 'healthy' sexuality is imagined along conventional lines, many people–trans people, queer people, kinky people, disabled people, older people, people who need assistive devices–may have no route forward to a socially acceptable sexuality. This framework also makes it harder to opt out of sexuality if we wish, since cis men and cis women in different ways are subject to imperatives to enjoy sex. And while the health framework may disproportionately disadvantage people with nonstandard sexualities, it is not clear that it is to the advantage even of conventional heterosexual men and women to understand their sexual flourishing and pleasure through a health paradigm that naturalizes specific sexual desires and performances. The heavy-handed application of a health framework makes it harder to see and value sexual pleasure and agency as dimensions of well-being and flourishing in their own right, regardless of their connection to health. Many of the social, political, and cultural forces that make good, pleasurable sex difficult have absolutely nothing to do with health.

21.6   Healthy Gender Identity

Transgender people have found themselves at the center of worldwide cultural warfare of late. While gender is not a medical category, and one might rightfully assume that no gender identities are any healthier or less healthy than any others, for trans folks, the very act of existing inserts them into conflicting and moralized visions of health.

Being trans is about having, or lacking, a particular gender. Gender is a complex and contested social phenomenon, but it is strange to think of it as any kind of medical condition. Yet for a long time, having–and indeed proving to medical professionals that you have–gender dysphoria has been taken as the privileged measure of whether you 'really count' as trans (Spade 2013). In turn, gender dysphoria is understood in health terms; it is the illness or disorder of feeling acutely that your body is an uncomfortable fit with your identity or sense of self. Trans people have typically earned whatever social legitimacy and respect they have managed to eke out by mobilizing and accepting the language of pathology and disorder to explain their identity.

Many trans folks do have gender dysphoria, but some do not. Placing gender dysphoria at the center of trans identity, as the measure of its authenticity, turns transness, even when it is 'accepted,' into an experience of ill health, deficiency, and sadness. The best that trans people can hope for, according to this picture, is to be 'fixed' so that they no longer feel dysphoric. Yet many trans folks have recently begun emphasizing that gender euphoria–the visceral experience of joy at how your body is expressing gender–is just as characteristic of and central to trans experience as is gender dysphoria (Ashley and Ells 2018; Dale 2021; Jacobsen and Devor 2022; Jones 2023). If we see being able to play with gender and express and build our identities through our bodies as a basic dimension of well-being, then euphoria will be just as important as dysphoria. But an understanding of trans identity that centers euphoria fits less well into a health and illness paradigm. Within a health framework, it becomes tempting to see the freedom to express a gender identity as important only to the extent that doing so cures ill health. Seeing transness as a special opportunity for various forms of euphoria has little hold on the general public's imagination; euphoria, within a health framework, is supererogatory.

The absorption of gender identity, and especially trans gender identity, into the domain of health has masked the ways in which gender dysphoria and euphoria are in fact quite common human experiences among trans and cis folks alike (Schall and Moses 2023). Many cis women experience discomfort with some bodily feature that feels dissonant with their conception of their femininity–too much body hair, too small breasts, whatever–and there are lucrative industries that have emerged to help relieve this dysphoria (waxing salons, push up bras, etc.), although they are not framed as health interventions when cis women use them. Cis women often take euphoric selfies when they feel they 'look cute' in ways that affirm the kind of femininity they want to express Cis men likewise often hide skinny chests, or strut in front of mirrors or take gym selfies when they feel masculine in a way that pleases them. More generally, our bodies cause us great aesthetic and existential pain and pleasure, and much of that is gendered. It does not usually occur to us to apply the language of health to such experiences when they belong to cis people, as we do in the case of trans folks.

Meanwhile, the issue of trans children's and adolescents' health has become polarized, and it has turned into a nexus for control over their lives and identities (for instance see Bazelon 2022).

On one side, there is intense distress over medical interventions that are open to teenagers and pre-teens, most often puberty blockers.[7] Those who oppose these interventions justify their concern on the grounds of their supposed health risks. However, the health risks are so far relatively speculative, and counterbalanced by (equally speculative) apparent benefits, including less need for more dramatic medical interventions later. Indeed, most articles cautioning against their use emphasize our *uncertainty* about their health effects rather than any known actual negative effects (e.g. Clayton et al. 2022; Mahfouda et al. 2017).[8] Puberty blockers have been used to treat children with precocious puberty since the 1980s, without alarming effects; only their use in trans youth is new (Radix and Silva 2014). Although we have little to no data about their long-term effects, the slowing of bone mineral density that apparently can accompany their use, which is the main direct health concern they raise, appears to be reversible (Jorgensen et al. 2022). We have a long history of being willing to prescribe drugs for children without solid evidence concerning their risks and benefits, because of the ethical complexities of studying drugs in paediatric populations. Perhaps this is poor practice, but many people do

---

[7] It is extremely rare to perform other gender affirming medical interventions on children, although some late teens do get "top surgery."

[8] A couple of critical reviews of the literature conclude that there is no strong evidence of harm but an all-around weak evidence base (Rew et al. 2021; Jorgensen et al. 2022).

seem to have a disproportionately heightened concern with trans kids' health when it comes to puberty blockers. On this side, the rhetoric seems to be that we should err on the side of withholding interventions that make space for gender affirmation in the name of children's health.

On the other side, protection of trans children's autonomy over their social presentation is often justified in terms of the health benefits of allowing this autonomy, with appeals to catastrophic consequences such as high suicide and depression rates in trans kids who are not supported or allowed to socially transition (see Rew et al. 2021). While suicide and depression are indeed serious health issues, it strikes me as significant that the discourse is one that is almost always framed in terms of health. We should not need to always appeal to catastrophic health consequences in order to defend the idea that we should let people of any age make their own broad-strokes choices about how they want to talk, dress, use their body, and be addressed, as a matter of basic respect.

Hence on the one side, the focus on care for transgender youth is on medical interventions and their health risks, while on the other, the focus is on the health risks of not affirming gender identity, whether socially or medically or both. Both sides seem to have replaced fundamental questions about how to respect people's bodily and social autonomy with health questions. What is ultimately a moral debate is funnelled on both sides through the discourse of health, which both infuses health with moral meaning, and restricts us to thinking about trans flourishing in terms of health risks, pathology, and fixing deficiency.

Importantly, I want to acknowledge that at this moment in history, there are important reasons for trans people to adopt a health framework in order to talk about their needs and well-being (Kukla 2015). There are indeed major health risks that come along with being trans right now, and framing trans identity as a health issue gives trans people access to social resources and social comprehension. Moreover, many, though far from all, trans people want access to medical interventions such as hormones or surgeries. Although cisgender people do not generally have to frame themselves as pathologically disordered or ill to get access to such things, trans folks do. But ultimately it seems like understanding gender freedom as primarily a health need distorts our understanding of gender identity and of people's basic autonomy rights, and encourages a picture of trans lives as pathological and burdened rather than as joyful.

21.7 Elite Capture of the Concept of Health

One way to get at the essence of my point in this paper is this: Health pragmatists have urged that we allow our concept of health to remain open-textured and available for strategic use by different communities and stakeholders. But when we allow this kind of fluidity and repurposing, we open the concept of health up to elite capture. Olúfẹ́mi Táíwò (2020, 2022) has recently influentially described elite capture as the recurring process by which the privileged elites within a group or community, by way of their heightened social and material capital, hijack the goals and values of the group, using the resources of the group to further their own interests, which they then portray as standing in for group interests. For example, within the gay rights movement, gender-conforming gays and lesbians with more money, education, and social privilege arguably managed to put gay marriage and access to conventional social settings and institutions at the top of the political agenda, even though these goods were not of interest to many. Less privileged gay folks often cared more about homelessness, access to health care, and vulnerability to violence, for instance (Táíwò 2020, 2022). But gay marriage and the aspiration to middle-class normalcy became the face of the gay rights movement. Similarly, privileged white women have often made lean-in style 'girl-boss' leadership and concerns with the glass ceiling in the world of high politics and high business the public face of feminism, leaving many women of color and poor women alienated from the movement.[9] Táíwò's main concern in his research is to show how identity politics, and political theorizing more generally, have been hijacked by elite capture. He argues, "Elites get outsize control over the ideas in circulation about identities by, more or less, the same methods and for the same reasons that they get control over everything else" (2020).

I want to suggest that the concept of health and its social use, both in discourse and in practice, are particularly vulnerable to elite capture. As I have argued, we live in a healthist regime in which health is presumed by almost everyone to be an overriding value, and in which both societies and individuals are presumed to be responsible for maximizing health and controlling health behaviours. Hence conceiving of something as a health issue or health need gives it pressing social weight. But at the same time, we have no clear shared conception of health or what it includes. This makes it especially easy for elites with social power over which ideas get currency—that is, people with education and social capital, who also tend to be white and ablebodied—to marshal the notion of health to their own ends, and to use health rhetoric to enforce their own conceptions of what sort of world they would like to inhabit, aesthetically and practically. The open texture of our notion of health, combined with its pragmatic power, makes it an exceptionally valuable tool over which to gain social control. Elites within and relative to any group are the ones best positioned to gain this social control.

This elite capture of health was perhaps most vivid in my discussion of food and eating. Elites generally enjoy leisurely strolls through farmers markets, baguettes, fresh vegetables, and family dinners at the dining room table. Not only have they succeeded in associating these activities and foods with health, but they have succeeded in entrenching a social and personal imperative that everyone shop and eat in this way. If people cannot or will not make these choices on their own, the line goes, we should create environments that compel these choices. Such initiatives present themselves as furthering social justice and collective well-being, but they

---

[9]This alienation is the cornerstone of third wave intersectional feminism. For good in-depth discussions of this dynamic, see Hay (2020) and Zakaria (2021).

impose an elite vision that does not necessarily track everyone's preferences or understanding of the good life.

I think that we see the elite capture of health in other domains that I have discussed as well. Our social visions of healthy sexuality are dominated by images of heterosexuals having vanilla sex in loving, monogamous relationships, with women finding sex life-affirming and men finding sex possible at every moment. Our social visions of healthy gender identity are of folks striving to have traditional binary gendered bodies and identities as best they can, rather than relishing in genderqueer joy and experimenting with gender as much as they like. The self-care industry packages expensive elitist pleasures such as spa days and destination vacations as health needs and indeed health imperatives.

Urban planners' dominant visions of 'healthy cities' are not just of walkable cities, but of city streets lined with coworking spaces, third wave coffee shops, and other signifiers of upper middle-class lifestyles, free of homeless shelters and methadone clinics that concretely protect the health of those most in need. More grandly, visions of a 'healthy society' and a 'healthy nation' generally presume that fat and disabled people are best off 'cured' out of existence rather than included and supported. Because these social visions are backed by and serve the purposes of elites, they end up being the ones best practically positioned to influence both social policy and people's reactive attitudes towards individuals who make 'unhealthy' choices or have 'unhealthy' lifestyles. We equate healthy living with white, wealthy, able-bodied, heterosexual lifestyles and aesthetics, and with lifestyles and behaviors that do not disrupt dominant cultural practices. Meanwhile the lifestyles and aesthetics of fat, queer, BIPOC, kinky, and disabled folks show up as inherently 'unhealthy' and in need of fixing and judgment.

This possibility for elite capture is opened by the twofold move of expanding health to include an unspecified number of dimensions of 'well-being,' combined with a pragmatist approach to health that marshals the concept to give an issue social heft and a claim on social resources. Well-being is a loose concept, and different people's conceptions of well-being differ dramatically. Tethering the already-slippery concept of health to the even more slippery concept of well-being and then giving it a great deal of social power lays the perfect groundwork for elite capture. Moreover, allowing local groups and communities more self-determination when it comes to their vision of health will not necessarily help, as social subgroups are intersectionally diverse and subject to internal elite capture (Táíwò 2022).

I do not take anything in this paper to be an argument against a pragmatist, pluralist conception of health. Indeed, as I said at the start, I believe we need such a conception. I do not think that it is possible to critique the elite capture of health by insisting that we use the concept of health only in a narrow 'essential' sense, because I do not believe that there is in fact an essence of health. Rather, I think that once we embrace a pragmatist, pluralist conception of health, we have no choice but to recognize that this will require us to be on our guard when it comes to oppressive and exclusionary uses of the concept. Overly broad and permissive conceptions of health are especially vulnerable to elite capture. Well-being is a particularly ideo-logically charged concept, and so the expansion of health to include well-being more generally makes it particularly prone to misuse. We should not assume that health promotion is always empowering or beneficial. Health is a pragmatic, pluralist concept, but in a healthist society it is also a dangerously potent and weaponizable one.

Acknowledgement I am grateful to the participants at the workshop on "health and disease as practical concepts" at Erasmus University in Rotterdam, as well as to the Political Philosophy workshop at Katholieke Universiteit Leuven, for excellent and helpful discussions of earlier drafts of this paper. I am Particularly grateful to Leen De Vreese for her helpful comments on the paper.

References

Afzal, Shoaib, Anne Tybjærg-Hansen, Gorm B. Jensen, and Børge G. Nordestgaard. 2016. Change in body mass index associated with lowest mortality in Denmark, 1976–2013. *JAMA, Journal of the American Medical Association* 315 (18): 1989–1996.

Allison, Kelly C., and Namni Goel. 2018. Timing of eating in adults across the weight spectrum: Metabolic factors and potential circadian mechanisms. *Physiology & Behavior* 192: 158–166.

Anastasiadis, Aristotelis G., Anne R. Davis, Mohamed A. Ghafar, Martin Burchardt, and Ridwan Shabsigh. 2002. The epidemiology and definition of female sexual disorders. *World Journal of Urology* 20: 74–78.

Ashley, F., and C. Ells. 2018. In favor of covering ethically important cosmetic surgeries: Facial feminization surgery for transgender people. *The American Journal of Bioethics* 18 (12): 23–25.

Barnes, Elizabeth. 2023. *Health problems – Philosophical puzzles about the nature of health.* Oxford University Press.

Barnhill, Anne, Katherine F. King, Nancy Kass, and Ruth Faden. 2014. The value of unhealthy eating and the ethics of healthy eating policies. *Kennedy Institute of Ethics Journal* 24 (3): 187–217.

Bazelon, Emily. 2022. The battle over gender therapy. *New York Times Magazine*, June 15, 2022.

Carpiano, Richard M. 2001. Passive medicalization: The case of Viagra and erectile dysfunction. *Sociological Spectrum* 21 (3): 441–450.

Carr, Willem Jacobus Petrus. 1990. *Soweto: Its creation, life, and decline.* South African Institute of Race Relations.

Clayton, Alison, William J. Malone, Patrick Clarke, Julia Mason, and Roberto D'Angelo. 2022. Commentary: The signal and the Noise—Questioning the benefits of puberty blockers for youth with gender dysphoria—A commentary on Rew et al. (2021). *Child and Adolescent Mental Health* 27 (3): 259–262.

Dale, L.K., ed. 2021. *Gender Euphoria: Stories of Joy from trans, non-binary and intersex writers.* Unbound.

Dean, Megan A. 2021. In defense of mindless eating. *Topoi* 40 (3): 507–516.

Hay, Carol. 2020. *Think like a feminist: The philosophy behind the revolution.* WW Norton & Company.

Henderson, Jillian T., George F. Sawaya, Maya Blum, Laura Stratton, and Cynthia C. Harper. 2010. Pelvic examinations and access to oral hormonal contraception. *Obstetrics and Gynecology* 116 (6): 1257.

Huber, M., J.A. Knottnerus, L. Green, et al. 2011. How should we define health? *British Medical Journal* 343: d4163.

Jacobsen, K., and A. Devor. 2022. Moving from gender Dysphoria to gender Euphoria: Trans experiences of positive gender-related emotions. *Bulletin of Applied Transgender Studies* 1 (1–2): 119–143.

Jones, Tiffany. 2023. Why be Euphorically Queer? An ecological model of Euphorias' influences & impacts. In *Euphorias in gender, sex and sexuality variations: Positive experiences*, by Tiffany Jones, 15-34. Palgrave Macillan Cham.

Jorgensen, Sarah C.J., Patrick K. Hunter, Lori Regenstreif, Joanne Sinai, and William J. Malone. 2022. Puberty blockers for gender dysphoric youth: A lack of sound science. *Journal of the American College of Clinical Pharmacy* 5 (9): 1005–1007.

Kaplan, Jonathan. 2019. Self-care as self-blame redux: Stress as personal and political. *Kennedy Institute of Ethics Journal* 29 (2): 97–123.

Keyes, Helen, Sarah Gradidge, Nicola Gibson, Annelie Harvey, Shyanne Roeloffs, Magdalena Zawisza, and Suzanna Forwood. 2022. Attending live sporting events predicts subjective wellbeing and reduces loneliness. *Frontiers in Public Health* 10: 989706. https://doi.org/10.3389/fpubh.2022.98970.

Kirkland, Anna. 2011. The environmental account of obesity: A case for feminist skepticism. *Signs: Journal of Women in Culture and Society* 36 (2): 463–485.

Kiszko, K.M., O.D. Martinez, C. Abrams, and B. Elbel. 2014. The influence of calorie labelling on food orders and consumption: A review of the literature. *Journal of Community Health* 39: 1248–1269.

Kukla, Rebecca. 2005. *Mass Hysteria: Medicine, culture, and mothers' bodies*. Rowman & Littlefield Publishers.

———. 2015. Medicalization, 'normal function,' and the definition of health. In *The Routledge companion to bioethics*, ed. John D. Arras, Elizabeth Fenton, and Rebecca Kukla, 515–530. London: Routledge.

———. 2016. *Failed medicalization and the cultural iconography of feminine sexuality. Philosophy of love and sex: Thinking through desire*, ed Sarah LaChance Adams, Christopher M. Davidson and Caroline R. Lundquist, 185–208. Rowman and Littlefield International Ltd.

Kukla, Quill R. 2022a. Healthism and the weaponization of health: A response to Valles. *Studies in the History and Philosophy of Science* 91: 316–319.

———. 2022b. What counts as a disease, and why does it matter? *The Journal of Philosophy of Disability* 2: 130–156.

Lenz, Matthias, Tanja Richter, and Ingrid Mühlhauser. 2009. The morbidity and mortality associated with overweight and obesity in adulthood: A systematic review. *Deutsches Ärzteblatt International* 106 (40): 641–648.

Lexchin, Joel. 2006. Bigger and better: How Pfizer redefined erectile dysfunction. *PLoS Medicine* 3 (4): e132. https://doi.org/10.1371/journal.pmed.0030132.

Mahfouda, Simone, Julia K. Moore, Aris Siafarikas, Florian D. Zepf, and Ashleigh Lin. 2017. Puberty suppression in transgender children and adolescents. *The Lancet Diabetes & Endocrinology* 5 (10): 816–826.

Metzl, J., A. Kirkland, and A.R. Kirkland, eds. 2010. *Against health: How health became the new morality*. New York: NYU Press.

Morris, Theresa, and Joan H. Robinson. 2017. Forced and coerced cesarean sections in the United States. *Contexts* 16 (2): 24–29.

Moynihan, Ray. 2003. The making of a disease: Female sexual dysfunction. *British Medical Journal* 326 (7379): 45–47.

Murkoff, Heidi. 2002. *What to expect when you're expecting*. New York: Workman Publishing.

Nuttall, Frank Q. 2015. Body mass index: Obesity, BMI, and health: A critical review. *Nutrition Today* 50 (3): 117–128.

Paltrow, Lynn M., and Jeanne Flavin. 2013. Arrests of and forced interventions on pregnant women in the United States, 1973–2005: Implications for women's legal status and public health. *Journal of Health Politics, Policy and Law* 38 (2): 299–343.

Radix, Anita, and Manel Silva. 2014. Beyond the guidelines: Challenges, controversies, and unanswered questions. *Pediatric Annals* 43 (6): e145–e150.

Rew, Lynn, Cara C. Young, Maria Monge, and Roxanne Bogucka. 2021. Puberty blockers for transgender and gender diverse youth—A critical review of the literature. *Child and Adolescent Mental Health* 26 (1): 3–14.

Schall, Theodore E., and Jacob D. Moses. 2023. Gender-affirming care for cisgender people. *Hastings Center Report* 53 (3): 15–24.

Segal, Judy Z. 2012. The sexualization of the medical. *Journal of Sex Research* 49 (4): 369–378.

Spade, D. 2013. Mutilating gender. In *The transgender studies reader*, ed. Susan Stryker and Stephen Whittle, 315–332. London: Routledge.

Táíwò, Olúfẹ́mi O. 2020. Identity politics and elite capture. *The Boston Review*, May 7. Available at https://www.bostonreview.net/articles/olufemi-o-taiwo-identity-politics-and-elite-capture/.

———. 2022. *Elite capture: How the powerful took over identity politics (and Everything Else)*. Haymarket Books.

Thoits, Peggy A. 2010. Stress and health: Major findings and policy implications. *Journal of Health and Social Behavior* 51 (1_suppl): S41–S53.

Valles, Sean A. 2018. *Philosophy of population health: Philosophy for a new public health era*. Routledge.

van der Linden, Rik, and Maartje Schermer. 2022. Health and disease as practical concepts: Exploring function in context-specific definitions. *Medicine, Health Care and Philosophy* 25 (1): 131–140.

Welch, Tala. 2011. Healthism and the bodies of women: Pleasure and discipline in the war against obesity. *Journal of Feminist Scholarship* 1: 33–48.

Williams, David R., Jourdyn A. Lawrence, and Brigette A. Davis. 2019. Racism and health: Evidence and needed research. *Annual Review of Public Health* 40: 105–125.

Zakaria, Rafia. 2021. *Against white feminism: Notes on disruption*. New York: WW Norton & Company.

Chapter 22
Pragmatism, Pluralism, Vigilance and Tools for Reflection: A Reply to Quill Kukla

Leen De Vreese

22.1 Introduction

In '*Healthism, Elite Capture, and the Pitfalls of an Expansive Concept of Health,*' Quill Kukla (Chap. 21, this volume) warns for the possible harms caused by a conception of health that is too open-textured and all-encompassing. Kukla admits that health is an ineliminably pluralist and pragmatist term—making it impossible to define its essential core. But they meanwhile emphasize that adopting a too broad and permissive conception of health might be harmful in its own right. The take-away message of Kukla's contribution is that, when we embrace a pragmatist, pluralist approach to the conception of health, we should be extra vigilant for the possible misuses of the term.

I can only agree with Kukla's point of view as clearly stated and well-defended in their contribution to this book. In former work, I adopted and defended a similar point of view regarding the notion of disease. I argued that, while we need to accept a pragmatist and pluralist point of view to get grips on the actual use of the notion, it is not the case that there can only be "variability and free choice" (cf. Schwartz 2007: 61). Rather we need philosophical analyses that can be used as tools for ongoing reflection on our actual conceptual practices. (De Vreese 2014, 2017, 2021).

Compared to the extensive debate on the notion of disease in philosophy (of medicine), the notion of health is rather undertheorized and getting much less attention. One of the reasons seems precisely to be this: while it is already very hard to pinpoint the meaning of the notion disease in a widely accepted way, it seems even harder to pinpoint the meaning of the notion health because the latter notion is actually used in an even broader and more encompassing sense. 'Health' does not

L. De Vreese (✉)
Centre for Logic and Philosophy of Science, Ghent University, Ghent, Belgium
e-mail: Leen.devreese@ugent.be

© The Author(s) 2024
M. Schermer, N. Binney (eds.), *A Pragmatic Approach to Conceptualization of Health and Disease*, Philosophy and Medicine 151,
https://doi.org/10.1007/978-3-031-62241-0_22

just mean 'free from disease'. 'Health'—in the ways we actually use the notion—seems to cover much more.[1]

My own pragmatic approach to analyzing the notion of disease has always been to start from the pluralism in the ways the concept is already used in different contexts, and to offer a philosophical framework that can be used as a tool to critically reflect on that pluralist use. I am convinced that such an approach is much more fruitful than one resulting from the reverse strategy often applied by philosophers in their (conceptual) armchair analyses, i.e. to start with critical reflection and deduce from that a (preferably monist) meaning in which the concept *should* be used accordingly. As I defended before (De Vreese 2014, 2017, 2021), the latter strategy does not seem to work. The ongoing disagreements among writers on this topic can be taken as proof of this. Hence, what I would like to defend here is that we need more than just a pluralistic, pragmatic view on the conception of health in combination with a vigilant attitude. What we need is a pragmatic approach to our (philosophical) analyses of health. While the former approach in the end still focusses on establishing some kind of standards (and/or who is going to set the standards) for deciding on the 'right' conception of health (be it monist or pluralist), the latter focuses rather on developing tools for reflection that can help in bringing an ongoing vigilance to life while respecting the actual pluralism and dynamics in our use of the concept.

The remainder of this comment will further pursue this line of thought. Given the possible problem(s) for the pragmatist, pluralist approach to the conception of health (as they have nicely been explored by Kukla), let us look at possible steps towards solutions. How can we proceed given the plurality in the use of the notion health and the possible pitfalls following from embracing that in our analyses? What do we, as medical humanity researchers, have to offer in that endeavor? In what follows, I will give some suggestions and raise some questions to be explored in view of such a project.

22.2 Health as a Multifaceted Term and Its Relation to Disease and Well-Being

First, given the very wide range of meanings of the concept of health as it is used in practice, it seems necessary to discern and characterize different kinds of 'health' and to be clear about what kind(s) of health one talks about when critically analyzing the notion. Otherwise, it is rather easy to conflate things and/or to talk past each other. That, e.g., in some uses of 'health' reference is made to domains of life far removed from the domain of medicine (as pointed out at several places in Kukla's argumentation), might not be problematic if we are just very clear about that. In

---

[1] For a recent contribution to the debate and an overview of the diversity in the use of the health notion in different scientific domains, see Scholl and Rattan (2022).

Kukla's paper, (indirect) references are already made to different (sub)types of health that we could distinguish such as medical health, physical health, biological health, mental health, emotional health, sexual health, social health. Some of these types of health seem indeed more closely related to the domain of medicine than others. This, however, should not be problematic in its own right, as long as we make these distinctions and are well aware of them. Also, more (sub)types of health might be thought of as worth discerning: e.g. spiritual health, phenomenological health, genetic health... just to name a few.[2] Once we have clarified useful distinctions, a next step that we can take is to analyze carefully how these different types of 'health' not only differ from each other, but are also related to each other and to other related notions such as disease, well-being... For now, I only point to some things that strike me as important here.

First, it seems to me that, while strictly 'medical health' (what we could define for now as an ideal state of the body that we try to maintain using medical means) is too narrow to be equated to 'well-being', the amalgam of all different kinds of health might come very close to what we refer to when using the concept well-being. Discerning medical health (and perhaps closely related types of health) from other types of health will hence also help in clarifying the relation between the conceptions of health and well-being.

Secondly, what unites all these different types of health seems to be the reference to some kind of balance. On the one hand, a 'within-type' positive balance between healthy and unhealthy elements of a specific kind[3] (e.g. healthy and unhealthy physical characteristics). On the other hand, an 'among-type' balance, for which we have to weigh the health values of different kinds of health against each other. To illustrate this with a simple example: choosing to eat a high-calorific piece of cake weekly on Sunday might be argued to be physically unhealthy, but might be considered to be a good choice nevertheless because it furthers the mental health of the person making that choice (e.g. as being part of someone's way of enjoying life). Hence, choices or habits that are e.g. medically or physically (un)healthy, might not necessarily be emotionally/mentally/... (un)healthy.

Additionally, although all or most kinds of 'health' might somehow have an influence on our physical and/or mental health state, unhealthy choices or characteristics (of whatever health kind) do not need to imply disease or disorder right away. This emphasizes the point made earlier that health and disease are not just two sides of the same coin.

The latter also relates to a further question to consider: are all kinds of 'health' concerned with fighting against disease as an end-goal? Or can different types of 'health' be related to different kinds of goals? For example, some kind(s) of 'health'

---

[2] I do not at all claim here to make an ideal and/or complete list of kinds of 'health' worth to be distinguished. I only want to point out the importance of these distinctions. Part of the task will precisely be to think further about useful kinds of 'health' to be discerned in function of a fruitful and enlightening debate.

[3] Meaning that the influence of the healthy aspects largely outweigh the unhealthy ones.

might be concerned primarily with enhancing the quality of life—whatever one's medical health state—, rather than fighting against disease per se or trying to prolong life.

Exploring these and related questions will bring us further in getting a better and more nuanced grip on the 'landscape' of the conception of health.

22.3 Health as a Value-Laden Term and the Need for Ongoing Reflection

Once we have a better grip on the different types of health, their relations and the intended goals in labelling them a 'health' matter, we can start using this as a framework for further reflection. Whenever talking in terms of health, we can take a step back and wonder (a) whether we have really good reasons for using the term 'health' to start with and if so, what type(s) of health we are precisely talking about (given the different meanings discerned), (b) what goals are implied, (c) how strong the relation is with strictly medical health matters, etc.

From there, further thought can also be given to what this implies for the enforcement of social imperatives regarding the 'health' matter(s) at hand. Do we have good (medical or other) reasons to enforce social imperatives? Or is there only a tenuous relationship between the social imperatives we want to enforce and these aspects of health we need to care about as a society? And, more generally, to what extent can we enforce social imperatives for which kind of 'health' matters?

Such reflections will not lead to final answers and values will always be involved in our deliberations. But we will at least have the tools at hand that can help in making explicit the reasons, goals and values that are involved. It will make us able to more thoughtfully weigh our words in practice and act accordingly. We might, for example, better be able to explicate why something which is labelled 'healthy' is not necessarily something we all should feel obliged to strive for. In this way, it can act against too simplistic healthist and elitist reflexes without having to alter existing concepts of health. We might be able to recognize that buying one's food at a farmer's market is a healthy choice, while recognizing that this does not imply that everyone should make this choice. Just as we might be able to recognize that going to the gym is healthy, without thereby stigmatizing people who cannot afford gym memberships or just make other choices.

22.4 Conclusion

Of course, this reply offers only a very sketchy proposal of topics for further elaboration. I hope it is nonetheless at least clear enough to convince the reader that the way to proceed is not to develop a final set of pragmatic concepts of health that should be applicable always and everywhere and meanwhile be restrictive enough in order not to be misused within a healthist paradigm. As humanities scholars, we would better develop tools that offer a framework for continuously, critically and pragmatically analyzing the diverse and evolving uses of the health notion by means of different (existing) concepts of health. This way, we can install a pragmatic set of practices within a reflective community which keeps the vigilance alive.

References

De Vreese, Leen. 2014. The concept of disease and our responsibility for children. In *Philosophical perspectives on classification and diagnosis in child and adolescent psychiatry*, ed. Lloyd Wells and Christian Perring, 35–55. Oxford: Oxford University Press.
———. 2017. How to proceed in the disease concept debate? A pragmatic approach. *Journal of Medicine and Philosophy* 42 (4): 424–446. https://doi.org/10.1093/jmp/jhx011.
———. 2021. Against the disorder/nondisorder dichotomy. In *Defining mental disorders: Jerome Wakefield and his critics*, ed. Luc Faucher and Denis Forest. Cambridge: M.I.T. Press.
Scholl, Jonathan, and Suresh I.S. Rattan, eds. 2022. *Explaining health across the sciences*. Cham: Springer Nature.
Schwartz, Peter H. 2007. Decision and discovery in defining 'disease'. In *Establishing medical reality. Essays in the metaphysics and epistemology of biomedical science*, ed. Harold Kincaid and Jennifer McKitrick, 47–63. Dordrecht: Springer.

Chapter 23
Epilogue: Towards a Toolbox for a Pragmatist Approach to Conceptualization of Health and Disease

Maartje Schermer (iD), **Rik van der Linden, Timo Bolt, and Nicholas Binney**

23.1 Introduction

One of the starting points of our pragmatist approach to the conceptualization of health and disease, as set out in the Chap. 2 to this volume, is the idea that only a contextualist, pluralist account of health and disease will do justice to the complexity of medicine and healthcare. Considering the various contributions to this volume seems to confirm this view. As Quill Kukla has argued elsewhere in defense of a pragmatist view: "the concept [of disease] shows up in deeply competing projects and is used for deeply different ends, and there is no consistent notion of disease that underlies these or ties them together" (Kukla 2022: 131).

The contributions in this volume indeed pertain to many different contexts and settings, are relevant to different domains of healthcare and reflect on different (types of) problematic situations. In this Epilogue, we will assess the 'lessons learned' from the individual chapters and the discussions during the workshop with regard to the possibilities for a pragmatist approach to health and disease. The approach we envision is, obviously, still work-in-progress. As stated in the Introduction, we do not aim for a new theory or definition of health and disease, but rather attempt to develop a research program—a way to define and approach problems related to health and disease as encountered in real-life practices. Therefore, in this Epilogue we will first consider what we have learned with regard to the problematic situations outlined in the Prologue (Chap. 2, this volume), and then address some issues and themes that came up during the workshop and that still need more work. Finally, we will also tentatively propose some guidance for further inquiry in the form of a toolbox.

M. Schermer (✉) · R. van der Linden · T. Bolt · N. Binney
Section Medical Ethics, Philosophy and History of Medicine, Erasmus MC University Medical
Center, Rotterdam, The Netherlands
e-mail: m.schermer@erasmusmc.nl

M. Schermer, N. Binney (eds.), *A Pragmatic Approach to Conceptualization
of Health and Disease*, Philosophy and Medicine 151,
https://doi.org/10.1007/978-3-031-62241-0_23

23.2 Insights into Some Problematic Situations

In the Prologue, we identified a number of (types of) problematic situations we encountered in the medical domain, which we considered to be—at least in part—caused by the ways in which health and disease are conceptualized. Most of these problematic situations and their underlying conceptual issues have been addressed in some way or other in the contributions to this volume. We briefly list them here and consider the proposed (conceptual) solutions.

First of all, the problem that patients who have symptoms and who suffer but in whom no pathology can be found are often not recognized as 'really' having a disease, is addressed head-on by Monica Greco (Chap. 17, this volume), supplemented by a reflection by Jenny Slatman (Chap. 18, this volume) on the problematically narrow understanding of embodiment in both medical practice and philosophy. The relation between signs and symptoms on the one hand and pathology on the other also plays a role in Mary Walker and Wendy Rogers' discussion of addiction (Chap. 15, this volume). Both chapters propose ways of re-conceptualizing disease, by questioning the assumed simple equations of pathology with disease, and of disease with various social expectations and normative judgements.

This problematic of symptoms without disease thus appears closely related to problems regarding the institutional designation of the sick role, which we mentioned as a final problematic in our Prologue. Societal attitudes and responses are often based one-on-one on the classification of a problem as a biomedical disease, as also discussed in the chapters by Walker and Rogers, Greco, and Gemma Blok. Walker and Rogers seek to remedy these simplistic and problematic equations by introducing the notion of disease as a vague cluster concept, while Blok (Chap. 16, this volume) makes a plea for understanding it as a "SUD-spectrum". Greco's tentative solution with regard to MUS (medically unexplained symptoms) lies in taking seriously the interactions between patient's illness experiences, physiology, and social structures and in focusing on a pragmatic co-construction of explanations for patients' symptoms.

Next, patients with some pathology or abnormal 'biomarkers' but no symptoms are often understood as diseased even if this classification doesn't necessarily benefit them. This problematic situation was discussed by Marianne Boenink and Lennart van der Molen (Chap. 11, this volume), who argue that we should not understand biomarkers as signaling pathology and hence as diagnostic tools, but rather as prognostic or "anticipatory" devices. This changes the whole "mode" of looking at diseases—from an ontological to a physiological mode, as they call it—emphasizing their development over time and stressing their internal heterogeneity. Elodie Giroux' chapter (13, this volume) in a way also addresses this same issue, by showing how disease and risk—or a pathological and an epidemiological mode of understanding health phenomena—are two distinctly different ways of conceptualizing ill health.

A third problematic we identified, that preventive medicine is often aimed at preventing pathology or pathophysiology as opposed to symptoms, seems related to the previous problem, and Giroux' distinction between epidemiological and pathological conceptualizations of disease is helpful in understanding this. Whereas a pathological model assumes that pathology will lead to symptoms and hence preventing pathology is necessary to prevent symptoms, the epidemiological model shows that not all pathologies in the end cause symptoms and that preventing symptoms may thus not necessarily require preventing all pathology. The shift towards "anticipatory medicine" that Boenink and Van der Molen envisage, similarly to the "personalized medicine" mentioned by Olaf Dekker (Chap. 14, this volume), opens the possibility of focusing more on the risk of developing symptoms, than on preventing pathological changes per se. Bjørn Hofmann, however, warns in his chapter (12, this volume) that biomarkers are often not used to anticipate symptoms and suffering. Biomarkers may well reconfigure the way disease, illness, sickness and even health are understood, but this may not place the needs of the patient at the center of medicine.

Fourth, the problem of overdiagnosis shows up in many contributions, from Nicholas Binney (Chap. 7, this volume)—where it is caused by a false presumption of homogeneity of disease categories in the case of the follicular variant of the papillary thyroid carcinoma (FVPTC)—to Walker and Rogers, Boenink and Van der Molen, and Kukla (Chap. 21, this volume). In the latter, it is not so much over-diagnosis of specific diseases but a more general risk of medicalizing all kinds of issues as being 'health-problems'. Next to the well-known risks of overdiagnosis, such as overtreatment, Kukla's chapter introduced the risk of "elite capture" and the reinforcement of social inequalities and existing power relations. That is to say: forms of overdiagnosis and medicalization may not hit different social groups equally. The problem of overdiagnosis appears very much related to some of the other problems addressed and no single (conceptual) solution is available.

The false presumption that patients with the same disease are homogeneous—our fifth problematic situation—is addressed by Binney and by Lara Keuck (Chap. 9, this volume) and Frank Wolters (Chap. 10, this volume). While Binney demonstrates that a historical approach is indispensable in understanding how researchers and physicians come to see diseases as homogeneous entities, Keuck proposes a new concept—scope validity—to remedy problems stemming from the increasingly recognized heterogeneity between animal models and patient groups with 'the same' disease.

The sixth problematic situation, that of defining 'health' in a helpful way is analyzed in relation to the new concept of positive health by Van der Linden and Schermer (Chap. 19, this volume), and concludes—among others—that empirical research is needed to evaluate whether new conceptualizations live up to their aims. Gili Yaron (Chap. 20, this volume) stresses that understanding the socio-political context and dynamics of new health concepts is also very important to understand their real-world effects. The contribution by Kukla argues that a pragmatic approach may risk embracing overly expansive notions of health which may have undesirable consequences. Leen De Vreese (Chap. 22, this volume) proposes some pragmatic strategies to help tackle these problems, like distinguishing between different 'kinds' of health.

Finally, some of the chapters—especially the reflections—address issues that are considered to be lacking, or underrepresented, in our sketch of problematic situations. Hub Zwart (Chap. 4, this volume), for example, argues that the voice of the patients is not heard enough—the subjective, experiential side of health and disease is underrepresented in most medical and medial-philosophical approaches. Timo Bolt (Chap. 8, this volume) argues for more historical inquiry in order to further understanding conceptualizations of health and disease, not only epistemically but also in the socio-political context. This call is supported by Yaron, arguing for more attention to the socio-political and material lives of concepts—not only in the past but also in the present. The contribution by Kukla might be taken as an example of such an approach.

23.3   Further Themes and Issues

In the discussions we had during the workshop, several themes emerged that were not directly related to 'problematic situations' but to more general complexities regarding conceptualization of health and disease.

23.3.1   Disease and Diseases

First, it appears important to clarify the often-overlooked distinction between two main projects of conceptualizing health and disease: one is about defining Disease as an umbrella concept, the other about how different diseases can be individuated and classified. The first project attempts to define what Disease is as distinct from non-disease (often called 'health'). The question is how to distinguish between the normal and the pathological, or how to classify thing as either belong or not belonging to the category of Disease. The second project is about understanding and determining how to name and classify and distinguish separate diseases or disease-entities. Here, the question is how to carve up the domain of Disease into different disease categories, how to individuate different diseases. An important observation here is that historically, the notion of specific disease 'entities' is a relatively recent idea (Rosenberg 2002, 2003), but one that appears to be presupposed in much of the classificatory project. It is, however, mostly implicitly present, staying in the background but determining much of our understanding, as an 'active element of knowledge' (Binney 2023) that we are hardly aware of. This idea of a dominant disease-entity model is also present in the chapters of Greco, Slatman and Boenink and Van der Molen, who discuss and question the biomedical model in which the disease entity—consisting of a unity of a pathophysiological leasion or dysfunction, an aetiology (cause), pathophysiology (mechanism), symptoms & signs and a natural course—is central. The contribution of Giroux forcefully shows how—subconsciously—sticking to this entity-model may prevent us from appreciating the radically different model that is implied in epidemiological approaches.

As discussed during the workshop, from an historical perspective, the first project—aiming to determine what counts as Disease—is challenged by states or conditions that were pathologized in the past but are no longer pathologized now, or the other way around. Drapetomania and masturbation are examples of the first, epilepsy of the second. Even when we agree that some condition is, was and has always been a Disease, there may still be discussion about changing classification or nosology—arguably an infection with the tubercle bacillus was always a Disease, but has been configured differently through time, e.g. as consumption or tuberculosis.

Classification and categorization thus play a very significant role in our understanding of health and disease concepts. From a pragmatist perspective, classification is an inherently pragmatic endeavor, intended to 'cut up' the world in ways that produce useful outcomes. In Fleckian terms: classifications and categorizations are some of the active elements of knowledge, necessary to attain passive resistance: facts about how these categories behave and relate to each other. Binney's chapter shows in detail how this might work out, and also why history is an indispensable element in understanding these processes of knowledge production—and hence the attained knowledge itself.

As Heiner Fangerau (Chap. 3, this volume) describes with a reference to Vaihinger's As-If philosophy: our classifications have an important function since they steer us in seeing things as 'the same', in noticing and highlighting similarities. However, this should not elude us into thinking they represent some fixed essence. These classifications and categories rather should be understood as useful fictions, or as Koch called them 'thought constructs'.

There exists, of course, a voluminous literature on classification and categorization in general, as well as in the health-domain, within the philosophy and the sociology of science (Bowker and Star 2000; Wadmann 2023; Lie and Greene 2020). Nevertheless, in the discussions during the workshop, it was not always clear how to understand these notions, and how to distinguish between, for example, classification and diagnosis, or between disease categories and disease entities. More clarity on this would enhance the discussion.

23.3.2 Context

Second, next to distinguishing these two different projects in the conceptualization of health and disease, it appears important to highlight the many different settings and contexts within the large domain of healthcare and 'the medical realm'. We must distinguish at least between medical research, clinical practice, and public health or population health practices. Within the domain of medical research, there are also important distinctions to be made between laboratory, epidemiological and clinical research, and between biomedically oriented research versus socio-psycho-behavioral research. Clinical practice is of course also by no means a homogenous area, ranging from general practice to, say, tertiary care oncology. The different questions and problematics addressed by different chapters in this volume reflect these differences between settings and highlight the importance of having a contextualized discussion about how we should understand and conceptualize health and disease. Questions about an appropriate understanding of a disease such as Alzheimer's in a laboratory research setting (as discussed in Keuck) are very different from the questions about appropriate understanding of the role of biomarkers in diagnosing or predicting Alzheimer's disease in a clinical setting (as described by Boenink and Van der Molen).

Moreover, we should recognize that notions of health and disease also play a role outside of the medical domain: in the courtroom, in the institutions of our social security system, in the public sphere and in the private sphere of self-understanding. In these contexts, concepts of health and disease often carry important normative and practical weight, and classifying something as disease or not can have far reaching consequences in social, economic, legal and psychological respects. Again, this impacts on the kind of problems raised and the kind of questions asked about the proper or desirable conceptualization of health and disease, as shows for example in the chapters by Kukla, Walker and Rogers, Blok, Van der Linden and Schermer, Yaron and Greco.

23.3.3 Concepts, Conceptions and Conceptualizations

This brings us to a third important issue: how exactly do we understand the notions of 'concept' and 'conceptualization'? And, related: how to understand the 'functions' of concepts? In the Prologue, we emphasized that one of the leading motivations behind our proposal for a pragmatist approach to conceptualization of health and disease is that conceptualizations have consequences. It matters how we categorize, name and 'see' things, for 'seeing as' often carries implicit calls for action and guides our behavior. We also stated that concepts should be seen as tools that have certain functions and can be helpful in producing desired effects—which is in line with recent ideas about conceptual engineering. However, when one compares the different contributions to this volume it will be apparent that 'concept' and 'conceptualization' are not unequivocal terms, as also became clear during our workshop discussions.

As pragmatists, we do not need to settle on one single correct concept of concepts, but we do need to make it clear what we are talking about in specific discussions. In the traditional philosophical view, concepts have been described as mental images/representations, and as logical or linguistic categories ('units of thought'). Concepts, in these views, have a classical definitional structure and can be described in terms of necessary and sufficient conditions, which are to be discovered via conceptual analysis. However, this view has received much criticism and in contemporary philosophical literature, concepts are more often described as prototypes (a set of typical features), exemplars (similarities with memorized objects), theory-theories (experimental and knowledge-based), or as multimodal devices (based on a

4E cognition framework). These contemporary views aim to account for fuzzy boundaries, unclear category membership, and/or for the embeddedness of concept acquisition in the world (see Nefdt 2021 for an overview).

In the field of conceptual engineering, many authors take a non-essentialist, pluralist and functionalist stance towards concepts, which implies that concepts may have various meanings, depending on the function that is at stake (Belleri 2021). It has also been proposed to take a broader approach, focusing on 'representational devices at large' rather than on concepts narrowly construed (Cappelen and Plunkett 2020). Recently, Westerblad has argued for an explicitly Deweyan concept of concepts, which sees concepts as 'rules for action', as a fruitful concept for conceptual engineering and philosophical inquiry more generally.[1] According to Westerbald, "concepts do not just guide thoughts through their determination of inferences, they also guide and determine action and make events and things meaningful" (2022: 11). Following this Deweyan concept of concepts, "implementation of concepts is a matter of changing people's actions, and not just their minds" (5). This idea is in line with the view taken by Koselleck in the field of conceptual history or 'Begriffsgeschichte' (Koselleck 2004). Koselleck defines concepts in two different ways. On the one hand, it indicates the meaning or the signified in the process of signification. Concepts are thus linked to words. On the other hand, concepts acquire an additional layer of meaning from their use; concepts determine "the space of experience and horizon of expectations" (Koselleck 2004: 81).

During the workshop, Kusch (Chap. 5, this volume) brought forward the idea of disease and health as—what is called in political philosophy—"essentially contested concepts" (Gallie 1964). One way to deal with such essentially contested concepts is to move the discussion towards 'conceptions' of the particular instantiations, or realizations of ideal and abstract notions. We might agree, for example, that Boorse's naturalist theory offers a specific conception of Disease, while Coopers' normativist definition captures another. Using the term 'conceptions' for different interpretations of the abstract notion of Disease is thus to some extent in line with pluralism but does not get away from contestation over the 'right' conception. The functionalism and contextualism that a pragmatist approach adds can focus disagreements over conceptions of disease to a more sharply defined context of use. We might all agree that certain conceptions of disease are fine for a research context, but unsuitable in a courtroom, for example.

Finally, the term 'concept' can also be used more loosely as a term for the models we use (e.g. the ontological vs the physiological model of disease, or the biomedical vs the biopsychosocial model); and it can also be used in an even more practical sense, as something we enact. In this latter sense, a concept is more like a social imaginary, a set of presumptions and beliefs that we express in our actions and in the

---

[1] Dewey's concept of concepts follows from their role in scientific inquiry. "Dewey thought of concepts as rules of actions or operations to be performed" (Westerblad 2022: 2). Concepts are thought of "in terms of expressing 'anticipated consequences of what will happen when certain operations are' performed" (Westerblad 2022: 8). They are not final or static, but integral tools in the dynamic activity of inquiry.

social and institutional organization of a practice.[2] One could even ask 'where' concepts are located. In individual minds, in a collective cognitive 'space', materialized in practices? Yet another notion related to this broad idea of a concept is 'cosmology' (Jewson 1976; Armstrong 1995). There remain some questions to be answered here: How should we understand such models, imaginaries and cosmologies in philosophical terms? Are they paradigms, thought styles, or different ways of 'seeing as'? How do they create or influence conceptions of health and disease (and vice versa)?

23.3.4 Functions and Functioning of Concepts

From a pragmatist perspective it is important to think about how concepts and conceptions function in different ways in different contexts. For example, as (regulative) ideals or instruments of power in the socio-political domain; and as scientific models, stipulative definitions or useful as-ifs in scientific practices. In the first case, philosophy could analyze the values and power relations inherent in concepts and analyze the 'social life of concepts' (Yaron, Chap. 20, this volume). In the second case, it is about the epistemic role of conceptions which can be more or less fruitful (cf Binney (Chap. 7), Keuck (Chap. 9), and Wolters (Chap. 10), this volume). In either case, however, we need to distinguish between the *function* and the *actual functioning* of concepts. There is a difference between what a concept of conceptualization it is *intended* to do, what its role or function is supposed to be within a certain practice (e.g.: gatekeeping, or providing access to the sick role, or distinguishing between groups with different prognosis); and the actual effects or *consequences that it turns out to have* in a practical context. The chapter by Van der Linden and Schermer shows how the functions a concept is supposed or intended to serve can change over time, and how a concept can become overburdened with too many expectations—our call for pluralism is strengthened by the observation that the single concept of Positive Health cannot meet all these expectations. As Kukla warns us of in their chapter, some conceptualizations of 'health' may actually turn out to be instruments of oppression rather than tools to promote well-being. The intended effect may not be achieved in practice due to the complexities of social and political life. Likewise, as Greco shows, conceptualizing MUS as a 'real' disease entity— intended to provide patients with more recognition and better care—may backfire when it reinforces the idea that only conditions with clear underlying pathophysiology are 'real' and worthy of medical attention. And while the concept FVPTC was

---

[2]This is somewhat similar to what Haslanger (2012) calls the 'operative concept'. However, enactment has a more material and practical aspect to it; it is not merely about the way we use words in practice, but about the way we do and organize things. Annemarie Mol's notion of 'doing' atherosclerosis, as discussed in Kusch, Chap. 9, this volume, may be a good example.

intended to better predict the prognosis and clinical behavior of certain thyroid tumors, it turned out differently (Binney, Chap. 7, this volume).

Here the distinction between political pragmatism and a speculative pragmatism as introduced by Greco (Chap. 17, this volume) can be helpful. In the first form of pragmatism, 'truth' hardly seems to matter. The strategy is to define or conceptualize health, pathology, disease in general or a specific disease in such a way that is most expedient to some purpose. Pushing to understand MUS as a 'real disease' based on scientifically unconvincing brain-scans is an example, as is defining 'health' as whatever some subgroup within a population wants it to be (cf Kukla, Chap. 21, this volume). This form of pragmatism begs the question whose interests should count here, or which goals or aims of defining something as health or disease are acceptable and which are not. For example: disease mongering for the purposes of financial gain by the pharmaceutical industry, versus claiming disease status in order to gain access to care for some underserved group. More fundamentally, it can be accused of being too superficial, not getting to the core of the problematic and offering only short term or superficial solutions.

Speculative pragmatism, described by Greco as being in line with James' radical empiricism, tries to dig deeper. It does care about 'truth' but takes an open attitude as to what it entails. It questions the underlying conceptualizations and thought patterns (e.g. the dominant biomedical disease model, monocausal explanation, or the dichotomies between health and disease, or mind and body). It traces the effects of such conceptualizations, and of proposed alternative or new conceptualization, not only in their direct appearance but also in their more subtle and nuanced effects. This may shift our understanding of the problematic situation at hand. For example, as discussed by Walker and Rogers, instead of claiming that addiction should (or should not) be regarded as a disease in the courtroom—as political pragmatism might do—speculative pragmatism may question the presumed link between disease and non-responsibility; or it may question the biomedical reductionism that ignores psycho-social causes of addiction.

23.3.5 Continuity and Change

Finally, an important theme throughout the discussions and one central to the chapters of Kusch, Fangerau and Binney, is that of change and continuity, difference and similarity. From a pragmatist point of view that is non-essentialist and puts the usefulness of concepts center stage, taking a historical perspective comes naturally. Understanding how disease concepts evolved, what their function and effects have been in the past and how they have changed in response to observations, discoveries and shifting attitudes opens up possibilities for influencing the directions in which concepts evolve in the present. At the same time, the historicity of disease and disease categories raises a great deal of questions about the 'nature' of diseases.

As we stated in the prologue, "Framing concepts of health and disease as pragmatic and historically contingent may raise concerns that these concepts are being reduced to whatever historical actors have found expedient to believe" (Binney et al., Chap. 2, this volume). We understand pragmatism as a position that rejects simplistic forms of realism and embraces contingency without slipping into pernicious and 'silly' forms of relativism.

In his contribution to this volume, Kusch discusses different forms of relativism in relation to retrospective diagnosis (Chap. 5, this volume). As his analysis makes clear, there are many forms of relativism available in the historical and philosophical literature on medicine, and these need not be "equal validity" relativisms. A relativism of distance may render different concepts of disease too remote, too alien, for physicians to actually pick up and use for pragmatic purposes today. A relativism of locality, however, may provide the necessary flexibility for physicians to employ different concepts in different contexts. As Kusch points out elsewhere, in the past actors were much more comfortable with relativism than actors tend to be today. Drawing on the philosophy of "as if" that Fangerau shows was championed in the early twentieth century, different concepts of disease might be understood as useful fictions and treated as if they were true, to see what the practical consequences of adopting such concepts in particular circumstances were. The tolerance of several such concepts of disease may constitute a relativism of locality. Even so, as Walker and Rogers remind us, any particular concept may help achieve the desired goals, or may hinder that achievement. As Ehni highlights, we must take care to prevent relativism of locality slipping into pernicious relativism. Attention must be paid to the empirical consequences of adopting certain concepts, and decisions to use one concept rather than another must be responsive to these empirical consequences and help to achieve the relevant goals. Combining relativist possibilities with pragmatic goals produces an epistemology that is neither an absolutist realism nor a pernicious sort of relativism.

Moreover, Binney (Chap. 7, this volume), suggests that such a pragmatic middle ground position actually requires an integrated historical and philosophical analysis and an epistemic role for history. He demonstrates how using Fleck's active and passive elements of knowledge as a tool for historical and epistemological analysis can achieve this.

So, instead of either trying to identify some time-independent 'essence' of Disease or the 'real' natural kinds of specific diseases, or concluding that different time-dependent concepts and definitions of disease are just incommensurable—turning historical analogies into a comparison of apples and oranges—we propose a different approach. We propose to analyze how different definitions, conceptions, and classifications are historically linked to each other. Rather than in historical comparisons in terms of differences and similarities between the past and the present, we are interested in a genealogical reconstruction of how a problematic situation has arisen in the present, as a result of the paths medicine did or did not take, or has abandoned, during its development. This approach fits well with recent calls for integration of historical perspectives in analytic philosophy (Dutilh Novaes 2015, 2020), for pragmatist interpretations of conceptual engineering (Westerblad 2022), and of using 'anamnesis' as historiographical method in the history of medicine (Tybjerg Forthcoming). In the next section, where we discuss our proposal for a toolbox, we will elaborate some more on such 'genealogical reconstruction'.

23.4   Towards a Toolbox

From the outset of our research project, we anticipated the results to come not so much in the form of a new definition or theory of health and disease, but rather in the form of a set of tools and methods and a proposal on how to approach issues regarding the conceptualization of health and disease. In this final section, we tentatively propose a 'toolbox' intended to give guidance to further inquiry into the concepts and conceptions of Disease, specific diseases and health, as they function in numerous practical contexts.

To start with, it is important to see that the notion of a toolbox can be understood in two ways. First, the term is a metaphor for a specific view on concepts and conceptual pluralism itself. It implies that the different conceptions of health and disease should be understood as tools that have specific functions and purposes in different contexts. As such, the metaphor points towards the practical use and usefulness of concepts, one of the tenets of pragmatism. Chang, for example notes that "pluralism in science is just as natural as wanting to have various types of tools in our toolbox or having different types of shoes in our cabinet to suit different occasions" (2012: 273). We present a non-exhaustive overview of possible disease concepts in the next section.

However, we also use the metaphor here to stand for a set of methods and (conceptual) strategies that may help us to solve 'problematic situations' related to health and disease concepts.[3] We agree with De Vreese (Chap. 22, this volume) that what we need is a pragmatic approach to our (philosophical) analyses of health and disease which focusses on developing tools for reflection. Based on our research and supplemented with insights from the final discussion at the workshop, we propose a 'working method' and a non-exhaustive set of (conceptual) strategies and method-ological approaches that we found can be useful in addressing problematic situations concerning the conceptualization of health or disease.

23.4.1   A Toolbox of Disease Concepts

It is worth providing something of an inventory of the wide array of concepts we have encountered in this project, in our reading and discussions with doctors and patients. Many are familiar from everyday discussions of medicine, or from philosophical, sociological and historical literature. Others are drawn from medical literature, medical practice, or face-to-face discussions. Table 23.1 provides an overview.

---

[3] In previous work one of the authors has for example propose strategies such as gradualization (e.g., conceptualization in terms of a spectrum instead of a dichotomy), common ground dialogue (i.e. open discussion about the purposes of conceptualization and definition among stakeholders), renaming and reframing, or multiplying (i.e. creating shared understandings by making plurality visible) as strategies to deal with ethical issues arising from technological development (Keulartz et al. 2004).

Table 23.1 A toolbox of disease concepts

Disease concepts	
Syndrome	Constellations of symptoms from which a patient suffers. These are often understood as inferior forms of classification to aetiological or pathological classifications, but can also be understood as highly sophisticated ways of focusing on the suffering of patients and their ability to live a fulfilling life (see Greco, Chap. 17, this volume).
Pathological lesion	Changes to the anatomy of the internal organs that are deemed pathological. Lesions are often understood as things that do not occur in the healthy body - health and disease are qualitatively different. Originally connected with lesions that could be seen at the postmortem examination, but new technologies have made a multitude of new types of lesion detectable (Dwivedi et al. 2017). The microscope made histological and cellular lesions, and eventually immunohistochemical lesions, detectable. In recent years, genetic technologies have made genetic mutations detectable, and these arguably are kinds of pathological lesion as well (although they may be lesions present in every cell, rather than of particular organs). Often these produce sub-categories of anatomically and histologically defined disease, especially in oncology (Zhao et al. 2019). However, entirely new categories can also arise, such as those used in basket trials based on molecular profiles rather than anatomical site (Park et al. 2019).
Deviation from physiological normality	In contrast to lesions, which are not present in healthy people, deviations from physiological normality are physiological parameters that are present in all healthy people (for example blood glucose concentration) that are too extreme (either too high or too low). Disease is quantitatively different from health, but not qualitatively different. This seems to have inspired prominent naturalistic accounts of disease (Boorse 1977). Biological functions are used as the physiological parameters that might deviate from normality. New technologies can also produce new ways of measuring physiology, and new disease concepts, as has happened with continuous glucose monitoring and gestational diabetes (Scott et al. 2020; Carreiro et al. 2018).
Aetiological disease entity	In an effort to produce classifications with greater homogeneity, aetiological disease entities, in which diseases were defined by one important cause, became popular in the late nineteenth century (Carter 2003). This simultaneously promoted the concept of homogeneous disease entities, with one specific treatment and uniform clinical course. Defining conditions using prognosis and response to treatment may become more important as the heterogeneity of patients is increasingly recognized. The emergence of such prognostic and therapeutically oriented categories can be seen in medical literature on 'dia-prognostic' testing, in therapy companion diagnostics, and in the development of conceptual schemes for assessing prognosis (Knottnerus 2002; Knottnerus et al. 2017; Huber et al. 2022; Kent et al. 2020).
Risk-model	As discussed by Giroux, Chap. 15, in this volume, risk-models use a wide variety of factors, including physiological parameters, lesions, symptoms, medical history, family history, diet, exercise

(continued)

Table 23.1 (continued)

	patterns and smoking habits, to assess the risk of unwanted future events. There are all sorts of things that a person might be at risk of, including future symptoms, future disease, or even future risk-states. As everyone is at some level of risk, and because people with recognizable pathology such as cancer might be at a lower risk of some unwanted future event than a person without recognizable pathology, the strict opposition between health and disease might break down on this view.
Systems/network model	Systems medicine is a buzzword used in a variety of contexts loosely connected with personalized medicine (Schleidgen et al. 2017). Conceptually, amongst the more novel innovations found is the attempt to cope with non-linearity of disease phenomena using concepts sourced from systems biology and network theory. These fields study complex, non-linear feedback loops, such that things that might be causes in a system produce effects that feed back to earlier positions in the causal chain to affect the putative causes (Higgins 2002; Berlin et al. 2017). Thus, it is become difficult to distinguish causes from effects. These disciplines seek to describe the networks of factors relevant to disease processes and to model their behavior. This can involve integrating genomic, epigenetic, proteomic and metabolomic levels of information to identify relevant networks of interacting factors. It is expected that these networks will not map onto existing disease categories, such that one network may have a role to play in multiple currently recognized disease states, and currently recognized disease states may involve several of these networks. Although this approach might constitute a new medical cosmology, it has yet to manifest in medical practice.
Biopsychosocial model	To counteract a tendency to understand disease as a biological process happening to the patient's body, the biopsychosocial model was put forward to emphasize the interaction of the patient's body with their psychological, and social environment (Bolton 2023; Bolton and Gillett 2019; Engel 1977; Kusnanto et al. 2018; see both Greco, Chap. 17 and Slatman, Chap. 18, in this volume). So, diabetes may not simply be thought of as an inability to regulate blood glucose requiring medication, but also as the result of a particular diet and lifestyle, which is itself the product a physical and social environment which influences material opportunities and psychological disposition. These other factors might serve as targets for intervention. Despite of widespread recognition of the BPS model, at least in clinical medicine, questions about the causal relation between the three spheres and their integration are considered to be problematic (de Haan 2021).
Symptomatic pathology	Disease is the combination of pathology and/or pathophysiology and the presence of symptoms (Tresker 2020).
Pathology with poor prognosis	Disease is the combination of pathology and/or pathophysiology and a poor prognosis. For example, see the ODx suggested by Rogers and Walker (2018). Both pathology with poor prognosis and symptomatic pathology might be thought of as practical diseases as found in the Boorse's biostatistical theory (1977); or harmful dysfunction as described by Wakefield (1992).

(continued)

Table 23.1 (continued)

Elements	
Biological function	What a bodily part or process ought to be doing. There are many different concepts of biological function, including the contribution made to survival and reproduction, and doing what the part or process evolved to do (Boorse 1977; Wakefield 1992).
Vital goals/capabilities	Something that is necessary for a person's minimal happiness in the long run, or necessary for the person's living a minimally decent life. These might be thought of as personally or socially contingent goals, or as capabilities applicable to all people (Nordenfelt 1995; Venkatapuram 2013).
Harm	Something is harmful if it is negatively valued.
Symptom	An element of suffering reported by the patients. Often contrasted with 'signs', which are reported by the physician.
Individual or group reference class	Normality can only be defined relative to some standard. This standard is usually set by a reference class of people: perhaps of the same age and sex as the patient (Boorse 1977). There is some debate in medicine about what the correct reference classes are. Some physicians and researchers argue for the use of *individualized* reference classes, in which this standard is set using measurement taken from the patient at an earlier point in time (Coskun et al. 2022; Zaninetti et al. 2015).
One dimensional or multidimensional	Deviation from physiological normality is usually conceived as a deviation along a *single* physiological axis. So, diabetes is about blood glucose levels. Disease might be understood as a position in a matrix composed of several axes. These axes might include things like islet autoimmunity, incretin activity, the ability of islet cells to produce insulin, obesity, fat distribution and insulin resistance (McCarthy 2017). This approach has been suggested for diabetes and is referred to as the "palette model" of disease, in contrast to the categorical "pigeon hole" model of disease (McCarthy 2017).
Homogeneous or heterogeneous entity	All patients with the same disease are sometimes taken to be the same as each other for all intents and purposes: people with the same disease have the same aetiology, pathology, symptoms, clinical course, prognosis, and response to treatment. When such homogeneity can be found it is very useful, but assuming that such homogeneity exists often causes significant problems in practice. Alternatively, the heterogeneity of patients might be recognized: patients with the same disease may not be the same, and have different aetiology, pathology, symptoms, clinical course, prognosis and response to treatment. It is important to recognize heterogeneity where it exists, but the failure to find homogeneity makes it very difficult to practice medicine. Shifts to focus medical attention on prognosis and response to therapy rather than diagnosis (see Boenink and Van der Molen, Chap. 11, this volume) are an effort to cope with patient heterogeneity following the recognition that knowing the diagnosis does not tell physicians and patients everything they want to know. New questions for researchers, such as 'what is the scope of this proposed intervention?' (see Keuck, Chap. 9, this volume), are a reflection of the same.

(continued)

Table 23.1 (continued)

Monocausal or multicausal	A disease might be though to have a single cause, or it might be thought to have multiple causes. The same set of causes might be necessary in each case of the disease, or different causes might produce the same disease. That set of causes might be specific for one disease, or may also cause other diseases. There are many subtly different ways to understand disease causation.
Linear development	Linear development assumes that a disease develops inevitably from early stages of the disease to later stages of the disease. Cancer screening programs have relied on this concept, as these seek to prevent later stages of disease by finding lesions identical to those found in the early stages of disease. Unfortunately, for many cancers, lesions identical to early stages of disease do not develop into later stages, and even can regress. This has led to the development of concepts such as the "indolent lesion of epithelial origin" (IDLE) (Esserman et al. 2014), and to the search for histological, genetic and molecular biomarkers that can distinguish those lesions with a good prognosis from those with a bad prognosis (See Boenink and Van der Molen, Chap. 11, this volume).
Multistage development	Disease progression might take place by the same physiological process becoming more and more extreme. Alternatively, different physiological processes may manifest at different steps in the development of the disease. So, a single mutation may cause the uncontrolled proliferation cells in a cancer seen throughout the entire disease process. Alternatively, different stages of the cancer might require different mutations (Vogelstein and Kinzler 1993; Berman 2018). Note that a multistage model of disease may or may not be linear.
Category or spectrum disorder	Diseases might be thought of as categorically distinct, both to other diseases and to health. Alternatively, health and other diseases might exist on a spectrum, admitting, for example, robust health, good health, mild disease, and severe disease.
Poor prognosis	The patient will suffer and/or die from the condition at some point in the future.
Response to treatment	The patient will have some form of recovery as a result of treatment at some point in the future.
Miscellaneous conditions	
Pre-diseases	Many diseases have associated pre-diseases, or precursor conditions (Boorse 2023). Pre-diseases are defined using the same physiological parameter used to define the fully fledged disease, at a lesser degree of abnormality. So, diabetes has prediabetes, hypertension and prehypertension, and osteoporosis has osteopenia. It is often assumed that people with pre-diseases are at high risk of developing symptomatic disease, but this is often not the case. Pre-diseases can be quite far removed from what are arguably of interest: symptoms. One might wonder if it would be better to use all available information to build a risk-model for the development of symptoms, for example, rather than to use pre-diseases to assess this risk more indirectly.
Target conditions	The concept of target conditions has emerged from test evaluation as a response to patient heterogeneity (Leeflang and Allerberger

(continued)

Table 23.1 (continued)

	2019; Irwig et al. 2002; Korevaar et al. 2019). Recognizing that the disease entity is not necessarily the object a test should be detecting, test evaluators needed another concept to describe the objects of medical interest—e.g. those African American men, over 65 years of age, who have a palpably enlarged prostate found at routine check-up, who will go on to suffer and/or die from prostate cancer.
Multidisease	Some tests are designed not to detect one disease, but many. For example, some tests for cancer are not targeted at one type or subtype of cancer, but rather at cancers in general (Kisiel et al. 2022; Putcha et al. 2021). This runs counter to the general trend of precision medicine, which aims to specificity smaller and smaller groups of patients. This is an interesting concept, but requires that commonalities be found across many heterogeneous groups of patients.
Minimal residual disease	Another concept found in cancer care. Should patient enter remission, there remains the suspicion that some residue of the disease remains, which might return at some point. New diagnostic technologies, such as 'liquid biopsy', allow detection of cells and cell-free DNA associated with cancerous growths, which are understood as the residue of the disease (cancer) (Honoré et al. 2021; Bou Zerdan et al. 2023). This raises interesting questions about why such 'residual' conditions should be seen as having cancer, and the effect this has on practice.
Phenotypes	The notion of a disease phenotype is occurring more and more frequently in medical literature. Often, this refers to a constellation of factors, which could be symptoms, lesions or physiological parameters, considered as the manifestations of a disease process. Using traditional methods as well as newer "omics" technologies, effort is often made to find an underlying causal mechanism, or endotype, for each phenotype (Chung and Adcock 2013; Ozdemir et al. 2018). The search for new phenotypes and endotypes is a search for new specific disease entities hidden within traditional disease categories, triggered by the recognition that traditional disease categories are too heterogeneous. Additional concepts such as genotype (combinations of genes), theratype (responses to therapy), and regiotype (geographical locations where there are different allergens) are also put forward to cope with patient heterogeneity, especially in asthma medicine. These concepts are innovative in so far as they facilitate the search for new entities whilst still employing the old ones. Another use of phenotype is closer to a cluster concept. When different phenotypes of a disease, such as polycystic ovarian syndrome, are described, there may be some disagreement about which phenotypes genuinely constitute the disease. Instead of arguing about this, physicians might specify which phenotypes they use when they work (Azziz 2021). For example, sometimes phenotypes A, B, and C might be used, but at other times, phenotypes A, B and D might be used. This conceptual innovation might help resolve ontological arguments about which set of phenotypes is universally correct.

(continued)

Table 23.1 (continued)

Disease as institutional fact	Within a legal or institutional context. Disease can be stipulatively defined, often by reference to medical professional standards or consensus definitions.
Illness	The patient's experience of their symptoms and suffering (Boyd 2000; Parsons 1951).
Sickness	The recognition by society that the patient has a disease (Parsons 1975; Burnham 2014).

Several authors have already provided their own inventory of concepts of health and disease (Haverkamp et al. 2018; Hofmann 2001). We have found the 'medical cosmologies' described in historical and sociological literature especially helpful (Armstrong 1995; Jewson 1976; Ackerknecht 2016). We have found it useful to divide the concepts we have encountered into three categories: concepts of disease, elements, and miscellaneous.

When listing different concepts of disease we have been quite broad, and include things that are not strictly diseases but are 'medical cosmologies' in the above sense. We do not focus on concepts such as the biostatistical theory and various normativist or hybrid theories (Ereshefsky 2009), as these concepts can often be thought of as particular instances of the ones we highlight here. For example, the biostatistical theory might be thought of as an instance of a deviation from physiological normality. In any case, these are already extensively discussed in the philosophical literature.

'Elements' refers to the various components from which the disease concepts are assembled. Elements such as 'biological functioning' are familiar to the philosophical literature, with some concepts of disease make use of this element whilst others do not. We focus on highlighting other elements that have received less attention. For example, some concepts of disease expect that all patients with a disease will be homogeneous, or that a disease will develop linearly, whilst others do not. We do not seek to provide a comprehensive inventory of elements here. We only seek to begin a project of exploring medical literature and practice for different elements that are used in different combinations to produce different concepts of disease.

Finally, we identify several miscellaneous concepts, that are not themselves diseases, but are still used in medical practice alongside disease concepts. Concepts such as 'illness' and 'sickness' have already received some attention (although they could probably do with more). We focus on others, such as 'phenotype', which have received less. Again, our goal is not to provide a comprehensive inventory, but rather to begin the process of exploring these sorts of concepts and considering their function and value to medical practice.

23.4.2 A Pragmatist Working Method

As part of developing a pragmatist approach to health and disease concepts, we have been discussing, thinking about, and experimenting with, different methodological tools, such as historical analysis, conceptual engineering and detailed case-analysis based on real-world problematic situations. In this way, over the course of our research project, the contours of a working method emerged. This four step method has important similarities with Dewey's five stages of inquiry, as described in Westerblad (2022):

1. Identify a problem from medical practice (or another health-related practice) in the present.
2. Produce an historical narrative that explains how the problem has arisen, taking into account the concepts used (and produced).
3. Reflect on how the problem might be addressed and what role (re-) conceptualization might play
4. Offer recommendations for how to proceed based on these insights.

The first step, diagnosing the problematic situation at hand, is a crucial one, since "a problem well put is half-solved" (Dewey 1910). It is important here to determine the facts of the case at hand, and to be critical and precise in formulating the issue. What is the problem, why is it a problem, and for whom? What is the role of conceptualization in this problematic situation?

There are several ways to go about the second, genealogical step which aims to produce understanding of how the problematic situation and the concepts related to it, arose. Binney's analysis of FVPTC (Chap. 7, this volume) which uses the Fleckian notions of active and passive elements of knowledge is an example of how this might work. By tracing a fluctuating network of active and passive elements back through time, it is possible to uncover why objects in the present, such as the FVPTC, are seen as a homogeneous kind. How else could the class of papillary thyroid carcinoma that has no papillary tissue in it have come to exist other than through a contingent historical process? As Fleck counseled:

> [I]t would never occur even to a modern research worker, equipped with a complete intellectual and material armory, to isolate all these multifarious aspects and sequelae of the disease from the totality of the cases he deals with or to segregate them from complications and lump them together. Only through organized cooperative research, supported by popular knowledge and continuing over several generations, might a unified picture emerge, for the development of the disease phenomena requires decades (Fleck 1979: 22).

Even though the FVPTC is now recognized as an error, there is still much to learn from this historical process. Contemporary tumor classifications are the product of the same historical processes. Even concepts of 'cancer' and 'malignant' were, are and continue to be conditioned by them. As Hanson says, "Only by seeing what sorts of things make a man fail to explain a phenomenon or fail to make a certain observation can we appreciate what is at work when he succeeds at these things" (Hanson 1958: 158).

Another example might be Lara Keuck's "historical and epistemological perspective" on Alzheimer's disease (Keuck 2021). The current picture of Alzheimer's Disease, Keuck argues, "represents a collage of different ways in which the disease had been conceptualized within the last century" (Keuck 2021: 19). What is interesting about this metaphor is that a collage does not form a single logical whole, but consists of various elements that coexist, sometimes in harmony, sometimes at odds with each other. This composition of the collage can only be understood historically, because its various elements have different origins and have been added in different times and contexts. Keuck's historical and epistemological perspective thus revisits the conceptual foundations of Alzheimer's Disease and helps to locate and evaluate current problematic ideas and promises about dementia prevention.

Still other approaches may be Aronowitz' (2008) analysis of path dependencies as a way of studying dynamic continuities in the way diseases get framed, or Feudtner's work on diabetes and the idea "disease transmutation". The latter tracks socially, medically and technologically mediated changes in the objects of medical concern—as well as in the experiences of patients—to create a dynamic understanding of the continuous transformations of diseases and disease concepts (Feudtner 1996, 2003).

Finally, as the chapters by Zwart (Chap. 4), Bolt (Chap. 8), Yaron (Chap. 20) and Kukla (Chap. 21) in this volume illustrate, it is important in this step to be aware of the perspective from which the narrative/genealogy is constructed. Whose point of view is taken, which interests are accounted for and to what extent should the broader socio-political and cultural-historical context be taken into consideration?

Once the problem and its history are clear, the third step urges one to reflect on how this historical narrative changes the way in which the problem is understood and how it might be addressed. One can analyze the intended functions and the unintended consequences the use of specific concepts or conceptualizations has had and consider whether re-conceptualization might help solve the problem. This requires a clear view of the aims or goals of the practice in question, the role or function of the concepts used, and reflection on the interests involved. In this step tools and methodologies from different disciplines next to history and philosophy of science—such as philosophy of language, conceptual engineering, STS, biomedicine, and social sciences—may come in to complement the picture. They may enrich our understanding of the socio-political context, existing interests and power relations, and moral implications of different conceptualizations. We think it is important to be aware that "political pragmatism"—simply adapting concepts to what appear to be convenient or serve short-term interests—may be too shallow and may backfire (cf Greco, Chap. 17, this volume).

Fourth and finally, these reflections should lead to recommendations for how to proceed based on the insights gained. Perhaps this step should be thought of as twofold: hypothesizing and proposing solutions, and testing them for their usefulness in actual practice (Westerblad 2022).

Strategies to be used in this step may include explication, precisification, specification and stipulation of disease concepts for specific functions in specific contexts. The proposal by Walker and Rogers (Chap. 15, this volume), to use a vague cluster concept of Disease in order to better understand addiction is an example, as is the proposal by Blok (Chap. 16, this volume) to move towards a spectrum model, instead of clinging to a dichotomous model of disease.

Introducing new concepts can also be helpful, as Keuck and Wolters show in their discussions of 'scope validity.' (Chap. 9 and 10, this volume). Importantly, we may look for new conceptual solutions emerging from the practice under consideration itself. For example, the concept of "symptoms in their own right" described in Greco's chapter (17) and the concept of Positive Health in Van der Linden and Schermer (Chap. 19, this volume), both originated in healthcare practice. Another example is that of "target condition" which has been proposed in the literature on medical test evaluation (Leeflang and Allerberger 2019). The conceptual toolbox (Table 23.1) may serve as a further source of solutions and of inspiration.

A strategy we did not pursue, but that might hold promise for future research, is to use more empirical methods, as is being done in the field of experimental philosophy or to work in closer collaboration with actual practice, like the philosophy of science in practice movement is advocating, and like bioethicists have been doing since the so-called 'empirical turn'. Strategies involving medical professionals and patients may also fit in here, such as consensus conference models (Moynihan et al. 2019; Doust et al. 2017). For sure, such strategies are not without their problems and do not offer an easy way to arrive at the 'right' conceptualizations of health and disease, but when these concepts have significant impact on people's lives, some form of democratic legitimation may be deemed important.

23.4.3 Concluding Remarks

The outline of a new, pragmatic approach to conceptualizing health and disease, which we described in the Prologue was intended to challenge and inspire the contributors to this volume to further develop their own ideas and views in a 'pragmatist spirit'. Both their contributions and the discussions during the workshop have enabled us to elaborate on the theoretical starting points of our proposed approach, as well as on the methods and tools one might use.

Of course, we acknowledge that there is still much work to be done to further develop and refine our approach, to address the many critical questions that might be asked, and to solve the theoretical and practical problems that may be encountered. However, we hope to have set a research agenda for future work that will engage philosophy and history of medicine with the problems encountered in medicine and other health related practices. As pragmatists, we believe the proof of the pudding is in the eating. Any scientific method or scholarly approach must prove its fruitfulness in practice. The contributions to this volume offer a rich palette of what a pragmatic approach to the conceptualization of health and disease might look like, and how it may help those working in different medical and healthcare-related contexts to improve their practice.

References

Ackerknecht, Erwin H. 2016. *A short history of medicine*. Baltimore: JHU Press.
Armstrong, David. 1995. The rise of surveillance medicine. *Sociology of Health & Illness* 17 (3): 393–404. https://doi.org/10.1111/1467-9566.ep10933329.

Aronowitz, Robert. 2008. Framing disease: An underappreciated mechanism for the social pattern-ing of health. *Social Science & Medicine (1982)* 67 (1): 1–9. https://doi.org/10.1016/j.socscimed.2008.02.017.

Azziz, Ricardo. 2021. How polycystic ovary syndrome came into its own. *F&S Science* 2 (1): 2–10. https://doi.org/10.1016/j.xfss.2020.12.007.

Belleri, Delia. 2021. On pluralism and conceptual engineering: Introduction and overview. *Inquiry* 0 (0): 1–19. https://doi.org/10.1080/0020174X.2021.1983457.

Berlin, Richard, Russell Gruen, and James Best. 2017. Systems medicine—Complexity within, simplicity without. *Journal of Healthcare Informatics Research* 1 (1): 119–137. https://doi.org/10.1007/s41666-017-0002-9.

Berman, Jules J. 2018. *Precision medicine and the reinvention of human disease*. London: Academic.

Binney, Nicholas. 2023. Ludwik Fleck's reasonable relativism about science. *Synthese* 201 (2): 40. https://doi.org/10.1007/s11229-022-04018-w.

Bolton, Derek. 2023. *A revitalized biopsychosocial model: Core theory, research paradigms, and clinical implications*, 1–8. *Psychological Medicine*, September. https://doi.org/10.1017/S0033291723002660.

Bolton, Derek, and Grant Gillett. 2019. The biopsychosocial model 40 years on. In *The biopsychosocial model of health and disease: New philosophical and scientific developments*, ed. Derek Bolton and Grant Gillett, 1–43. Cham: Springer. https://doi.org/10.1007/978-3-030-11899-0_1.

Boorse, Christopher. 1977. Health as a theoretical concept. *Philosophy of Science* 44 (4): 542–573.
———. 2023. Boundaries of disease: Disease and risk. *Medical Research Archives* 11 (4). https://doi.org/10.18103/mra.v11i4.3599.

Bou Zerdan, Maroun, Joseph Kassab, Ludovic Saba, Elio Haroun, Morgan Bou Zerdan, Sabine Allam, Lewis Nasr, et al. 2023. Liquid biopsies and minimal residual disease in lymphoid malignancies. *Frontiers in Oncology* 13 (May): 1173701. https://doi.org/10.3389/fonc.2023.1173701.

Bowker, Geoffrey C., and Susan Leigh Star. 2000. *Sorting things out: Classification and its consequences*. The MIT Press. https://doi.org/10.7551/mitpress/6352.001.0001.

Boyd, Kenneth M. 2000. Disease, illness, sickness, health, healing and wholeness: Exploring some elusive concepts. *Medical Humanities* 26 (1): 9–17. https://doi.org/10.1136/mh.26.1.9.

Burnham, John C. 2014. Why sociologists abandoned the sick role concept. *History of the Human Sciences* 27 (1): 70–87. https://doi.org/10.1177/0952695113507572.

Cappelen, Herman, and David Plunkett. 2020. A guided tour of conceptual engineering and conceptual ethics. In *Conceptual engineering and conceptual ethics*, ed. Herman Cappelen, David Plunkett, and Alexis Burgess, 1–26. Oxford University Press. https://philarchive.org/rec/CAPAGT.

Carreiro, Marina P., Anelise I. Nogueira, and Antonio Ribeiro-Oliveira. 2018. Controversies and advances in gestational diabetes—An update in the era of continuous glucose monitoring. *Journal of Clinical Medicine* 7 (2): 11. https://doi.org/10.3390/jcm7020011.

Carter, K. Codell. 2003. *The rise of causal concepts of disease: Case histories*. London: Routledge. https://doi.org/10.4324/9781315237305.

Chang, Hasok. 2012. *Is water H₂O? Evidence, realism and pluralism*, Boston Studies in the Philosophy and History of Science. Springer.

Chung, Kian Fan, and Ian M. Adcock. 2013. How variability in clinical phenotypes should guide research into disease mechanisms in Asthma. *Annals of the American Thoracic Society* 10 (Suppl): S109–S117. https://doi.org/10.1513/AnnalsATS.201304-087AW.

Coskun, Abdurrahman, Sverre Sandberg, Ibrahim Unsal, Mustafa Serteser, and Aasne K. Aarsand. 2022. Personalized reference intervals: From theory to practice. *Critical Reviews in Clinical Laboratory Sciences* 0 (0): 1–16. https://doi.org/10.1080/10408363.2022.2070905.

Dewey, John. 1910. *How we think*. Washington, DC: Heath & Company.

Doust, Jenny, Per O. Vandvik, Amir Qaseem, Reem A. Mustafa, Andrea R. Horvath, Allen Frances, Lubna Al-Ansary, et al. 2017. Guidance for modifying the definition of diseases: A checklist. *JAMA Internal Medicine* 177 (7): 1020–1025. https://doi.org/10.1001/jamainternmed.2017. 1302.

Dutilh Novaes, Catarina. 2015. Conceptual genealogy for analytic philosophy. In *Beyond the analytic-continental divide*, ed. Jeffrey A. Bell, Andrew Culrofello, and Paul M. Livingston, 75–110. Routledge.

———. 2020. Carnap meets Foucault: Conceptual engineering and genealogical investigations. *Inquiry* 0 (0): 1–27. https://doi.org/10.1080/0020174X.2020.1860122.

Dwivedi, Shailendra, Purvi Purohit, Radhieka Misra, Puneet Pareek, Apul Goel, Sanjay Khattri, Kamlesh Kumar Pant, Sanjeev Misra, and Praveen Sharma. 2017. Diseases and molecular diagnostics: A step closer to precision medicine. *Indian Journal of Clinical Biochemistry* 32 (4): 374–398. https://doi.org/10.1007/s12291-017-0688-8.

Engel, G.L. 1977. The need for a new medical model: A challenge for biomedicine. *Science (New York, N.Y.)* 196 (4286): 129–136. https://doi.org/10.1126/science.847460.

Ereshefsky, Marc. 2009. Defining "Health" and "Disease". *Studies in History and Philosophy of Science Part C: Studies in History and Philosophy of Biological and Biomedical Sciences* 40 (3): 221–227. https://doi.org/10.1016/j.shpsc.2009.06.005.

Esserman, Laura J., Ian M. Thompson, Brian Reid, Peter Nelson, David F. Ransohoff, H. Gilbert Welch, Shelley Hwang, et al. 2014. Addressing overdiagnosis and overtreatment in cancer: A prescription for change. *The Lancet Oncology* 15 (6): e234–e242. https://doi.org/10.1016/S1470-2045(13)70598-9.

Feudtner, Christopher. 1996. A disease in motion: Diabetes history and the new paradigm of transmuted disease. *Perspectives in Biology and Medicine* 39 (2): 158–170. https://doi.org/10. 1353/pbm.1996.0027.

———. 2003. *Bittersweet: Diabetes, insulin, and the transformation of illness.* Chapel Hill: University of North Carolina Press.

Fleck, Ludwik. 1979. *Genesis and development of a scientific fact.* Chicago: University of Chicago Press.

Gallie, Walter Bryce. 1964. Essentially contested concepts. In *Philosophy and the historical understanding*, ed. WB Gallie, 157–191. Chatto & Windus: London.

Haan, Sanneke de. 2021. Bio-Psycho-Social interaction: An enactive perspective. *International Review of Psychiatry* 33 (5): 471–477. https://doi.org/10.1080/09540261.2020.1830753.

Hanson, Norwood Russell. 1958. *Patterns of discovery: An inquiry into the conceptual foundations of science.* Cambridge: University Press.

Haslanger, Sally. 2012. *Resisting reality: Social construction and social critique.* Oxford: Oxford University Press.

Haverkamp, Beatrijs, Bernice Bovenkerk, and Marcel F. Verweij. 2018. A practice-oriented review of health concepts. *The Journal of Medicine and Philosophy: A Forum for Bioethics and Philosophy of Medicine* 43 (4): 381–401. https://doi.org/10.1093/jmp/jhy011.

Higgins, John P. 2002. Nonlinear systems in medicine. *The Yale Journal of Biology and Medicine* 75 (5–6): 247–260.

Hofmann, Bjørn. 2001. Complexity of the concept of disease as shown through rival theoretical frameworks. *Theoretical Medicine and Bioethics* 22 (3): 211–236. https://doi.org/10.1023/A:1011416302494.

Honoré, Natasha, Rachel Galot, Cédric van Marcke, Nisha Limaye, and Jean-Pascal Machiels. 2021. Liquid biopsy to detect minimal residual disease: Methodology and impact. *Cancers* 13 (21): 5364. https://doi.org/10.3390/cancers13215364.

Huber, Cynthia, Tim Friede, Julia Stingl, and Norbert Benda. 2022. Classification of companion diagnostics: A new framework for biomarker-driven patient selection. *Therapeutic Innovation & Regulatory Science* 56 (2): 244–254. https://doi.org/10.1007/s43441-021-00352-2.

Irwig, Les, Patrick Bossuyt, Paul Glasziou, Constantine Gatsonis, and Jeroen Lijmer. 2002. Designing studies to ensure that estimates of test accuracy are transferable. *BMJ* 324 (7338): 669–671. https://doi.org/10.1136/bmj.324.7338.669.

Jewson, N.D. 1976. The disappearance of the sick-man from medical cosmology, 1770–1870. *Sociology* 10 (2): 225–244. https://doi.org/10.1177/003803857601000202.

Kent, Peter, Carol Cancelliere, J. Eleanor Boyle, David Cassidy, and Alice Kongsted. 2020. A conceptual framework for prognostic research. *BMC Medical Research Methodology* 20 (1): 172. https://doi.org/10.1186/s12874-020-01050-7.

Keuck, Lara. 2021. A WINDOW TO ACT? Revisiting the conceptual foundations of Alzheimer's disease in dementia prevention. In *Preventing Dementia?* NED-New edition, 1, 7:19–39. Critical Perspectives on a New Paradigm of Preparing for Old Age, ed. Annette Leibing and Silke Schicktanz. Berghahn Books. https://doi.org/10.2307/jj.7079965.5.

Keulartz, Jozef, Maartje Schermer, Michiel Korthals, and Tsjalling Swierstra. 2004. Ethics in technological culture: A programmatic proposal for a pragmatist approach. *Science, Technology & Human Values* 29 (1): 3–29. https://doi.org/10.1177/0162243903259188.

Kisiel, John B., Nickolas Papadopoulos, Minetta C. Liu, David Crosby, Sudhir Srivastava, and Ernest T. Hawk. 2022. Multicancer early detection test: Preclinical, translational, and clinical evidence–generation plan and provocative questions. *Cancer* 128 (S4): 861–874. https://doi.org/10.1002/cncr.33912.

Knottnerus, J. André. 2002. Challenges in dia-prognostic research. *Journal of Epidemiology & Community Health* 56 (5): 340–341. https://doi.org/10.1136/jech.56.5.340.

Knottnerus, J. André, Peter Tugwell, Andrea C. Tricco, and Jessie McGowan. 2017. From testing to diagnostic strategies and dia-prognostic research. *Journal of Clinical Epidemiology* 92 (December): 1–3. https://doi.org/10.1016/j.jclinepi.2017.11.010.

Korevaar, Daniël A., Gowri Gopalakrishna, Jérémie F. Cohen, and Patrick M. Bossuyt. 2019. Targeted test evaluation: A framework for designing diagnostic accuracy studies with clear study hypotheses. *Diagnostic and Prognostic Research* 3 (1): 22. https://doi.org/10.1186/s41512-019-0069-2.

Koselleck, Reinhart. 2004. *Futures past: On the semantics of historical time*. Translated by with an introduction by Keith Tribe. Columbia University Press.

Kukla, Quill R. 2022. What counts as a disease, and why does it matter? *The Journal of Philosophy of Disability* 2 (November): 130–156. https://doi.org/10.5840/jpd20226613.

Kusnanto, Hari, Dwi Agustian, and Dany Hilmanto. 2018. Biopsychosocial model of illnesses in primary care: A hermeneutic literature review. *Journal of Family Medicine and Primary Care* 7 (3): 497–500. https://doi.org/10.4103/jfmpc.jfmpc_145_17.

Leeflang, M.M.G., and F. Allerberger. 2019. How to: Evaluate a diagnostic test. *Clinical Microbiology and Infection* 25 (1): 54–59. https://doi.org/10.1016/j.cmi.2018.06.011.

Lie, Anne Kveim, and Jeremy A. Greene. 2020. From Ariadne's thread to the Labyrinth itself: Nosology and the infrastructure of modern medicine. *The New England Journal of Medicine* 382 (13): 1273–1277. https://doi.org/10.1056/NEJMms1913140.

McCarthy, Mark I. 2017. Painting a new picture of personalised medicine for diabetes. *Diabetologia* 60 (5): 793–799. https://doi.org/10.1007/s00125-017-4210-x.

Moynihan, Ray, John Brodersen, Iona Heath, Minna Johansson, Thomas Kuehlein, Sergio Minué-Lorenzo, Halfdan Petursson, et al. 2019. Reforming disease definitions: A new primary care led, people-centred approach. *BMJ Evidence-Based Medicine* 24 (5): 170–173. https://doi.org/10.1136/bmjebm-2018-111148.

Nefdt, Ryan M. 2021. Concepts and conceptual engineering: Answering Cappelen's challenge. *Inquiry* 0 (0): 1–29. https://doi.org/10.1080/0020174X.2021.1926316.

Nordenfelt, L.Y. 1995. *On the nature of health: An action-theoretic approach*. Dordrecht: Springer.

Ozdemir, Cevdet, Umut Can Kucuksezer, Mubeccel Akdis, and Cezmi A. Akdis. 2018. The concepts of Asthma endotypes and phenotypes to guide current and novel treatment strategies. *Expert Review of Respiratory Medicine* 12 (9): 733–743. https://doi.org/10.1080/17476348.2018.1505507.

Park, Jay J.H., Ellie Siden, Michael J. Zoratti, Louis Dron, Ofir Harari, Joel Singer, Richard T. Lester, Kristian Thorlund, and Edward J. Mills. 2019. Systematic review of basket trials, umbrella trials, and platform trials: A landscape analysis of master protocols. *Trials* 20 (1): 572. https://doi.org/10.1186/s13063-019-3664-1.

Parsons, Talcott. 1951. Illness and the role of the physician: A sociological perspective. *American Journal of Orthopsychiatry* 21 (3): 452–460. https://doi.org/10.1111/j.1939-0025.1951. tb00003.x.

———. 1975. The sick role and the role of the physician reconsidered. *The Milbank Memorial Fund Quarterly. Health and Society* 53 (3): 257–278. https://doi.org/10.2307/3349493.

Putcha, Girish, Alberto Gutierrez, and Steven Skates. 2021. Multicancer screening: One size does not fit all. *JCO Precision Oncology* 5 (November): 574–576. https://doi.org/10.1200/PO.20. 00488.

Rogers, Wendy A., and Mary J. Walker. 2018. Précising definitions as a way to combat overdiagnosis. *Journal of Evaluation in Clinical Practice* 24 (5): 1019–1025. https://doi.org/10.1111/ jep.12909.

Rosenberg, Charles E. 2002. The tyranny of diagnosis: Specific entities and individual experience. *The Milbank Quarterly* 80 (2): 237–260. https://doi.org/10.1111/1468-0009.t01-1-00003.

———. 2003. What is disease? In memory of Owsei Temkin. *Bulletin of the History of Medicine* 77 (3): 491–505.

Schleidgen, Sebastian, Sandra Fernau, Henrike Fleischer, Christoph Schickhardt, Ann-Kristin Oßa, and Eva C. Winkler. 2017. Applying systems biology to biomedical research and health care: A précising definition of systems medicine. *BMC Health Services Research* 17 (1): 761. https:// doi.org/10.1186/s12913-017-2688-z.

Scott, Eleanor M., Denice S. Feig, Helen R. Murphy, Graham R. Law, and The CONCEPTT Collaborative Group. 2020. Continuous glucose monitoring in pregnancy: Importance of analyzing temporal profiles to understand clinical outcomes. *Diabetes Care* 43 (6): 1178–1184. https://doi.org/10.2337/dc19-2527.

Tresker, Steven. 2020. A typology of clinical conditions. *Studies in History and Philosophy of Science Part C: Studies in History and Philosophy of Biological and Biomedical Sciences* 83 (October): 101291. https://doi.org/10.1016/j.shpsc.2020.101291.

Tybjerg, Karin. Forthcoming. Medical Anamnesis: Collecting and recollecting the past in medicine. *Centaurus*.

Venkatapuram, Sridhar. 2013. Health, vital goals, and central human capabilities. *Bioethics* 27 (5): 271–279. https://doi.org/10.1111/j.1467-8519.2011.01953.x.

Vogelstein, Bert, and Kenneth W. Kinzler. 1993. The multistep nature of cancer. *Trends in Genetics* 9 (4): 138–141. https://doi.org/10.1016/0168-9525(93)90209-Z.

Wadmann, Sarah. 2023. Disease classification: A framework for analysis of contemporary developments in precision medicine. *SSM – Qualitative Research in Health* 3 (June): 100217. https:// doi.org/10.1016/j.ssmqr.2023.100217.

Wakefield, J.C. 1992. Disorder as harmful dysfunction: A conceptual critique of DSM-III-R's definition of mental disorder. *Psychological Review* 99 (2): 232–247. https://doi.org/10.1037/ 0033-295x.99.2.232.

Westerblad, Oscar. 2022. Deweyan conceptual engineering: Reconstruction, concepts, and philosophical inquiry. *Inquiry* 0 (0): 1–24. https://doi.org/10.1080/0020174X.2022.2118163.

Zaninetti, Carlo, Ginevra Biino, Patrizia Noris, Federica Melazzini, Elisa Civaschi, and Carlo L. Balduini. 2015. Personalized reference intervals for platelet count reduce the number of subjects with unexplained Thrombocytopenia. *Haematologica* 100 (9): e338–e340. https://doi. org/10.3324/haematol.2015.127597.

Zhao, Lan, Victor H.F. Lee, Michael K. Ng, Hong Yan, and Maarten F. Bijlsma. 2019. Molecular subtyping of cancer: Current status and moving toward clinical applications. *Briefings in Bioinformatics* 20 (2): 572–584. https://doi.org/10.1093/bib/bby026.